电子科学与技术专业英语

光电信息技术分册

总主编　张爱红
主　编　周彦平

哈爾濱工業大學出版社

内 容 简 介

本书由光学基础部分、半导体激光器、光电处理方法、光电新技术以及论文范例五部分组成。全书涵盖了光电信息技术领域的基本内容，同时又介绍了该领域一些较新的发展。

本书可作为光电子信息技术等相关专业的本科生专业英语教材，也可供相关专业的工程技术人员使用。

图书在版编目(CIP)数据

电子科学与技术专业英语·光电信息技术分册/张爱红主编.
—2版.—哈尔滨：哈尔滨工业大学出版社，2008.6(2025.1 重印)
ISBN 978-7-5603-1833-2

Ⅰ.电…　Ⅱ.张…　Ⅲ.①电子技术-英语-高等学校-教材
②光电子技术-英语-高等学校-教材　Ⅳ.H31

中国版本图书馆 CIP 数据核字(2008)第 031350 号

策划编辑　张晓京
责任编辑　杨明蕾
封面设计　卞秉利
出版发行　哈尔滨工业大学出版社
社　　址　哈尔滨市南岗区复华四道街 10 号　邮编 150006
传　　真　0451-86414749
网　　址　http://hitpress.hit.edu.cn
印　　刷　哈尔滨圣铂印刷有限公司
开　　本　880 mm×1230 mm　1/32　印张 14.25　字数 380 千字
版　　次　2004 年 3 月第 1 版　2008 年 6 月第 2 版
　　　　　2025 年 1 月第 10 次印刷
书　　号　ISBN 978-7-5603-1833-2
定　　价　38.00 元(本册)

前　言

21 世纪是一个国际化的高科技时代,国际化的信息时代,作为国际间交流的重要载体——英语,其作用显得更为重要。专业英语是大学英语教学的重要组成部分,其目的在于强化巩固基础英语并进行实践应用,从而掌握科技英语技能,使学生能够熟练阅读国外相关文献,掌握国内外本专业发展前沿的最新动态,并具有一定的科技写作能力。为此我们编写这套专业英语教材,以满足高等院校电子科学与技术和电子信息科学与技术专业及相关专业的本科生专业英语教学的需要和从事上述专业的工程技术人员学习英语的要求。全套教材分三册,包括光电子技术分册、微电子技术分册和光电信息技术分册。

本书是光电信息技术分册,由周彦平教授担任主编。

全书分为五个部分:光学基础部分、半导体激光器、光电处理方法、光电新技术以及论文范例。第一部分包括:第一章几何光学;第二章波动光学;第三章光学仪器;第四章电磁基本理论。第二部分包括:第五章半导体激光器的理论和应用;第六章量子激光器、分布式反馈激光器及垂直空腔表面发射激光器等先进的半导体激光器理论。第三部分包括:第七章光辐射的探测;第八章量子光学、量子噪声和压缩的经典处理方法。第四部分即第 9 章阐述全息技术和光信息存储技术。最后一部分为第 10 章,介绍三篇论文范例,使读者对英语论文写作的格式有总体了解。为了有利于教学和阅读理解，帮助学生更好地理解原文，每章后面给出了主要专业词汇及难句注释，并在每部分后面列出了有关参考文献。

本书在编写过程中得到了光学工程专业许多老师和同学的帮助，张伟教授和许士文教授在内容选择方面提出了许多宝贵意见；俞建杰、董先向、李刚、卢春莲、孙金霞、黄振永等硕士研究生在文章的专业词汇及难句注释方面做了很多工作，光学工程专业2002级硕士研究生对本书的主要章节进行了试读，最后由博士研究生舒锐对全书进行了统稿和整理，在此向他们表示感谢。编者对给予本书提出各种建议和意见的其他同志一并表示衷心的感谢。

由于编者水平有限，书中难免有不妥之处，衷心希望广大读者批评指正。

编　者

2004年3月

CONTENTS

目　录

PART ONE
BASIC PARTS

1

Geometrical Optics

1.1 Models of Light: Rays and Waves

Although we usually consider light to be a wave phenomenon, that is not the only possible view. The idea that light is composed of a stream of particles is also consistent with the observation that a beam of light from a flashlight or a laser seems to travel in straight lines except when it encounters an interface between two optical media (Figure 1.1). We call this straight-line path a ray of light. The ray model allows us to explain in simplest terms the formation of images by lenses and mirrors.

Figure 1.1 A ray of light from a laser. The light seems to travel in straight lines unless it encounters an interface between two optical media.

However, light displays particle like characteristics when it interacts with matter, as it does, for example, when sunlight falls on a leaf and photosynthesis takes place. The apparent conflict over this waveparticle duality can not be understood with just the theories of Newton or Maxwell, but is resolved by the twentieth-century theory of quantum mechanics.

The realization that light was the same as the electromagnetic radiation predicted by Maxwell's equations united the studies of electromagnetism and optics. The range of wavelengths that comprise visible light makes up only a small portion of the entire electromagnetic spectrum(Figure 1.2), which spans many orders of magnitude. There are no sharp dividing lines separating the various regions; there is just a continuous blending from one region to the next.

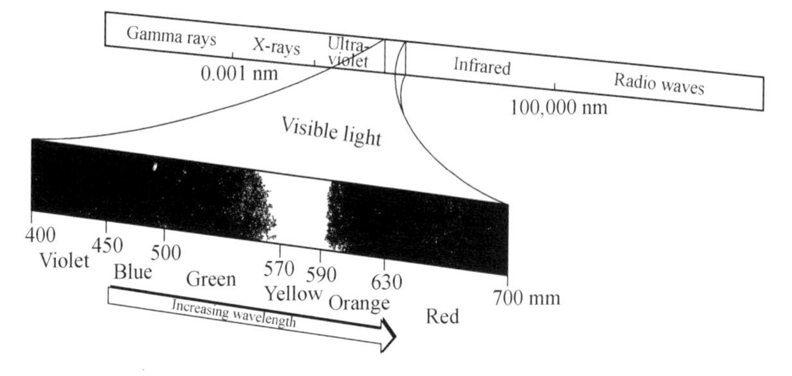

Figure 1.2 The electromagnetic spectrum. There are no sharp divisions between types of electromagnetic waves, nor are there sharp boundaries between different colors of visible light.

When sunlight is spread into a spectrum, we see the characteristic band of visible colors from red to violet. Beyond the violet edge of the visible spectrum are frequencies of radiation that exceed that of the violet. We use the name ultraviolet to describe this invisible extension of the spectrum. Beyond the red end of the visible spectrum lie frequencies below those we can see. These wavelengths make up the infrared region of the spectrum.

The fundamental behavior of all components of the electromagnetic spectrum is the same. They differ only in their wavelengths and frequencies and in the kinds of devices that can be used to generate and detect them. The behavior of all electromagnetic waves can be predicted from Maxwell's equations and a knowledge of the composition and shape of the lenses, reflectors, and other components involved. For example, the design of microwave antennas (Figure 1.3) follows the same underlying principles as does the design of telescope mirrors.

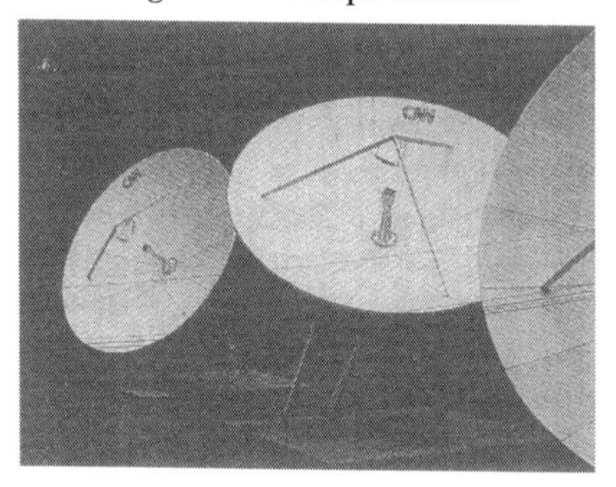

Figure 1.3 Microwave antennas are curved reflectors used to focus microwaves.

1.2 Reflection and Refraction

Suppose we shine a beam of light at a mirror (Figure 1.4). The light strikes the mirror at a point P and is then reflected. What is the direction of the reflected beam? You may know the answer even if you have not formulated it mathematically: The incident and reflected beams make equal angles with the mirror.

In optics it is an established convention to measure angles with respect to the normal, which is a line perpendicular to the surface, as indicated in Figure 1.4. The angle between the incoming ray and the normal is called the angle of incidence, θ_i. The angle between the outgoing ray and the normal is the angle of reflection, θ_r. Both angles are measured positive from the normal. When these angles are measured, the angle of reflection is always found to equal the angle of incidence. Thus, the observed law of reflection is that the angle of reflection is equal to the angle of incidence, or

$$\boxed{\theta_r = \theta_i} \tag{1.1}$$

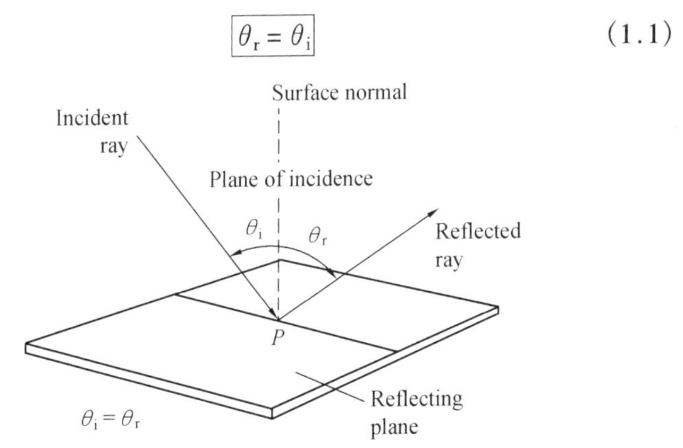

Figure 1.4 Reflection of light from a flat mirror. The incident and reflected beams make equal angles with the normal. The plane of incidence is defined by the incident and reflected rays.

The normal, the incident ray, and the reflected ray all lie in a plane perpendicular to the reflecting surface, known as the plane of incidence. The light ray does not "turn" out of the plane of incidence as it is reflected.

The law of reflection applies to both flat and curved surfaces. For a curved surface, the angle of reflection is determined by the angle of incidence between the incident ray and the surface normal at the point where the incident ray strikes the surface.

Reflection from a smooth mirrorlike surface is called specular reflection [Figure 1.5(a)]. When you look at a mirror, you do not usually see the mirror surface itself; you see the specularly reflected image of other objects instead. For example, you may look at a mirror and see an image of yourself or someone else. What happens when light is reflected from an object whose surface is not perfectly smooth? In that case, the light is diffusely reflected, with different parts of the incident light beam scattered in different directions according to the law of reflection [Figure 1.5(b)].

The light reflected at each little region of the surface is reflected at an angle equal to the local angle of incidence. Most objects that we see are made visible by the diffuse reflection of light from their surfaces. The paper in this book reflects light diffusely so that you can see it from any angle. To see a light reflected in a mirror, however, requires that you be in just the right place so that the specularly reflected light can reach your eye. This effect is used at Boston's Logan Airport to create interesting patterns of multiple images formed by a wall of plane mirrors intentionally set at slightly different angles (Figure 1.6).

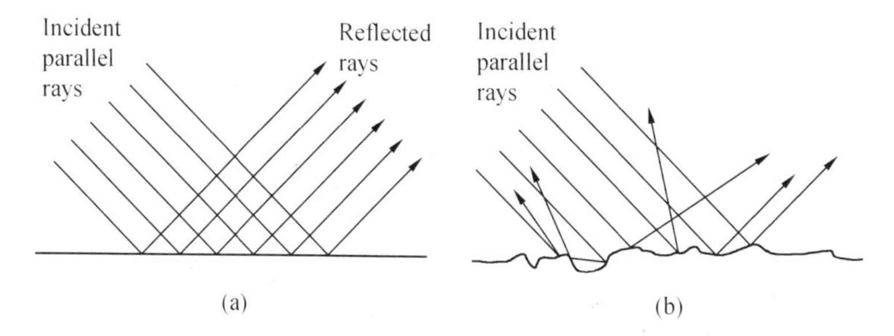

Figure 1.5 (a) Specular reflection from a smooth surface. (b) Diffuse reflection from a rough surface.

Figure 1.6 A wall of plane mirrors set at slightly different angles reflects multiple image, of the authors.

Sometimes both diffuse and specular reflection occur simultaneously from the same surface. For example, when sunlight shines on a car, you can see the car from any direction around it. That is a diffuse reflection. If the car is highly polished, you may also see the image of distant objects reflected from its surface. That is specular reflection.

Master the Concept

Reflection

Question: How do the day/night rearview mirrors commonly found in automobiles work?

Answer: The day/night mirror is made from glass that is wedge shaped and silvered on the back side. The mirror is mounted so that the base of the wedge is at the top. When light strikes the mirror, a small portion (4%) is specularly reflected from the air-glass interface and the rest is transmitted through the glass to the back surface when it is strongly reflected. Because the front and back surfaces are not parallel, the light reflected from the back surface emerges in a direction that is several degrees above the front surface reflection. In normal daytime use, the mirror is positioned so that the strong reflection of light from the rear window is reflected to the driver's eyes. At night the mirror is tilted slightly upward so that the light reflected from the silver backing is directed over the drivers head. In that position only the dimmer light from the front surface reflection reaches the driver's eyes.

1.3 Total internal Reflection

A particularly interesting situation arises when the light incident on an interface comes from within the optically denser medium — that is, from the material with the greater index of refraction. In this case, n_1 is greater than n_2. Consequently, in accord with Snell's law, θ_2 is larger than θ_1. As θ_1 increases, θ_2 must also increase, but it reaches a limit at 90° [Figure 1.7(a)]. At this point, the refracted ray runs along the interface. The incidence angle corresponding to a 90° angle of refraction is called the critical angle θ_c and is found from Snell's law, where

$$n_1 \sin \theta_c = n_2 \sin 90°,$$

$$\boxed{\sin \theta_c = \frac{n_2}{n_1}} \tag{1.2}$$

When the angle of incidence is smaller than θ_c, the light is transmitted at some θ_2 between 0 and 90°. But for angles of incidence greater than θ_c. Snell's law is no longer valid and experiments show that no light penetrates into the less-dense medium. Instead, when the angle of incidence exceeds the critical angle, all the incident light is reflected back into the denser medium by the interface between two normally transparent materials. We call this phenomenon total internal reflection. When light is reflected in by total internal reflection, no light is lost at the reflecting surface. By comparison, some light is lost in reflection from even the best silvered mirrors, which typically reflect only about 90% of the incident light. The light from the ruler in [Figure(1.7(b)] is totally reflected from the back of the prism to create the mirror image.

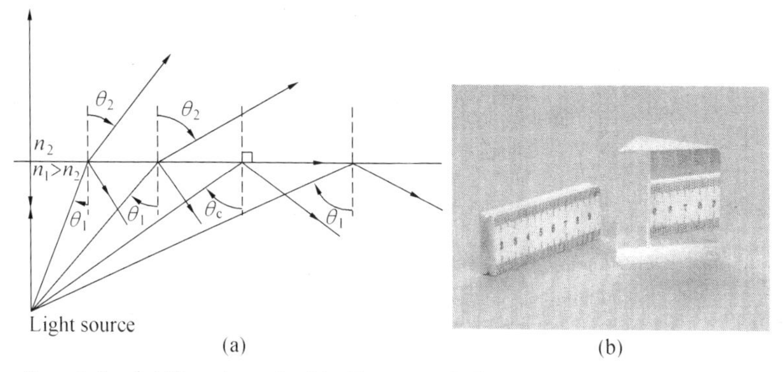

Figure 1.7 (a) When the angle of incidence exceeds the critical angle θ_c, light is totally internally reflected. (b) The image of a ruler is totally reflected from the back surface of a triangular prism.

1.4 Fiber Optics

Total internal reflection can work for us in light pipes. If we direct light into one end of a long rod of glass or plastic, the light is totally internally reflected by the walls, bouncing back and forth until it emerges at the far end [Figure 1.8(a)]. If we bend the light pipe into a particular shape, the light follows the shape of the pipe and emerges only at the end.

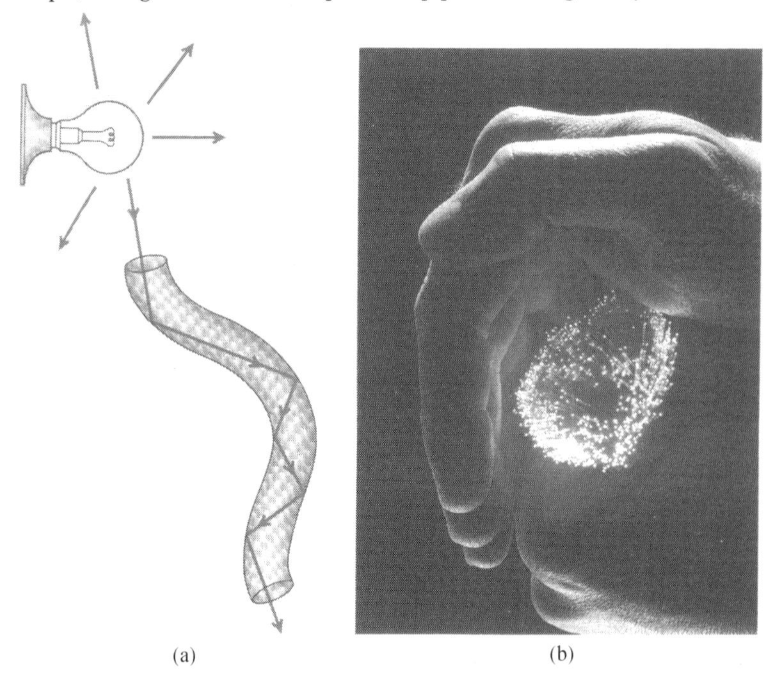

(a) (b)

Figure 1.8 (a) Light rays trapped by total internal reflection are channeled along the length of a light pipe. (b) Light can be seen emerging from the ends of tiny optical fibers.

Sometimes light pipes are made of very thin optical fibers, which may be grouped into a bundle [Figure 1.8(b)]. If fiber ends are polished and their spatial arrangement is the same at both ends, an image may be transmitted from one end of the fiber bundle to the other [Figure 1.9 (a)]. A fiber bundle that transmits an image is called a coherent bundle, or an image conduit. Bundles of fibers that do not have exactly the same alignment at both ends transmit light but not images. They are known as incoherent bundles or light guides. Because their fibers are so flexible, fiber optics light guides and coherent bundles are used in instruments designed to permit direct visual observation of otherwise inaccessible objects. Prime examples are endoscopes[Figure 1.9(b)], which are used by physicians examine the interior of a patient's body.

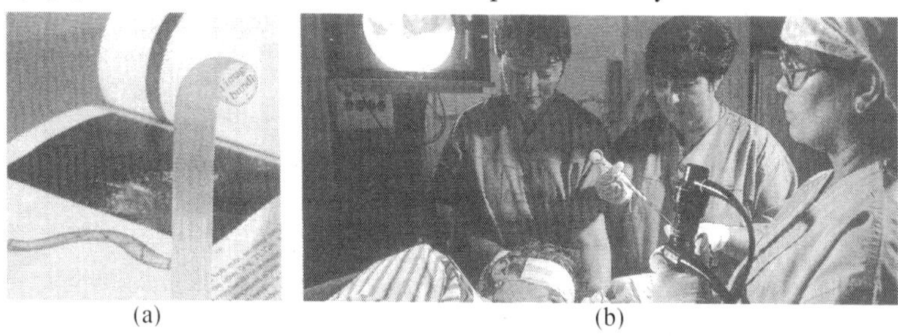

(a) (b)

Figure 1.9 (a) An image conduit composed of a coherent bundle of optical fibers transmits an image from one end to the other. (b) Physicians using a fiber optic endoscope examine the inside of a patient's body by viewing a television image.

Optical fibers are also used in communication systems to transmit modulated light beams. Because of the high frequency of the light waves, these fibers can carry more information in the same space than the metal wires that they replace. The fibers are less expensive to produce than are copper wires, and they are resistant to noise caused by stray electromagnetic signals.

1.5 Thin Lenses

An optical lens is a piece of glass or other transparent material used to direct or control rays of light. Usually the lens surfaces are spherical, although other shapes (parabolic, cylindrical, etc.) are not uncommon. The refraction of light at the surface of a lens depends on its shape, its index of refraction, and the nature of the medium surrounding it (usually air), in accordance with Snell's law. Because lenses can be used to produce images of objects, they form the basis of most optical instruments, from cameras and projectors to microscopes and telescopes. In fact, the eye itself contains a lens and functions as an optical system.

There are only a few distinct ways of combining flat, convex, and concave surfaces to form a single lens (Figure 1.10). If you hold a single lens like one of those in the figure at a moderate distance in front of your eye and look at something through it, you can make several observations. Consider first, lenses that are thicker in the center than at their edges, such as those in [Figure 1.10(a)]. These are called positive or converging lenses, because they refract incident parallel rays so that they converge on a focal point located on the opposite side of the lens. A distant object viewed through such a lens appears smaller and inverted if the lens is held some distance from the eye[Figure 1.10(b)]. An object held close to the same lens appears erect and enlarged[Figure 1.10(c)]. Now look at lenses that are thinner in the center than at their edges, such as those in [Figure 1.10(d)]. These are called negative or diverging lenses, because they refract incident parallel rays so that they appear to diverge from a focal point located on the incident side of the lens. Any object viewed through such a lens always appears erect and smaller than when viewed with the unaided eye. The image remains erect no matter how far the object is from the lens or how far the lens is from the eye [Figure 1.10(e)].

Let's take a converging lens and allow light from the sun to fall on it. Because the sun is so far away, the rays striking the lens are essentially parallel to one another. When these light rays strike the lens parallel to its axis of symmetry, the rays converge to a point called the focal point of the lens[Figure 1.11]. Thus, we can say that any ray incident on a converging lens and parallel to its optical axis (that is, the symmetry axis of the lens) passes through the focal point upon leaving the lens. A lens

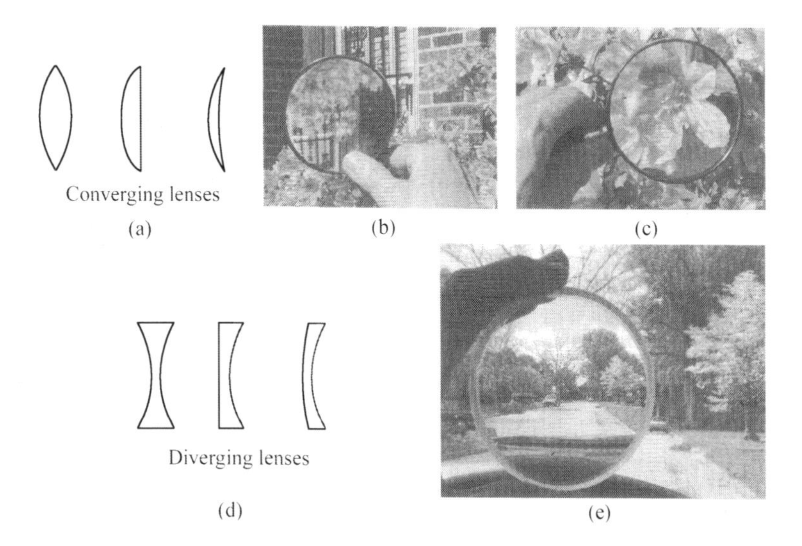

Figure 1.10 (a) Cross-sectional views of the most common types of thin converging lenses having spherical surfaces. (b) View through a converging lens of a distant object. (c) View through a converging lens of an object close to the lens. (d) Gross-sectional views of thin diverging lenses. (e) View Through a diverging lens.

whose thickness is small in comparison with its focal length is called a thin lens. The distance from the center of the thin lens to the focal point is the focal length of the lens. Each lens has its own particular focal length determined by the curvature of its surfaces and the index of refraction of the material from which it is made.

There is a symmetry to the passage of light through a thin lens. Any light ray entering the converging lens from the left parallel to the optical axis[Figure 1.11(a)]passes through the focal point F on the right after it leaves the lens. Conversely, any ray coming from the focal point on the right and striking the lens emerges parallel to the axis[Figure 1.11(b)]. Parallel light entering the same lens from the right converges to a point F' on the left of the lens[Figure 1.11(c)]. For a thin lens, the distances f and f' from the center of the lens to F and F' are the same, even though the lens surfaces are not identical. Thus, we consider a thin lens to have

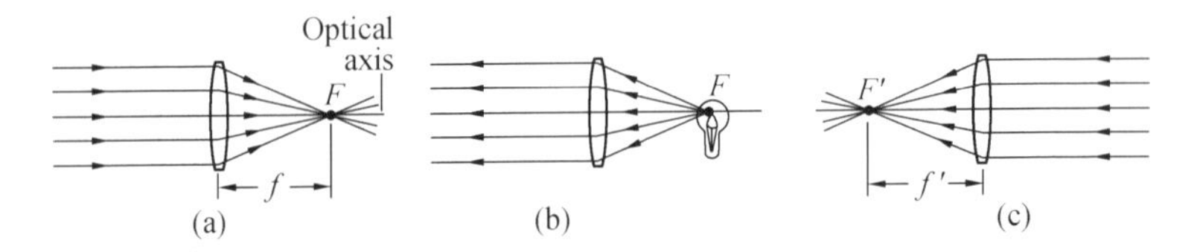

Figure 1.11 (a) Parallel light incident from the left on a converging lens is brought to a focus at the focal point F. (b) Light emerging from this focal point passes through the lens and emerges parallel to the optical axis. (c) Parallel light incident from the right converges to a focal point F' on the left side of the lens. For a thin lens, these two focal points are equidistant from the lens.

two identical focal lengths, one on either side of the lens.① In the remainder of this chapter, we will consider all lenses to be thin lenses. In addition, we will assume that the incident light rays are nearly parallel to the lens axis (the paraxial approximation). Later, in Section 1.9, we discuss some of the ways in which real lenses differ from our idealized thin lens.

Parallel light incident on a concave lens parallel to its optical axis diverges [Figure 1.12(a)]. After passing through the lens, the light behaves as if it came from a point source located at F. Thus f is the focal length for this lens. [Figure 1.12(b)] shows that the same lens refracts light directed toward the focal point on the other side so that the rays emerge parallel to the optical axis after passing through the lens.

An ordinary piece of flat glass, such as a window with parallel faces, passes light without changing its direction. Although the direction of the ray does not change, the ray is displaced laterally [Figure 1.13(a)]. At the center of a lens, the front and back surfaces are parallel, so to a very good approximation, a light ray striking the center of the lens goes through the lens undeviated. If the incoming ray does not make too large an angle with the axis and if the lens is thin, the offset in the ray is negligible

① For a thick lens — that is, one for which the thickness is not small in comparison with the focal length — it is possible for f and f' to be different.

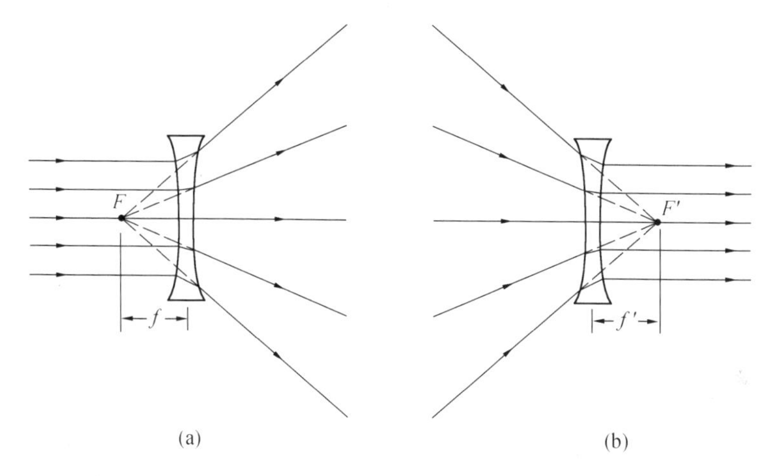

Figure 1.12 (a) Parallel light incident on a concave lens diverges as if it came from the point F. (b) Light converging toward the point F′ diverges so that it emerges parallel after passing through the concave lens.

[Figure 1.13(b)].

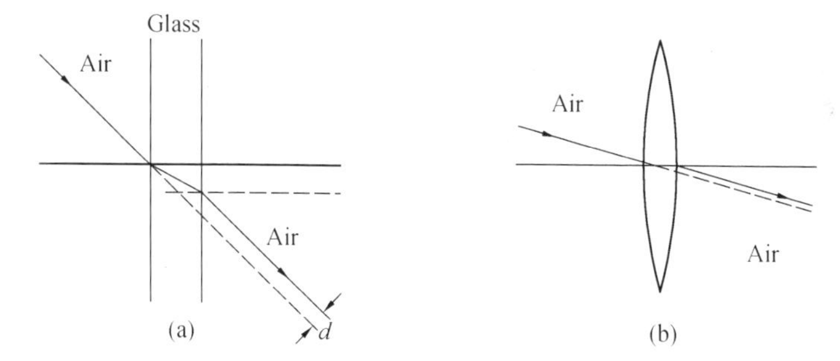

Figure 1.13 (a) A ray of light incident on a parallel-sided slab of glass is displaced an amount d, proportional to the thickness of the glass. The direction of the emerging beam is parallel to the incident beam. (b) A ray of light incident on the center of a thin lens is essentially undeviated if the angle of incidence is small.

1.6 Locating Images by Ray Tracing

The major usefulness of lenses is their ability to form images of an object. The object may be self-luminous, giving off its own light (like the sun or a light bulb), or it may reflect the light that falls on it (like an apple or a page of this book). In either case, an image of the object is formed where light rays that come from points on the object intersect or at the points from which the rays appear to originate. When the light rays actually intersect at the image, we call this a real image. A screen placed at a real image point would show the image in the same way that pictures appear on a movie screen. A virtual image is formed at a point where the light appears to converge, or from which it appears to come. If you place a screen at the position of a virtual image, no image is observed on the screen, for the light rays do not actually intersect there. An example of images in a mirror may help to explain virtual images. If you stand one meter in front of a mirror, your image appears to lie one meter behind the mirror. However, you will not find an image on a screen placed at the image position one meter behind the mirror.

An object, such as a pencil, reflects light rays in all directions from all points of the pencil. We illustrate this in Figure 1.14 by using arrows from the tip of the pencil to represent light rays. However, to locate the image of the pencil formed by a thin lens, you don't need to trace the light rays from all over the object. Instead, you can locate the image by tracing three particular rays from a point on the object. [Figure 1.14(a)] shows these rays refracted by a converging lens; [Figure 1.14(b)] shows them for a diverging lens. In each case the focal length of the lens and the distance from the object to the center of the lens (the object distance) determine the distance of the image from the lens (the image distance).

Although we can use these three specific rays to locate the position of the image, the actual formation of the image is much more complex. The light from each point on the object converges to form the corresponding point on the image. Rays from a single point on the object strike the entire lens surface. Conversely, any small area of the lens surface is struck by rays coming from every point on the object. All of the rays striking each

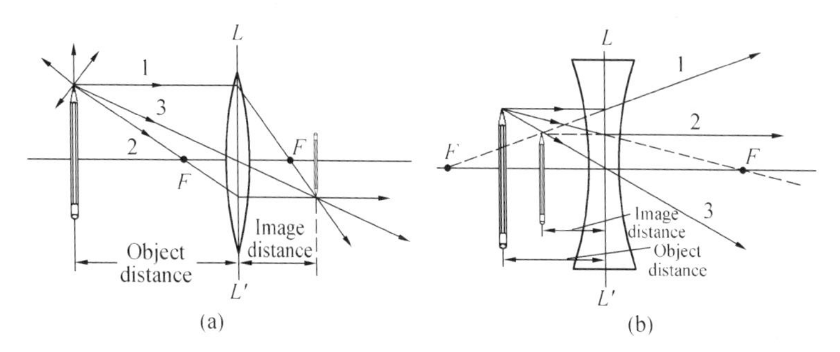

Figure 1.14 Graphical technique for locating the image position with (a) a converging lens and (b) a diverging lens. Dashed lines indicate where light appears to come from (or go to).

small area are focused to form the image. Thus, the image is composed of contributions of light that pass through every part of the lens. For this reason, blocking out part of the lens does not cast a shadow on the image, but simply reduces the brightness of the image by restricting the amount of light that gets to it.

When the rays are close to and nearly parallel to the axis (the paraxial approximation), a small planar object perpendicular to the optical axis is imaged by a thin lens in to a planar region that is also perpendicular to that axis. Thus, all points in the object plane are represented as points in an image plane. When the object is three-dimensional, like the pencil in Figure 1.14(a), points located in different object planes do not focus in a single image plane. For instructional purposes we restrict our discussion to planar objects and their planar images. Thus, we normally represent objects and images with simple arrows perpendicular to the optical axis.

Let us state the graphical procedure to follow for locating the image from a given lens when you know the object's position. In the next section, we will describe how to calculate the image height and location algebraically from this graphical ray diagram. You should refer to both parts of Figure 1.14 as you read these steps.

A. Select an appropriate scale (meters, centimeters, etc.) and mark the position of the lens on the optical axis (also referred to as the principal axis).

B. Draw a line LL' through the center of the lens and perpendicular to the optical axis, as shown. This line will replace the actual lens surfaces that refract the light.

C. Mark the focal points F of the lens on the optical axis and locate the object on the axis, all to the same scale. (In problems of this sort, you will ordinarily be given the focal length f of the lens.)

D. Draw the following rays. Any two of these rays are sufficient to locate the image, but you should always draw the third ray as a check.

1. Draw ray 1 parallel to the optical axis from the object to the line LL' (the lens). For a converging lens, extend this ray from LL' through the focal point on the side opposite the incident light. For a diverging lens, extend the ray from LL' as though the ray came from the focal point on the same side as the incident light.

2. Draw ray 2 from the object to the line LL', passing through a focal point. For a converging lens, draw ray 2 through the focal point on the same side as the incident light; for a diverging lens, draw ray 2 in the direction of the focal point on the opposite side. In both cases, continue the ray from LL' parallel to the optical axis.

3. Draw ray 3 from the object to line LL' at the center of the lens and continue it undeviated.

E. If the rays converge on the side opposite the incident light, the image is real and is located at the intersection of these three rays. If the rays diverge. the image is virtual and is located where these three rays appear to originate. You can measure the position and image height directly from the scale drawing.

You could make a ray diagram for each point on the object to fully locate the image. However, if all of the object is at approximately the same distance from the lens, then you need only consider the image of one point of the object and the rest of the object is thereby located. Again, note that for a given object distance, the size and position of the image are completely determined by the focal length of the lens. Once you have determined the height of the image, you can determine the lateral or linear magnification, the ratio of the image height to the object height.

1.7 The Thin-Lens Equation

We now derive an algebraic expression relating the object distance, the image distance, and the focal length of the lens, using a geometrical ray diagram for a general case. The resulting mathematical expression will allow you to calculate distances and focal lengths more precisely than can be done by ray tracing. However, applying this equation requires careful treatment of positive and negative distances from the lens, which we will discuss after presenting the basic derivation. You should always sketch a ray diagram when solving problems, even when you are using the algebraic method.

A general ray diagram for a convex lens is drawn in Figure 1.15. We designate the object distance by o, the image distance by i, and the focal length by f. The focal points are labeled F and F'. The height of the object is labeled h_0 and that of the image h_i. We use the geometry of the similar triangles APH and GPI to get

$$\frac{HA}{GI} = \frac{AP}{PG}$$

From the similar triangles CPF and IGF, we have

$$\frac{CP}{GI} = \frac{PF}{GF}$$

But $HA = CP$, so

$$\frac{AP}{PG} = \frac{PF}{GF}$$

From the figure we see that $AP = o$, $PG = i$, $PF = f$, and $GF = i - f$, so that this equation becomes

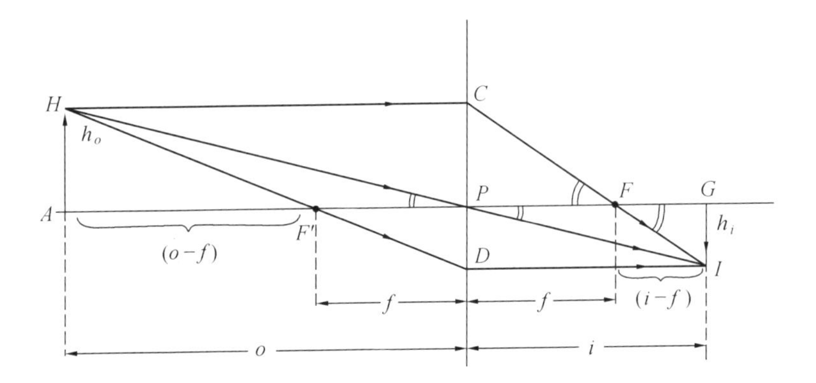

Figure 1.15 Geometry for derivation of the thinlens formula.

$$\frac{o}{i}=\frac{f}{i-f}$$

which, upon rearranging, becomes

$$\boxed{\frac{1}{i}+\frac{1}{o}=\frac{1}{f}} \tag{1.3}$$

This equation is called the thin-lens equation in Gaussian form.① Recall that by a thin lens we mean one for which the focal length is much greater than the thickness of the lens. For such a lens, Eq. (1.5) is sufficient to specify the position of the image.

The lateral of linear magnification of an object is given by the ratio of the image height to the height of the orginal object:

$$m=\frac{h_i}{h_o} \tag{1.4}$$

From the similar triangles *HAP* and *IGP* in Figure 1.15, we see that

$$\frac{h_i}{h_o}=\frac{i}{o} \tag{1.5}$$

so the lateral magnification becomes

① The behavior of a thin lens is described using other characteristic distances in the Newtonian form of the thin-lens equation. The Gaussian form is more common.

$$\boxed{m = -\frac{i}{o}} \tag{1.6}$$

The minus sign is included because, as we will see, it is convenient to have a positive magnification for an erect image and a negative magnification for an inverted (upside-down) image.

In the preceding derivation, we examined the special case of a converging lens with the object placed outside the focal point. However, the thin-lens equation holds for both converging and diverging thin lenses and for any object distance, provided the following sign conventions are maintained:

1. The sign of f is positive for a converging lens and negative for a diverging lens.

2. The sign of o is positive if the object is on the same side of the lens as the incident light (real object), negative if the object is on the other side (virtual object).①

3. The sign of i is positive (real) if the image is on the side of the lens opposite the incident light, negative (virtual) if the image is on the same side.

4. Object and image heights are positive if above the optical axis, negative if below it.

These conventions are consistent with the results of geometrical ray tracing. In fact, the use of Eq. (1.5) and the use of ray tracing techniques always give the same results. In general, it is helpful to make a scale drawing to clarify the problem in your mind and then use the equation for numerical precision. A few examples will serve to illustrate these principles and conventions.

① Virtual objects occur in compound optical systems like those described in Chapter 3

1.8 Spherical Mirrors

Many optical instruments use curved mirrors as image-forming devices. Curved mirrors, like lenses, are used to focus light and create images. However, mirrors work by reflection of light, rather than by refraction, so their practical applications are different.

Curved mirrors are often spherical in shape. A concave mirror reflects light from the inner surface of a sphere; a convex mirror reflects light from the outer surface. Figure 1.16 depicts a beam of parallel light reflecting from a concave spherical mirror and converging to a single point, called the focus of the mirror. However, if the diameter of the incident beam is a large fraction of the diameter of the mirror [Figure 1.17], the reflected rays do not all converge to the same point, and the image is somewhat blurred. This effect, called spherical aberration, is discussed in more detail in Section 1.9. For this reason we limit our discussion of spherical mirrors to those cases where the diameter of the incident light beam is small compared with the radius of curvature of the surface.

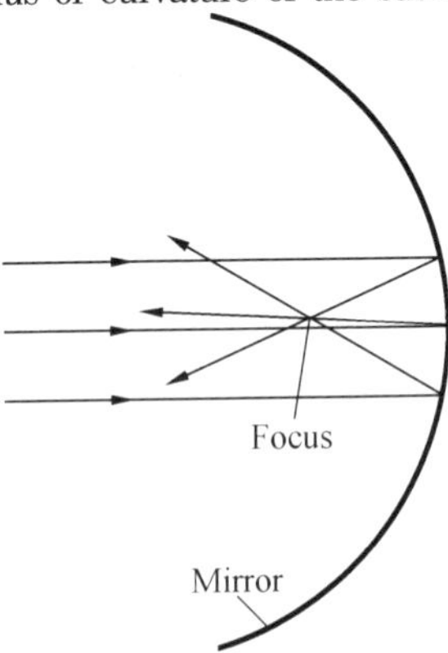

Figure 1.16 The reflection of light rays from a curved surface obeys the law of reflection at each point. When parallel light rays strike near the center of a concave spherical mirror, they are reflected to a common point, called the focus.

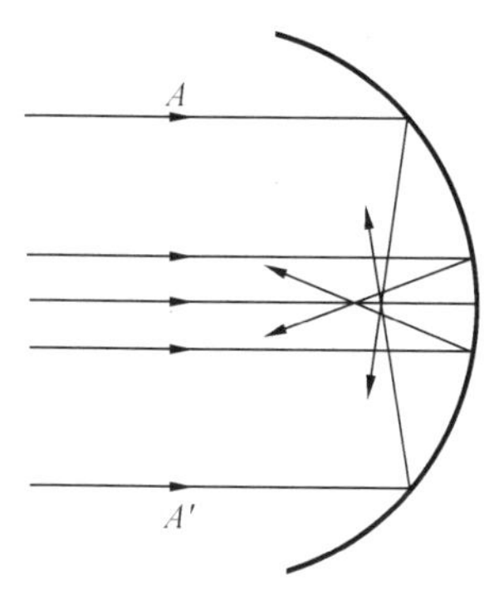

Figure 1.17 Parallel light rays A and A' strike the spherical mirror at so great a distance from the central axis that they are not reflected to the common focus of the rays incident close to the axis.

Figure 1.18 shows a concave mirror illuminated by a beam of light parallel to the optical axis. A single ray parallel to the axis strikes the mirror surface at M and is reflected through the axis at the focal point F. A line drawn from the center of curvature C to the mirror at M forms the base of an isosceles triangle CMF, with equal sides CF and FM. In the limit of small angle θ, the distances CF and FM are equal to one-half the radius R of the surface. The focal length f of the concave spherical mirror is then

$$\boxed{f = \frac{R}{2}} \tag{1.7}$$

where R is the radius of curvature of the mirror.

A concave mirror focuses a parallel beam of light in a manner analogous to the way a converging lens does. On the other hand, if the mirror surface is convex, the reflected light diverges, in a manner analogous to the spreading of a beam by a diverging lens. A convex mirror makes the reflected beams appear to come from a focal point F inside the mirror surface [Figure 1.19]. For a small angle θ, this focal point is halfway between the reflecting surface and the center of curvature, so again $f = R/2$.

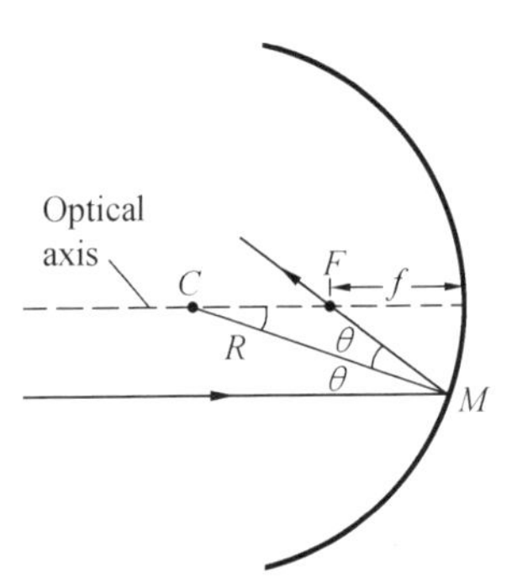

Figure 1.18 Geometrical construction showing that rays near the axis of a spherical mirror converge through a point F located a distance R/2 from the mirror.

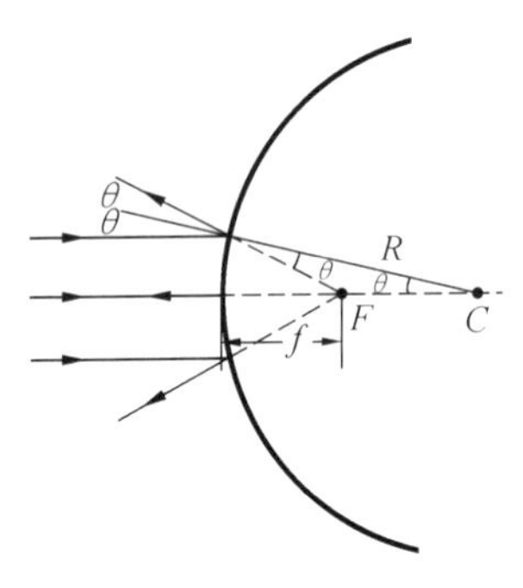

Figure 1.19 Parallel light rays diverge upon reflection from a convex spherical mirror. The diverging rays appear to come from a virtual source point located at $F = R/2$ inside the mirror.

The geometrical procedure for locating the image produced by a spherical mirror is similar to that used for lenses. As you read these steps, follow along with the aid of Figure 1.20.

A. Select an appropriate scale and mark the positions of the object and the mirror on the optical axis, as shown.

B. Draw a line MM' through the center of the mirror and perpendicular to the optical axis.

C. Mark the focal point F of the mirror at $f = R/2$ on the optical axis and locate the object on the axis, all to the same scale.

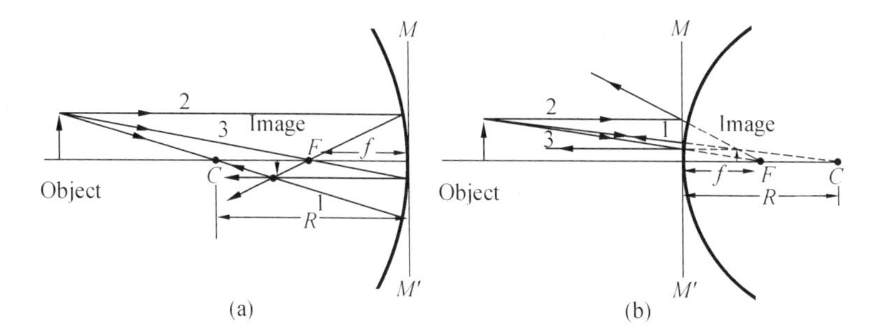

Figure 1.20 (a) Graphical technique for locating the image position for a concave mirror. (b) Graphical technique for locating the image position for a convex mirror.

D. Draw the following rays to locate the image. As with lenses, any two of these reflected rays are sufficient to locate the image position at their intersection, but you should always draw the third ray as a check.

1. Ray 1 is drawn from the object through the center of curvature. It reflects back along itself upon striking the mirror.

2. Ray 2 is drawn parallel to the optical axis from the object to the mirror (MM'), where it is reflected through the focal point of a concave mirror or from the focal point of a convex mirror.

3. Ray 3 is drawn from the object to the mirror(MM') along a path through or toward the focal point and is reflected parallel to the optical axis.

1.9 Lens Aberrations

Up to now, our discussion of lenses has not taken into account some of the optical imperfections that are inherent in single lenses made of uniform material with spherical surfaces. These failures of a lens to give a perfect image are known as aberrations. They have nothing to do with the composition of a lens or the smoothness of its surfaces, but arise naturally from the geometry of image formation. They may be reduced to an acceptable level, but in general, the reduction of one type of aberration tends to increase another.

If we carefully reexamine our original observation that the sun's rays all converge to a focal point after passing through a simple lens, we see that this is only approximately so, [Figure 1.21(a)] shows incident rays parallel to the optical axis of a converging lens. Careful measurement of the refraction angles shows that incident rays farther from the optical axis intersect slightly closer to the lens than the focal point. This effect is called spherical aberration. [Figure 1.21(b)] shows incident rays at an angle to the optical axis. Again, careful measurement shows that the refracted rays do not meet at one point. But create a trailing blur away from the optical axis. This effect is called coma. We see that what we said earlier about thin lenses is strictly true only in the so-called paraxial approximation — that is, only for incoming rays that are not far from the axis and that are parallel to the axis. These limitations explain why the center of an image is often relatively clear while the edges are considerably more distorted.

Aberrations can be reduced, but not removed entirely, by suitably combining different spherical surfaces with appropriate radii. However, to reduce aberrations of this kind to an acceptable level for optical instruments such as microscopes, binoculars, or cameras, each lens must be made of several thin lenses in combination. Figure 1.22 shows such a lens designed for a camera. The design of these lenses is quite complicated in practice, but the basic principle used is still that of refraction as determined by Snell's law. Using computers, lens designers trace ray paths through complicated optical systems in a short time. Snell's law is applied at each boundary between materials with different indices of refraction. The number of individual lenses, their composition, and the shapes of the surfaces are chosen so as to reduce the distortion. The application of computers to lens-design problems has resulted in improved camera lenses and other optical systems, along with reductions in their

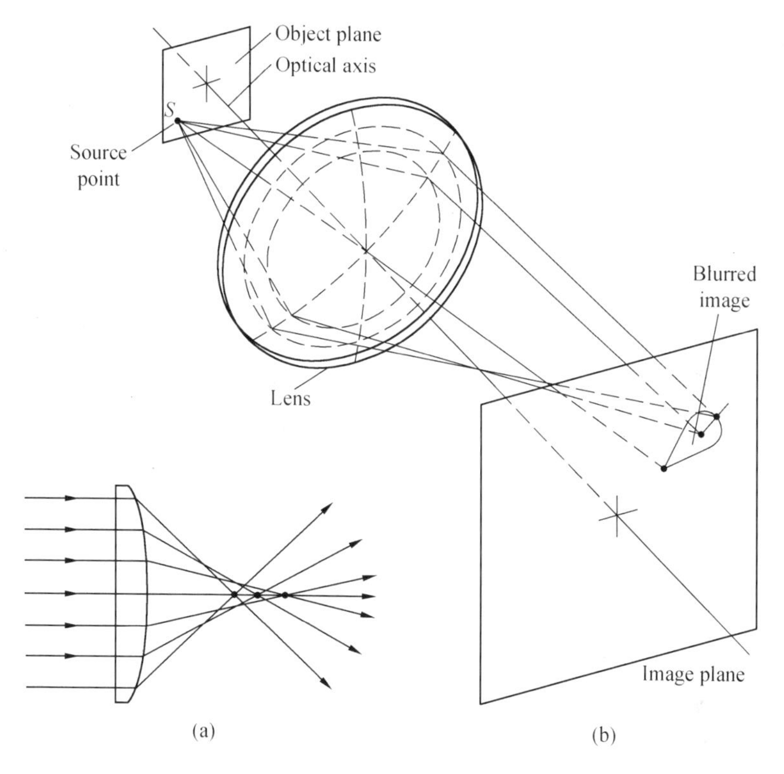

Figure 1.21 (a) Rays coming in parallel to the optical axis do not all intersect at the same point on the axis. Incident rays farther from the optical axis intersect closer to the lens. This effect is called spherical aberration. (b) Rays incident at an angle to the optical axis of the lens fail to intersect at a point, an effect called coma.

cost.

Along with the aberrations mentioned above, there is an additional complication: The index of refraction varies slightly for light of different colors — that is, different wavelengths. This variation is most easily seen in the action of a prism [Figure 1.23]. Incoming white light is a mixture of all colors. When white light strikes the prism surface, it is refracted according to Snell's law, but since the indices of refraction are slightly different for the various colors, each color refracts through a different angle. This spreading of light according to wavelength is called dispersion. A similar thing happens at the second surface, resulting in a spectrum of colors. Because the surfaces of a single lens are not parallel, it behaves like a prism. (Figure 1.24) shows this behavior in somewhat exaggerated form, in an effect called chromatic aberration. It can not be removed by geometrical shaping of a single lens.

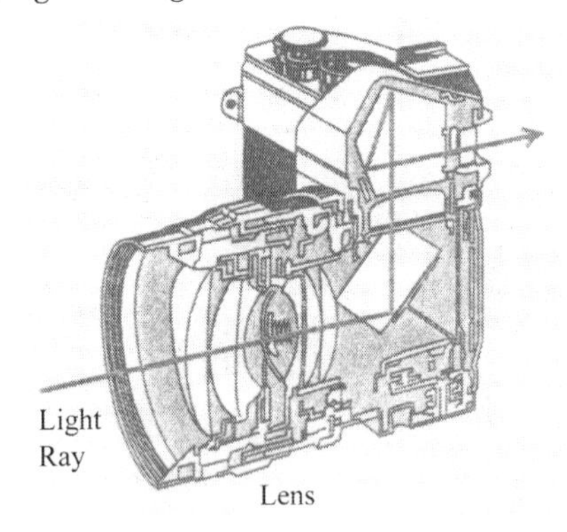

Figure 1.22 Cross section of a single-lens reflex camera with a modern photographic lens. The combination of several lenses helps correct for aberrations.

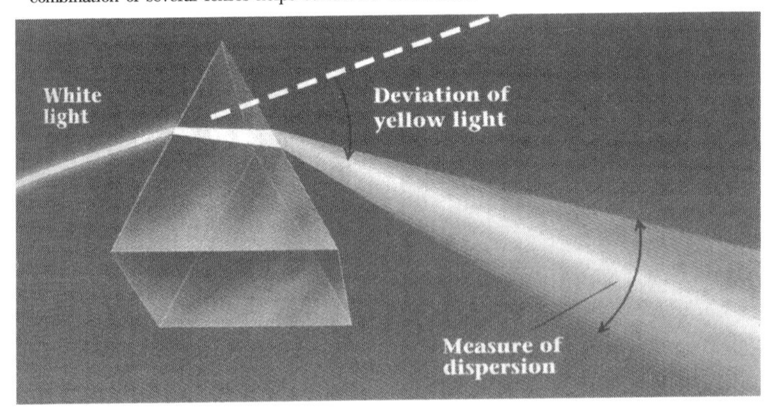

Figure 1.23 Dispersion of light into a spectrum by a prism. The angular spread of the emerging beam has been exaggerated. In a typical glass prism, it is much less than shown here.

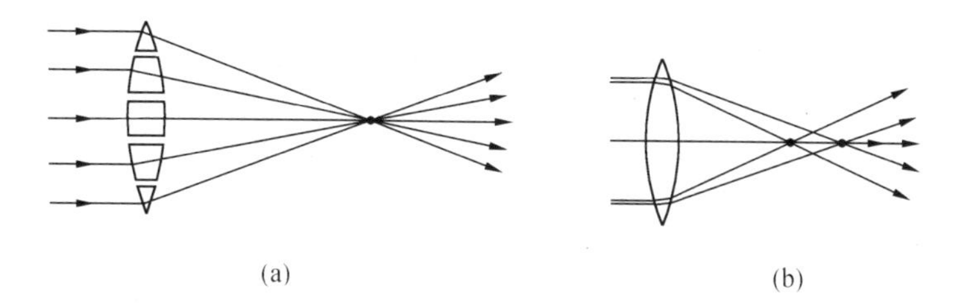

(a) (b)

Figure 1.24 (a) A lens can be considered to be made of sections of prisms. (b) Dispersion of the light by a lens causes a spreading of the focus, with blue light converging nearer to the lens than red light.

Chromatic aberration can be troublesome in any optical instrument that uses lenses. To overcome this problem in telescopes, Isaac Newton invented the reflecting telescope, which uses a focusing mirror instead of an objective lens. No dispersion occurs when light is reflected from the mirror, so chromatic aberration is eliminated from that element.

As we have seen, modern camera lenses are compound lenses made with groups of simple lenses used together. By choosing materials with indices of refraction that depend in different ways on the color of the light, lens designers can ensure that positive and negative lens elements compensate for each other by having equal but opposite chromatic effects. Such lenses are said to be achromatic, because they exhibit only minimal chromatic aberration. The first achromatic lens was made by John Dollond, a London optician, who succeeded in building an achromatic telescope in 1758.

We can take a complicated grouping of lenses like that in (Figure 1.22) and consider the whole as if it were a single lens with a single focal point on either side. Clearly, such a lens is not literally a thin lens. Often, the lens is almost as thick as its stated focal length. However, the reason for all the aberration corrections is to make the lens behave just as we said an ideal simple lens does. We can therefore deal with such compound lenses as single lenses and use the formulas developed for thin lenses to a good approximation.

Figure 1.25 illustrates lens aberrations and their correction. It shows the same scene photographed through a simple uncorrected lens and through a well corrected lens. The blurring in the first photograph [Figure 1.25(a)] is primarily the result of spherical aberration, and the inaccuracy in the color is due to chromatic aberration. A color photograph taken through the well-corrected lens[Figure 1.25(b)] shows no visible color errors.

(a) (b)

Figure 1.25 (a) A photograph taken using a single thin lens, a plano-convex 50-mm lens. (b) The same scene photographed through a well-corrected camera lens of the same focal length (50mm).

In recent years, lensmakers have begun to depart from the traditional manufacturing techniques of grinding lenses to spherical surfaces. Newer methods allow the production of lenses with one or both surfaces neither planar nor spherical. The resulting lenses, known as aspherics, have less aberration that an equivalent spherical lens. Alternatively, they can be made with shorter focal lengths than those possible with spherical lenses of equal diameter and equal sperical aberration. Lenses of this type are now being used in cameras and as condenser lenses in projectors.

Summary

Useful Concepts

- The law of reflection is

$$\theta_r = \theta_i$$

- Snell's law of refraction is

$$n_1 \sin \theta_1 = n_2 \sin \theta_2$$

where n_1 is the index of refraction of material 1 and n_2 is the index of refraction of material 2. The index of refraction of a medium is the ratio of the speed of light in vacuum c to the speed of light in the medium v

$$n = \frac{c}{v}$$

- The critical angle for total internal reflection within an optically denser medium of refractive index n_1 is

$$\sin \theta_c = \frac{n_2}{n_1}$$

- The thin-lens(and spherical mirror) equation is

$$\frac{1}{i} + \frac{1}{o} = \frac{1}{f}$$

- The lateral magnification due to a thin lens is given by

$$m = -\frac{i}{o}$$

- The focal length of a spherical mirror is

$$f = \frac{R}{2}$$

- The location of the image of a point formed by a lens can be found from a scale diagram by drawing at least two of the following rays:

1. A ray parallel to the optical axis from the object to the lens. For a converging lens, extend the ray through the focal point on the opposite side. For a diverging lens, extend the ray from the lens as though the ray came from the focal point on the same side.

2. A ray from the object to the lens passing through a focal point. For a converging lens, draw the ray through the focal point on the same side as the incident light; for a diverging lens, draw the ray in the direction of the focal point on the opposite side. Continue the ray parallel to the optical axis after passing through the lens.

3. A ray from the object through the center of the lens, which continues undeviated.

- The intersection of these rays (or their extensions) locates the image of the object point.

- The location of the image of a point formed by a mirror can be found from a scale diagram by drawing at least two of the following rays:

1. A ray from the object through the center of curvature that is reflected back along itself upon striking the mirror.

2. A ray parallel to the optical axis from the object to the mirror, where it is reflected through the focal point of a concave mirror or from the focal point of a convex mirror.

3. A ray from the object to the mirror along a path passing through or toward the focal point that is reflected parallel to the optical axis.

- The intersection of these rays (or their extensions) locates the image of the object point.

Important Terms

You should be able to write the definition or meaning of each of the following:

ray
ultraviolet
infrared
angle of incidence
angle of reflection
plane of incidence
law of reflection
law of refraction
index of refraction
diverging lens
thin lens
focal length
real image
virtual image
object distance
image distance
lateral magnification
thin-lens equation

critical angle
total internal reflection
converging lens
aberrations
dispersion

New Words and Expressions

geometrical optics	*n*.几何光学
Ptolemy	*n*.托勒密(地心说的创立者)
Corpuscle	*n*.血球,< 物理 > 微粒
Hooke	*n*.虎克,物理学家
Hungens	*n*.惠更斯,物理学家
Fresnel	*n*.菲涅尔,物理学家
Thomas Young	*n*.杨式干涉实验发明人
Kepler	*n*.开普勒
Newton	*n*.牛顿
Maxwell	*n*.麦克斯韦
Hertz	*n*.赫兹
photosynthesis	*n*.光合作用
duality	*n*.二元性
ultraviolet	*adj*.紫外线的,紫外的　*n*.紫外线辐射
infrared	*adj*.红外线的　*n*.红外线
perpendicular	*adj*.垂直的,正交的　*n*.垂线
angle of incidence	*n*.入射角
Jupiter	*n*.木星
half-silvered mirror	*n*.半面镀眼镜
cesim-beam atomic clocks	*n*.铯原子钟
specular reflection	*n*.镜面反射
diffuse reflection	*n*.漫反射
rearview	*n*.(车辆上的)后视镜
index of refraction	*n*.折射系数
reversible	*adj*.可逆的
critical angle	*n*.[物][空]临界角

prism *n*.[物]棱镜,[数]棱柱

periscope *n*.潜望镜,展望镜

fiber optics *n*.光纤

coherent bundle *n*.相干光束

image conduit *n*.传像管

light guide *n*.光导,光控制,光学纤维

parabolic *adj*.比喻的,寓言似的,抛物线的

converging lens *n*.会聚透镜

diverging lens *n*.发散透镜

unaided eye *n*.肉眼

thin lens *n*.薄透镜

focal length *n*.焦距

optical axis *n*.光轴

convex *adj*.表面弯曲如球的外侧,凸起的

concave *adj*.凹的,凹入的 *n*.凹,凹面

curvature *n*.弯曲,曲率

object distance *n*.物距

image distance *n*.像距

algebraical *adj*.(= algebraic)代数学的,代数的

linear magnification *n*.线性放大率

derivation *n*.引出,来历,出处

thin-lens equation *n*.薄透镜方程

linear magnification *n*.线性放大率

convention *n*.大会,协定,习俗,惯例

dime *n*. <美>一角硬币

image-forming devices *n*.成像设备

spherical aberration *n*.球差

isosceles *adj*.二等边的

analogous *adj*.类似的,相似的,可比拟的

parabolic mirror *n*.抛物柱面(反射)镜

aberrations	*n*.失常
paraxial approximation	*n*.旁轴近似
radii	*n*.半径
binocular	*n*.双筒望远镜
chromatic aberration	*n*.色差
troublesome	*n*.麻烦的,讨厌的,棘手的
grinding	*adj*.磨的,摩擦的,碾的
aspherics	[复]*n*.非球面镜头
condenser	*n*.冷凝器,电容器

NOTES

1. The outstanding achievement of this theory was the prediction of electromagnetic radiation, culminating in the realization that light is a form of this radiation. 这个理论最显著的成就是电磁辐射的预测,光波是辐射的一种形式的认识达到了顶峰。

2. Although we now know that light behaves like an electromagnetic wave, as described by Maxwell's equations, most of the fundamental laws of optics were discovered before the time of Maxwell. 虽然我们现在知道光波的行为跟电磁波类似,正如麦克斯韦方程中所描述的,大多数光波基本定律在麦克斯韦时代之前就已经被发现了。

3. Perhaps the best way to think of it is that waves and particles are both simplified models of reality, and that light is a complicated phenomenon that doesn't quite fit either model alone. 也许解释它的最好方法是波和粒子都是简化的实际模型,而光是一种不能单独符合其中任何一种模型的复杂现象。

4. The range of wavelengths that comprise visible light makes up only a small portion of the entire electromagnetic spectrum, which spans many orders of magnitude. There are no sharp dividing lines separating the various regions; there is just a continuous blending from one region to the next. 包含可见光的波长范围仅仅是整个电磁频谱中的一小部分,它跨越了许多种顺序。没有严格的分界线区分不同的区域,从一

个区域到下一个只有一个连续的混合。

5. The behavior of all electromagnetic waves can be predicted from Maxwell's equations and a knowledge of the composition and shape of the lenses, reflectors, and other components involved. 所有电磁波的行为都可以由麦克斯韦方程来预测，它是一种包括透镜、反射镜以及其他一些相关元件的形状及它们的组合的知识。

6. As it happens, the speed of light is far too great and human reactions are much too slow for this experiment to succeed. However, the principle is sound and forms a basis for a number of successful determinations of the speed of light. 当实验开始时，由于光波的速度太快而人类对这样试验的反应太慢而不能取得成功。然而，光波的定律是完全的，它构成了许多成功的光波速度判定的基础。

7. Because the chief limitation in these measurements was the uncertainty in the length of the meter based on the previous standards, this value of the speed of light has been adopted and the meter redefined in the terms of the speed of light. 因为在这些测量中的主要限制是基于先前的标准米的长度不确定造成的，光速的这种优势已经被利用，而且用光速重新定义了米的长度。

8. Instead, when the angle of incidence exceeds the critical angle, all the incident light is reflected back into the denser medium by the interface between two normally transparent materials. 当入射角超过临界角时，在两种透明材料的分界面上，所有的入射光都被反射回光密介质。

9. The refraction of light at the surface of a lens depends on its shape, its index of refraction, and the nature of the medium surrounding it (unusally air), in accordance with Snell's law. 光在透镜表面的折射取决于它的形状，它的折射系数，以及周围介质(通常是空气)的性质，这跟斯涅尔定律是一致的。

10. Thus, we can say that any ray incident on a converging lens and parallel to its optical axis (that is, the symmetry axis of the lens) passes through the focal point upon leaving the lens. 这样，我们可以说，平行

于光轴入射到会聚透镜的任一光束,出射后经过透镜的焦点。

11. Each lens has its own particular focal length determined by the curvature of its surfaces and the index of refraction of the material from which it is made.每个透镜都有自己的焦距,这由透镜的表面曲率和组成透镜材料的折射系数决定。

12. We illustrate this in Figure 1.18 by using arrows from the tip of the pencil to represent light ray. However, to locate the image of the pencil formed by a thin lens, you don't need to trace the light rays from allover the object.我们用图解的方式在图 1.18 中用箭头从铅笔的尖端来描述光线。然而,将铅笔通过薄透镜后的像定位,并不需要描绘从物体发出的每一束光线。

13. When the rays are close to and nearly parallel to the axis (the paraxial approximation), a small planar object perpendicular to the optical axis is imaged by a thin lens into a planar region that is also perpendicular to that axis.当光线靠近光轴并且近似平行于它时(旁轴近似),垂直于光轴的物平面经薄透镜成像,其像也位于垂直于光轴的平面内。

14. However, applying this equation requires careful treatment of positive and negative distances from the lens, which we will discuss after presenting the basic derivation.然而,应用这个方程需要我们小心区分距离的正负,我们将会在呈现基本引出之后讨论这些问题。

15. The minus sign is included because, as we will see, it is convenient to have a positive magnification for an erect image and a negative magnification for an inverted (upside-down) image.正如我们将看到的一样,负号被引用进来,这对正立图像对应一个正放大率和倒立图像对应负放大率的表示方式是很方便的。

16. A parallel beam incident along the optical axis of a parabolic mirror is imaged to a point, without the complications of spherical aberration mentioned earlier.一束沿光轴平行入射的光线经抛物反射镜后成像于一点,它没有先前所提及的球差。

17. We see that what we said earlier about thin lenses is strictly true only in the so-called paraxial approximation — that is, only for incoming rays that are not far from the axis and that are parallel to the axis. 我们应该看到先前所说的有关薄透镜只在旁轴近似的条件下才严格成立——也就是说,仅在入射光线离轴不远并且平行于轴才成立。

18. However, to reduce aberrations of this kind of an acceptable level for optical instruments such as microscopes, binoculars, or cameras, each lens must be made of several thin lenses in combination. 然而,为了减小这种由于光学仪器所带来的在允许之内的像差,例如显微镜,双筒望远镜,或者照相机,每一个镜头必须由几个混合的薄透镜组成。

19. The design of these lenses is quite complicated in practice, but the basic principle used is still that of refraction as determined by Snell's law. 这些透镜的设计在实际工程中是相当复杂的,但是运用的基本原理仍然是由斯涅尔公式所决定的折射定律。

20. By choosing materials with indices of refraction that depend in different ways on the color of the light, lens designers can ensure that positive and negative lens elements compensate for each other by having equal but opposite chromatic effects. 对不同折射率材料的选择依靠光线在不同方式中的颜色,透镜设计者可以通过使正负透镜相等但有相反的色效益来确保正负透镜相互补偿。

2

Wave Optics

2.1 Huygens' Principle

The first thing necessary to discuss a wave theory of light is a technique for describing wave theory of light is a technique for describing wave motion. A simple technique was developed by the Dutch scientist Christiaan Huygens to explain reflection and refraction. We can understand his principle in terms of simple water waves. Suppose we toss a rock into a pond. When the rock hits the water, it generates circular waves that spread out from the point of impact [Figure 2.1]. As we watch the wavefront expand, we are actually following a series of points of constant phase. Every point along the wave crest all around the circle has the same phase, and as the wave expands, this circle of constant phase expands. If the wave encounters a barrier, the wave is reflected; but if the barrier has an opening, part of the wave passes through.

Huygens recognized that it is possible to determine how the wave crest advances by considering each point along the wave crest to be a source point for tiny, expanding, circular wavelets, which expand with the speed of the wave. The contour of the advancing wave is the envelope tangent to these wavelets. In turn, this envelope generates source points for determining a later position of the wave. Figure 2.2 illustrates Huygens' constructions for describing circular waves and plane waves. The radius of each wavelet, equal to the distance between successive wavefronts, is taken to be one wavelength. This technique for describing the motion of waves, called **Huygens' principle**, is valid for all types of waves.

Figure 2.1 Circular ripples spread from a disturbance at the surface of a pond.

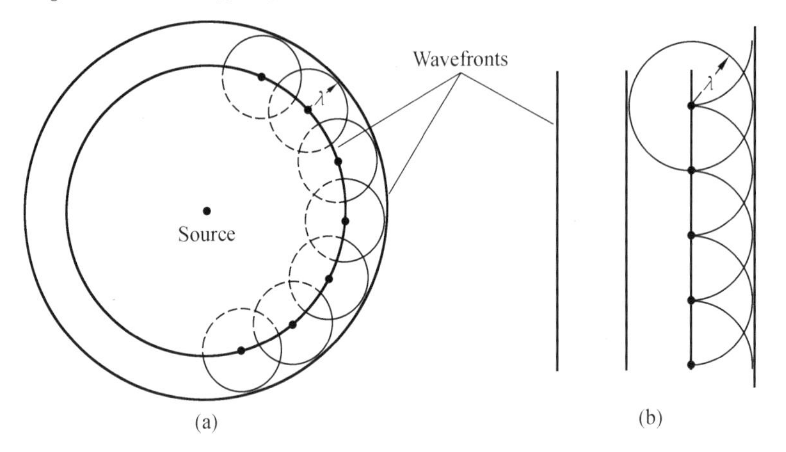

Figure 2.2 Huygens' construction of (a) a circular wave and (b) a plane wave. The radius of each wavelet, equal to the distance between successive wavefronts, is usually taken to be one wavelength.

We can extend Huygens' principle beyond two-dimensional waves to include three-dimensional waves, such as sound and light. The points of constant phase define a surface called the **wave surface** or **wavefront.**

We can determine the shape of the wave surface as the wave advances by applying Huygens' principle as just described. Because the wavefront at any point is perpendicular to the direction in which the wave is advancing, the wavefront is perpendicular to the ray that describes the path of the light.

2.2 Reflection and Refraction of Light Waves

We can use Huygens' principle to derive the laws of reflection and refraction. Recall that the law of reflection states that the angle of incidence and the angle of reflection are equal. This rule comes straight from everyday observations, as we discussed in Chapter 1. However, we can derive it geometrically with Huygens' principle. Imagine a beam of light incident at an angle θ_i on a smooth surface [Figure 2.3(a)]. The wavefront WF is perpendicular to the normal and AF is perpendicular to the incident ray.

According to Huygens' principle, we can locate the next successive wavefront by considering the expansion of spherical wavelets from the wavefront WF. In Figure 2.3, wavelets along WF strike the surface and reflect upward. The wavelet from W reaches the surface first. As it reflects up by one wavelength, the wavelet from F extends one wavelength toward the surface [Figure 2.3(b)]. Then, one by one, other wavelets along WF strike the surface and reflect up. Now, the right triangle AOC in [Figure 2.3(c)] is congruent to the right triangle OAF of [Figure 2.3(a)] because first, they share the common side AO, and second, $AC = FO$, since the wave speed remains the same. This means that the angle AOC is the same as OAF. Hence, the angle of reflection θ_r is identical to the angle of incidence θ_i. This statement is the law of reflection given earlier in Chapter 1:

$$\theta_i = \theta_r \tag{2.1}$$

We can also derive the law of refraction from Huygens' principle. In

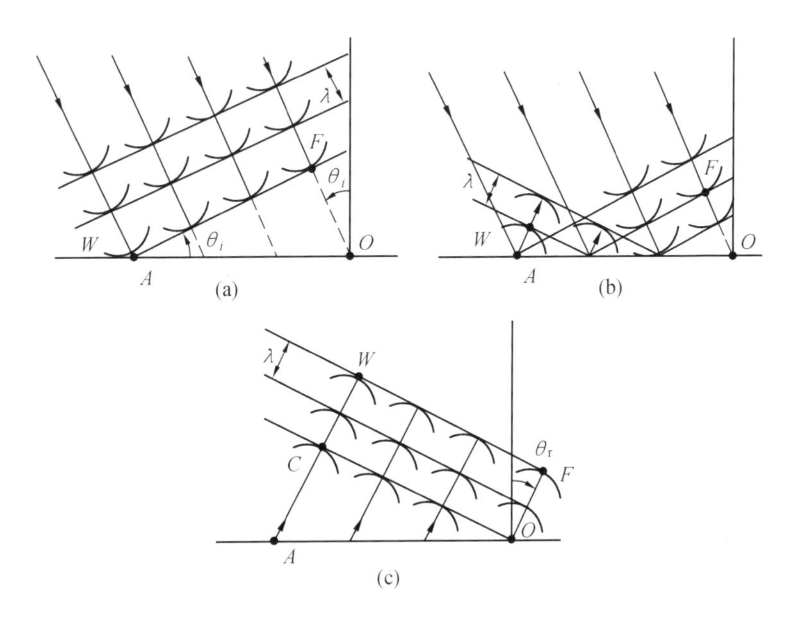

Figure 2.3 Reflection of a plane wave by a smooth (mirror) surface as described with Huygens' waves. Illustrations (a), (b), and (c) show the progress of the wavefronts with time.

Chapter 1, we discussed Snell's law, the relationship between the indices of refraction and the angles of a light ray as it passes from a medium of one index to that of another. Let's now look at a wave derivation of Snell's law. Consider a Huygens wavefront incident upon an interface between medium 1 (say air) with index of refraction 1 and medium 2 (perhaps glass) with a larger index of refraction n_2. The resulting behavior is like that shown in Figure 2.4. The wave speed in air is v_i (incident light), and the wave speed in the glass is v_t (transmitted light). Since we assumed that n_2 is greater than n_1, v_i is greater than v_t. As the wave enters into medium 2 at point A, its speed is reduced, and thus its wavelength is shortened [Figure 2.4(b)]. The wavefront transmitted in the lower medium consequently travels at a different angle with respect to the normal.

We observe that the two triangles *AOF* [Figure 2.4(a)] and *AOC*

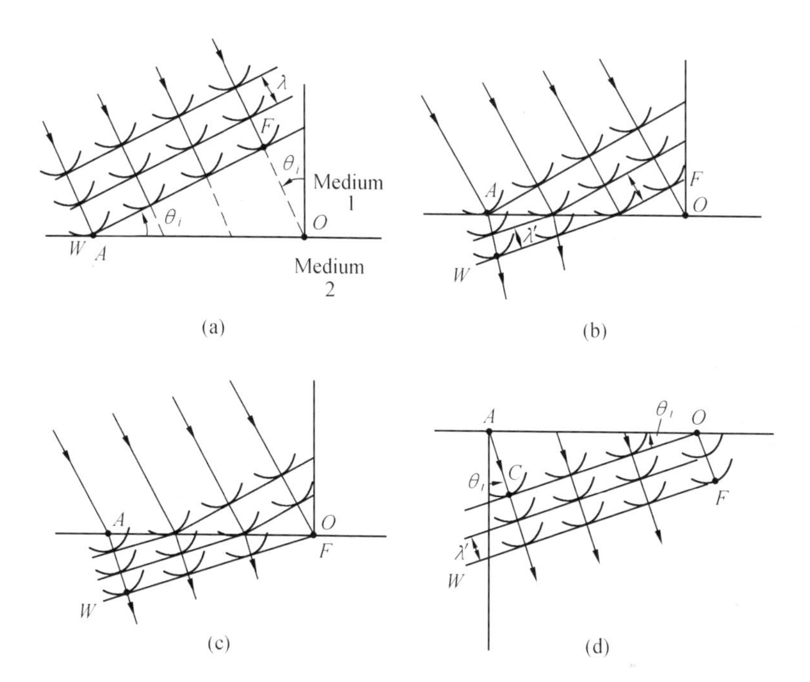

Figure 2.4 Refraction of a plane wave at a flat interface between two transparent media, with n_2 greater than n_1. Illustrations (a), (b), (c), and (d) show the progress of the wavefronts with time.

[Figure 2.4(d)] have one side (AO) in common. Further, we see that

$$\sin\theta_{\mathrm{i}} = \frac{OF}{AO} \text{ and } \sin\theta_{\mathrm{t}} = \frac{AC}{AO}$$

Upon dividing, we find

$$\frac{\sin\theta_{\mathrm{i}}}{\sin\theta_{\mathrm{t}}} = \frac{OF}{AC}$$

However, OF and AC are proportional to the speeds v_{i} and v_{t} of the wave. In a time t while the incident wave moves a distance $v_{\mathrm{i}}\mathrm{t} = OF$ [Figure 2.4(a)], the transmitted wave moves a distance $v_{\mathrm{t}}t = AC$ [Figure 2.4(d)]. So we obtain

$$\frac{\sin\theta_{\mathrm{i}}}{\sin\theta_{\mathrm{t}}} = \frac{v_{\mathrm{i}}t}{v_{\mathrm{t}}t} = \frac{v_{\mathrm{i}}}{v_{\mathrm{n}}}$$

Then using the definition of the index of refraction, we get

$$\frac{\sin \theta_i}{\sin \theta_t} = \frac{c/n_i}{c/n_t} = \frac{n_t}{n_i}$$

This equation is the law of refraction given in Chapter 22 as

$$n_i \sin \theta_i = n_t \sin \theta_t \tag{2.2}$$

Thus, we have shown here that the observed laws of geometrical optics — the law of reflection and Snell's law — follow naturally from the assumption that light is a wave.

We have already seen that the velocity of light in a medium is not the same as the free-space velocity c, but is given by $v = c/n$. When a light wave passes from one material into another, the frequency remains constant.① The wavelength inside a material of index of refraction n is smaller than the free-space wavelength because, the velocity of a wave is the product of its wavelength and frequency, $v = f\lambda$. Thus if the velocity chnges, the wavelength changes proportionately. The new wavelengti inside the material of index n is

$$\lambda' = \lambda / n \tag{2.3}$$

We will see the effects of the wavelength λ' when we consider interference in thin films.

2.3 Interference of Light

In 1807, Thomas Young (1773 ~ 1829) published his *Lectures on Natural Philosophy*, containing the description of an optical experiment now referred to as Young's double-slit experiment now referred to as Young's double-slit experiment. This demonstration of interference effects firmly established the wave theory of light on experimental grounds and

① In this respects, light waves are like sound waves. Remember, the frequency of sound waves from a guitar or violin is determined by the vibrational frequency of the string. As the waves pass into the air, their frequency does not change, even though the speed of sound waves in air is different from the speed of waves in the string.

provided a straightforward means for measuring the wavelengths.

To understand why Thomas Young's double-slit experiment was crucial to a wave theory of light, we first need to examine the interference effects of two inphase wave sources. Then we will show that these same effects were produced by Young's experimental setup, implying that light is a wave. Finally, we will follow Young's calculations in determining the wavelength of visible light.

We observed that a wave disturbance due to two or more sources can usually be taken as the algebraic sum of the individual waves. If the individual sources vibrate with different frequencies, there is nothing special about the resulting disturbance. But if two or more sources vibrate at the same frequency and with a constant relative phase, interesting interference effects occur.

Consider a little bob vibrating up and down on the surface of a body of water. Its motion causes circular water waves to spread out from the bob (Figure 2.5)

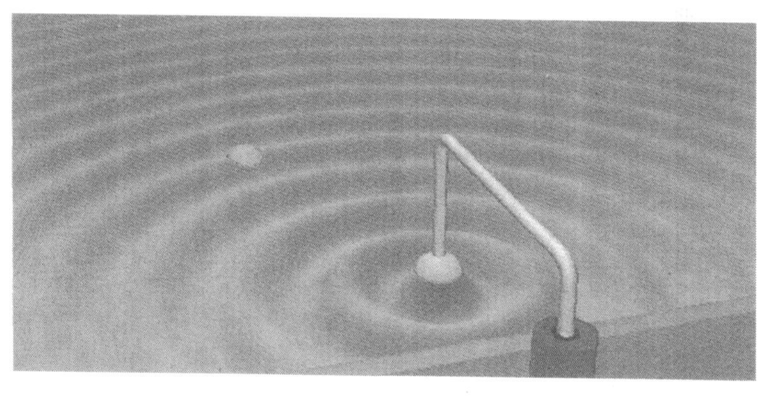

Figure 2.5 A computer-generated image of circular waves spreading out from the point of contact of an oscillating bob. A small yellow sphere is seen floating on the surface.

Now if two bobs are made to vibrate with the same frequency, each of them causes circular water waves to spread out from the point of contact. The waves from these two sources interfere with one another. In some directions, the waves combine constructively, making waves of larger amplitude. In other directions, they combine destructively, so that there is little or no wave amplitude. A pattern develops, as seen in Figure 2.6. The wedge-shaped areas of sharp contrast indicate the crests and troughs of strongly reinforced waves. However, in some radial directions, waves from the two sources arrive exactly out of phase. Since the resulting amplitude is zero, we see no contrasting lines representing the crests and troughs; instead we see a smooth region indicating no wave motion.

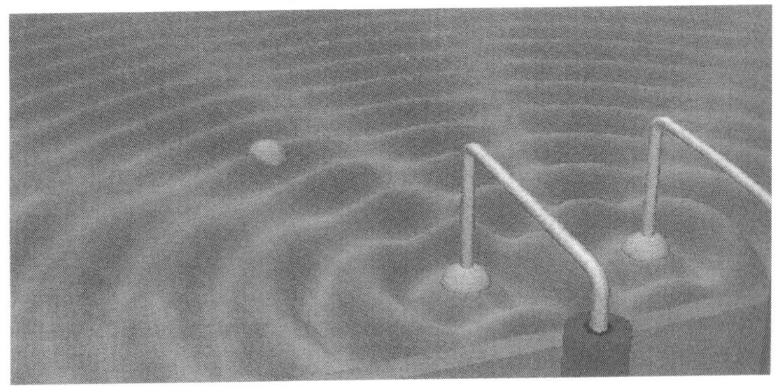

Figure 2.6 A pattern of constructive and destructive interference, produced by two in-phase sets of circular water waves

The condition for **constructive interference** is that waves from two sources with the same frequency arrive at the same point together with the same phase. The result is an amplitude that is greater than the amplitude of either wave alone. We can see from [Figure 2.7(a)] that the resultant wave at the position of the floating sphere has maximum amplitude because the two contributing wave crests (and subsequently the troughs) arrive at the same time. If the two sources oscillate with exactly the same phase, then the condition for maximum constructive interference is that the path lengths of the two waves must be identical or else differ by an integer multiple of the wavelength [Figure 2.7(b)]; that is,

$$D_1 - D_2 = \Delta D = m\lambda, \quad m = 0, 1, 2, 3, \ldots$$

(constructive interference)

where D_1 and D_2 are the path lengths of the waves from their source to the point P.

When the two waves arrive exactly out of phase, **destructive interference** occurs and the resulting wave is diminished.① If the two sources have the same amplitude, destructive interference results in zero net amplitude. This occurs when the path lengths differ by half a wavelength or by any odd half-integer multiple of a wavelength; that is,

$$\Delta D = (m + \frac{1}{2})\lambda, m = 0,1,2,3,\ldots$$

(destructive interference).

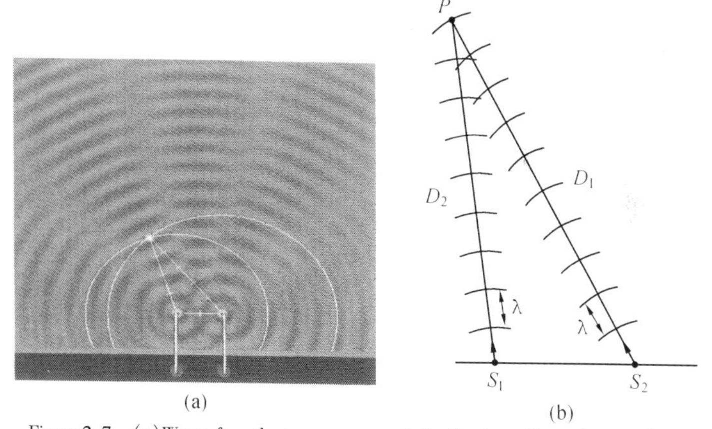

Figure 2.7 (a) Waves from the two sources reach the floating yellow sphere in phase when their path lengths differ by an integer multiple of the wavelength. (b) Waves from sources S_1 and S_2 reach point P in phase when $D_1 - D_2 = m\lambda$.

If the distances D_1 and D_2 are quite large compared with the separation d between the sources, we can express the conditions for constructive and destructive interference in terms of the angle θ shown in

---

① The conditions for constructive and destructive interference depend on the relative phase of the waves. They apply, in general, to all types of waves, no matter what the source of the phase difference.

Figure 2.8. When the distance L between the sources and the plane containing the observation point is much greater than d, the two paths D_1 and D_2 are nearly parallel. We can approximate the difference between them by $\Delta D \cong d \sin \theta$. This result leads to two new equations for determining the maxima and minima of the resultant wave at P:

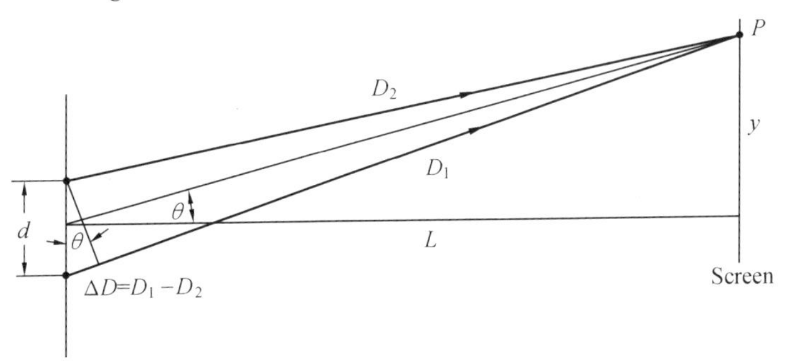

Figure 2.8 Geometrical construction for describing the interference pattern of two sources when the path length is much greater that the distance between the sources.

$$m\lambda = d \sin \theta, \quad m = 0,1,2,3,\ldots \quad \text{(maxima)} \tag{2.4a}$$

$$\left(m + \frac{1}{2}\right)\lambda = d \sin \theta, \quad m = 0,1,2,3,\ldots \quad \text{(minima)} \tag{2.4b}$$

Equation (2.4a,b) enable us to relate the directions of the maxima and minima with the wavelength and the source separation d. These equations are true for all waves, not just water waves. In particular, they accurately describe the interference of two sound waves and, as we shall see, two light waves.

Notice that when the separation d gets larger in Eqs. (2.4a,b), the angles θ corresponding to the respective maxima and minima get smaller, as seen in [Figure 2.9(a)]. Similarly, when the separation between the wave sources gets smaller, the angles of the corresponding maxima and minima increase [Figure 2.9(b)].

In Figure 2.8, the maximum (or minimum) occurs at a point P above the center line. When D_2 is greater than D_1, the maximum (or

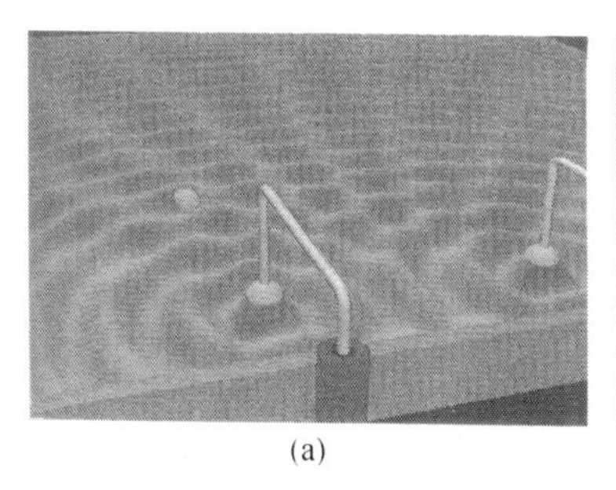

(a)

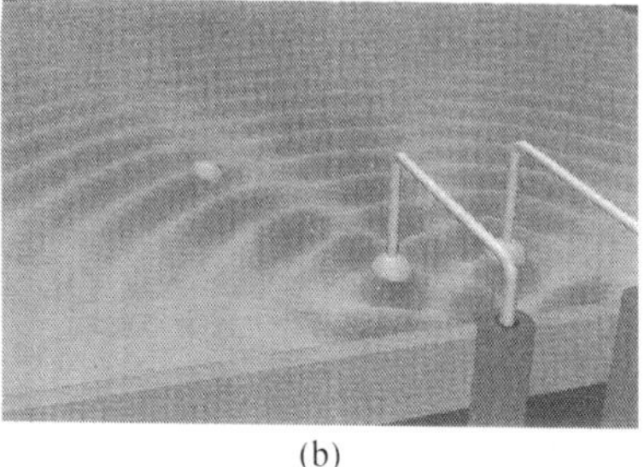

(b)

Figure 2.9 The angle between successive maxima (a) decreases when the separation between the bobs is increased and (b) increases when the separation is decreased.

minimum) lies below the center line. Thus, the pattern of maxima and minima is symmetric about the center line (Figure 2.6). A central maximum occurs along the center line for $m=0$ in Eq. (2.4a). For this reason, it is also referred to as the zero-order maximum. The next maxima on either side of the central beam are called the first order maxima and correspond to $m = \pm 1$. Similarly, the other peaks are labeled by their order. For example, the second-order maximum is the second maximum away from the center line and corresponds to $m = \pm 2$ in Eq. (2.4a). Similarly, the minima given by Eq. (2.4b) are labeled in order of their position from the central maximum. A first minimum occurs to either side of the central maximum. A second minimum occurs beyond the first, and so on. Note, however, that the first minimum corresponds to $m=0$ in Eq. (2.4b)

The interference pattern just described also arises when a single wave strikes a barrier pierced by two narrow slits. The wave passes through the slits, which act as if they were new sources of waves (Figure 2.10). The waves emerging from the two slits must have the same frequency because they were generated from the same initial wave. Because the incident wave crests strike both slits at the same time, the waves passing through the two slits also have the same phase. The slits thus act like two sources of identical frequency and phase, a condition known as **coherence.** The emerging waves produce an interference pattern exactly like the one described for two sources. This type of interference occurs for all kinds of waves and is known as double-slit interference.

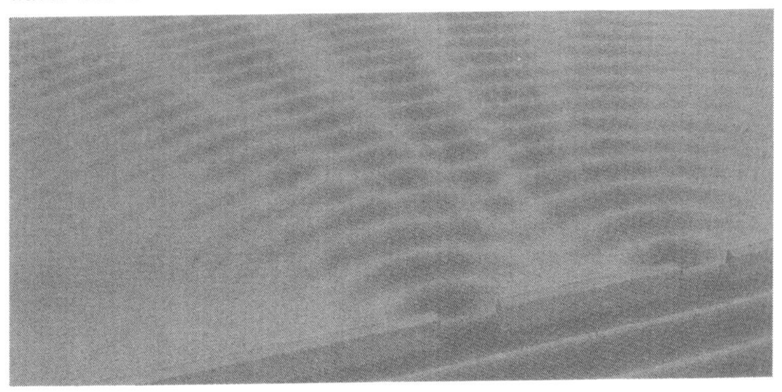

Figure 2.10 Interference pattern of water waves caused when a plane wave (bottom) passes through a pair of slits. Note the similarity with Figure 2.6.

Thomas Young's experimental setup [Figure 2.11(a)] allowed sunlight emerging from a small aperture to strike two very small slits. When light from the two slits fell upon a screen, dark stripes appeared, dividing the area into regularly spaced light and dark portions. [Figure 2.11(b)] shows the appearance of a typical double-slit pattern, made with the red light of a helium-neon laser.

Young recognized that for interference to occur, the light falling on the two slits must be coherent. The purpose of the first aperture was to ensure that the only light striking the two narrow slits came from the same source and was thus coherent,. Figure 2.12 shows the wave diagram drawn by Young to explain the origin of the light and dark bands of the interference pattern. Young reasoned, as we did above, that the maxima (bright bands) occurred when the path lengths from the two slits to the screen differed by integral multiples of the wavelength, and that the minima(dark bands) occurred when the paths differed by an odd number of half-wavelengths. The situation is exactly the same as that described in Figure 2.8.

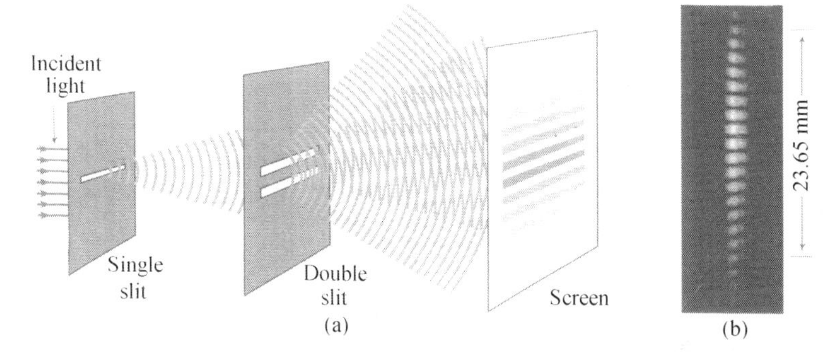

Figure 2.11 (a) The arrangement for Young's double-slit experiment. Sunlight passing through the first slit is coherent and falls on two slits close to each other. Light passing beyond the two slits produces an interference pattern on a screen. (b) The double-slit pattern of a small-diameter beam of light from a helium-neon laser. The center-to-center separation of the two slits was 0.113mm. The film was located 29.1cm from the slits. The reference marks seen along the edge of the film were spaced 23.65mm apart. From these data, try to compute the wavelength of the laser light.

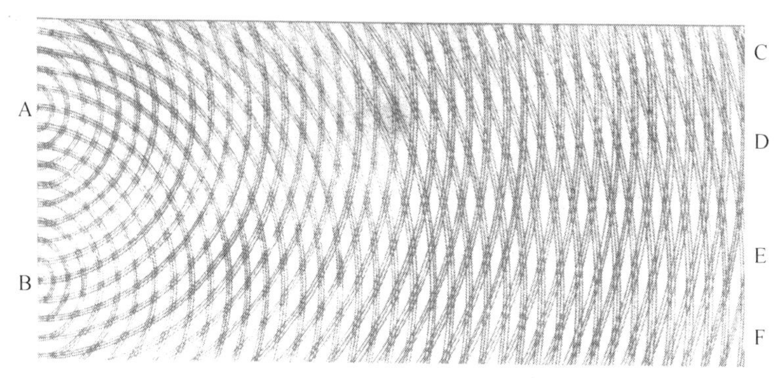

Figure 2.12 Diagram of the interference of light waves emerging from two points A and B as drawn by Thomas Young. Points C, D, E, and F indicate regions of destructive interference.

Once Young had worked out this explanation of the double-slit interference pattern, he realized that by measuring the spacing between the slits, the distance L to the screen, and the positions y of the maxima and minima, he could determine the wavelength of light. Young used an analysis similar to our use of Eq. (2.4) to find that the wavelengths of light range from about 400 nm in the extreme violet to about 700 nm in the extreme red. (Yong's actual values were given as one 60-thousandth of an inch to one 36-thousandth of an inch.)

Master the Concept

Interference of Waves

Question: Interference is a general property of all waves and occurs with light waves, Water waves, and sound waves. When listening to music on a stereo system with two speakers, why do you not experience places in the room where you hear no sound (because of destructive interference)?

Answer: You have already seen interference in standing sound waves in pipes, although the term *interference* was not used. The equations for interference due to two in-phase sources of the same frequency hold for sound waves. You can demonstrate this interference with two speakers held a half-meter apart and a common source (audio oscillator). As you move about in front of the speakers, you will notice strong variations in the sound intensity in accord with the double-slit equations.

The music that you hear from the stereo contains many wavelengths, not just one. Thus, a location for destructive interference of one wave length will not be destructive for other wavelengths. This observation is also true for points of constructive interference. Even though you can detect interference effects with two speakers and a single frequency, you do not normally observe them when listening to music. In addition, waves reflected from the walls and scattered by objects within the room contribute to what you hear.

2.4 Interference in Thin Films

We can observe the interference of light in ways other than the double-slit experiment. For example, interference also occurs when light is reflected from or transmitted by a thin film of transparent material. This interference is responsible for the colors of soap bubbles and oil slicks. As we will show, it is also the basis for the nonreflecting coatings commonly used on binoculars and photographic lenses.

(I) A Thin Film of Index *n* Surrounded by a Medium of Lower Index

Figure 2.13 shows a diagram of monochromatic light incident from above on a thin transparent dielectric film of thickness t and index of refraction n that is greater than that of the surrounding medium. At the first interface, the light is partly transmitted (refracted) and partly reflected. At the second interface, a portion of the transmitted light is reflected and follows the path indicated in the figure. Thus, an incident light beam produces two coherent reflecting beams, one from each surface of the thin film.

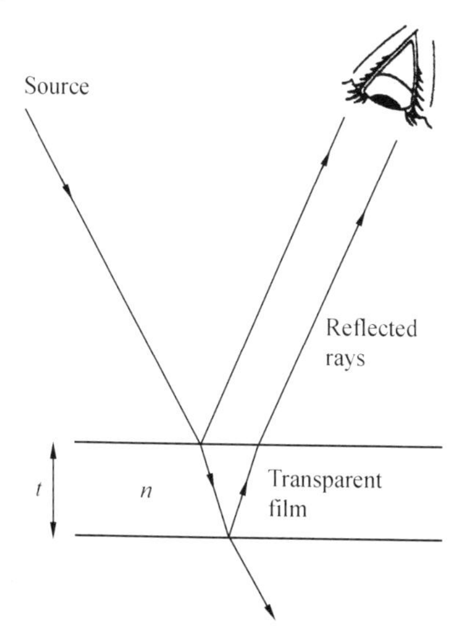

Figure 2.13 Monochromatic light incident on a thin transparent film is reflected from both the top and bottom surfaces.

If light strikes the film at nearly normal incidence, the two reflected beams may interfere constructively or destructively, depending on whether they are in phase or out of phase. The path length of the beam reflected from the second surface is $2t$ greater than that of the beam reflected from the first surface. If there are no other effects, the beams will interfere constructively when this path difference equals a whole number of wavelengths, for reasons similar to those discussed in connection with the double-slit experiment. However, there is another effect at work here. When light of wavelength λ is transmitted from a medium of lesser index of refraction to one of greater index, a phase change of 180°, corresponding to $\frac{1}{2}\lambda$, takes place upon reflection. The situation is analogous to the case of pulses reflecting back from the boundary between two ropes of different density (Figure 2.14). A pulse from a lighter rope to a heavier rope reverses its phase upon reflection, whereas a pulse from a heavier rope to a lighter one does not change phase on reflection. Similarly, light going from a medium of smaller index of refraction (air) to one of larger index (oil or water) changes phase upon reflection, while light going in the opposite direction does not change phase.

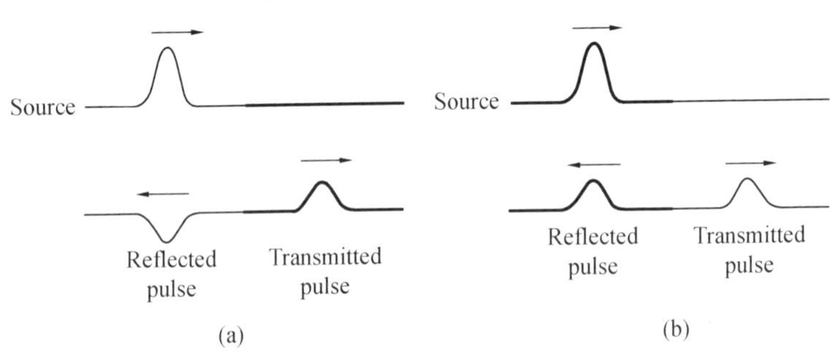

Figure 2.14 Reflection of a pulse wave at the boundary between ropes of different linear density. (a) Pulse incident in the less-dense rope changes phase upon reflection. (b) Pulse incident in the denser rope does not change phase.

Applying these ideas of phase changes to our thin film, we see that the light ray reflected from the top surface of the film undergoes a phase change of 180°. The transmitted ray does not experience any phase change during refraction at the upper surface of the film, nor does it undergo any phase change as it reflects from the bottom surface of the film. There is, however, a phase change associated with the path traveled through the film. The total effect taking place here, combining the phase change upon reflection with the path difference through the film, introduces an extra difference of $\lambda/2$ into our previous conditions for constructive and destructive interference.

Inside the film, where the index of refraction is n, the wavelength λ' is smaller than the incident wavelength λ by a factor of $1/n$. For constructive interference of light at normal incidence reflected from a thin film, the path length in the film must be an odd half-integer multiple of the wavelength λ'. Thus, the condition for constructive interference is

$$\left(m+\frac{1}{2}\right)\lambda = 2nt, \quad m = 0,1,2,3,\ldots \quad \text{(maxima)} \qquad (2.5a)$$

Similarly, minima in the reflected intensity occur for

$$m\lambda = 2nt, \quad m = 0,1,2,3,\ldots \quad \text{(minima)} \qquad (2.5b)$$

An interesting case occurs when the thickness of the film changes along its length, giving rise to alternating regions of constructive and destructive interference. For example, for a soap film suspended in a loop (Figure 2.15), the upper portion of the film is thinner than the lower portion. When the light striking the soap film of varying thickness is white, the various wavelengths of light constructively interfere at different places in the film, leading to a separation of the colors of white light. This thin-film phenomenon is also responsible for the rainbow colors visible on oil slicks. Some fish have scales with thin-film coatings that produce colors in the same way.

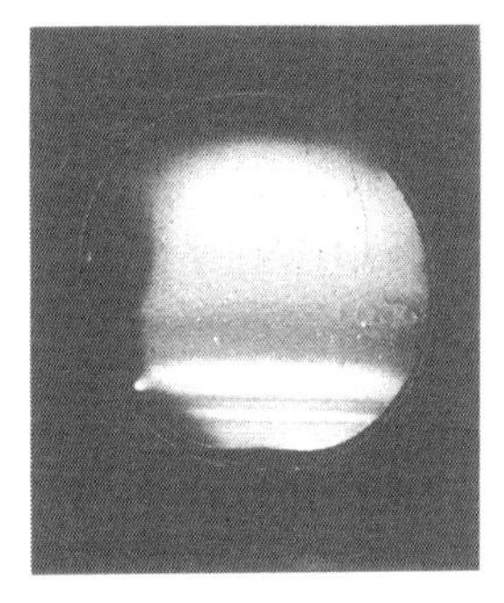

Figure 2.15 Interference pattern of a soap film suspended in a loop of wire.

(Ⅱ) A Thin Film of Index *n* Surrounded by a Medium of Higher Index

The same considerations of interference also hold for light reflected from a thin film of air (or thickness t) between two media of higher index, such as two glass plates (Figure 2.16). Light reflected from the first air-glass interface does not have a change in phase (higher refractive index to lower). However, light reflected from the second air-glass interface does undergo a phase change (lower refractive index to higher). The resulting equations are the same as Eqs. (2.5a, b), where n is the index of refraction of air.

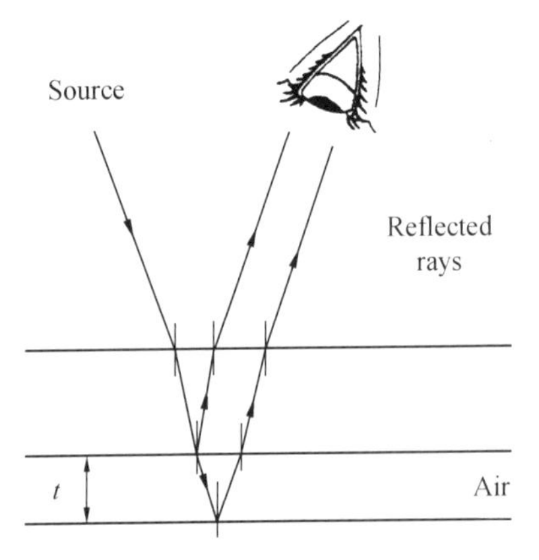

Figure 2.16 Reflection of light by a thin film of air between two glass plates.

We can see this effect when a piece of glass with a slightly curved bottom (such as a lens with a large radius of curvature) is placed on a flat glass plate and illuminated from overhead by a point source of light. When you look from above, you see a series of concentric rings. These rings, known as Newton's rings, arise from the interference between light reflected at the curved surface and light reflected from the underlying flat surface. Their appearance can be used to judge the flatness of the plate or the sphericity of the lens surface. If the source is not monochromatic but provides white light instead, the rings will be colored.

(Ⅲ) A Thin Film of Index *n* Surrounded by a Medium of Lower Index on One Side and a Medium of Higher Index on the Other Side

Interference of reflected light occurs in many thin-film situations involving different indices of refraction. For example, a thin film of oil ($n = 1.36$) on a glass plate ($n = 1.55$) involves two phase changes, one for reflection from each interface. The conditions for constructive and destructive interference are the same as Eqs. (2.5), only now the locations of maxima and minima are reversed. Rather than try to remember which combination of equations fits which thin-film situation, simply remember to count up the phase changes that occur for light reflected from each interface between media of different refractive indices. Then write down the conditions for constructive interference (path difference equals integer number of wavelengths) and destructive interference (path difference equals odd half-integer number of wavelengths) and add a factor of one halfwavelength for each difference in phase change between reflected rays.

One of the most important applications of interference in thin films occurs when we place a thin film in contact with a third medium of still-greater index of refraction (Figure 2.17). As we have already explained, a phase change occurs upon reflection at both surfaces, since both surfaces are low-index to high-index boundaries. As a result, the condition for constructive interference in the reflected beams is

$$m\lambda = 2n_2 t, \quad m = 1,2,3,\ldots \quad \text{(maxima)} \qquad (2.6a)$$

where n_2 is the index of refraction of the thin film.

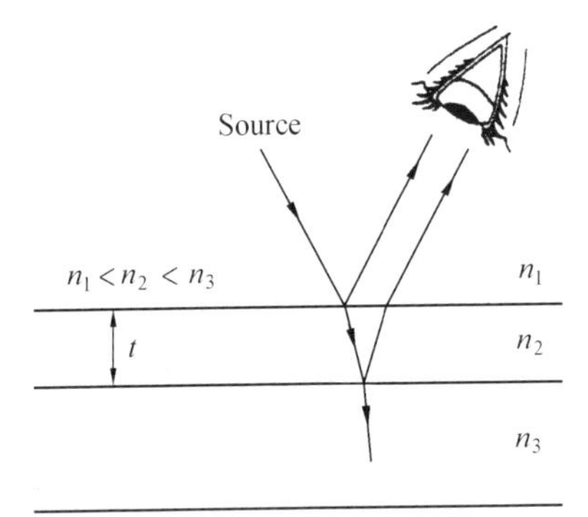

Figure 2.17 Reflection of light from a thin transparent coating on a material of higher index of refraction.

When monochromatic light strikes the surface at normal incidence, a phase difference occurs between the two reflected beams as a result of the difference in their paths. In this case, destructive interference occurs for

$$(m + \frac{1}{2})\lambda = 2n_2 t, \quad m = 1,2,3,\ldots \quad \text{(minima)} \qquad (2.6b)$$

When the thickness of the film is $\lambda/4n_2$, corresponding to $m = 0$, there is no reflection at that wavelength. So, if we coat a piece of glass, such as a lens, with a thin film that has the right thickness and an index of refraction intermediate between those of air and glass, we can minimize reflection from the glass. Such a film is called an **antireflection coating.** Because reflections are reduced, more light is transmitted through lenses that have antireflection coatings.

2.5 Diffraction by a Single Slit

We are all familiar with shadows thrown on a wall by an object, such as a hand, that blocks part of a beam of light, and we know that the shadow has approximately the same geometric shape as the object. Young's double-slit experiment shows that light does not travel past objects in simple straight lines, but instead spreads out in wavefronts that can interfere with each other. This spreading out of light passing through a small aperture or around a sharp edge is called **diffraction.** Diffractive spreading is exactly what we would expect from Huygens principle and, as we saw, was used by Young to illuminate his two slits.

Light always spreads out as it travels, but diffractive effects become noticeable only when light travels through a small enough aperture or past a sharp edge. Figure 2.18 shows a plane wave passing through (a) a wide slit and (b) a narrower slit. You can see that once the slit width approaches the dimension of the wavelength, the waves spread out as if from a point source. Figure 2.18 shows another example: the shadow of a ball bearing. Since the ball bearing has circular symmetry, light diffracted around its edge interferes constructively at the very center of the shadow. This bright spot in the center of a shadow was first predicted in 1819 by the French physicist Simeon Poisson as a necessary consequence of Augustin Fresnel's wave theory of light. Poisson believed the prediction — and the theory — ridiculous; but in fact, Francois Arago showed experimentally that the bright spot did exist, thus

Figure 2.18 Shadow of a ball bearing illuminated with laser light.

supporting rather than disproving the wave theory.

The pattern produced by a plane light wave illuminating a single slit depends on the size of the slit relative to the wavelength λ. When the slit is very wide compared to λ, the pattern closely resembles the geometrical shadow of the slit. As in the case of the water waves in Figure 2.18, the pattern spreads out as the width of the slit is narrowed (Figure 2.19).

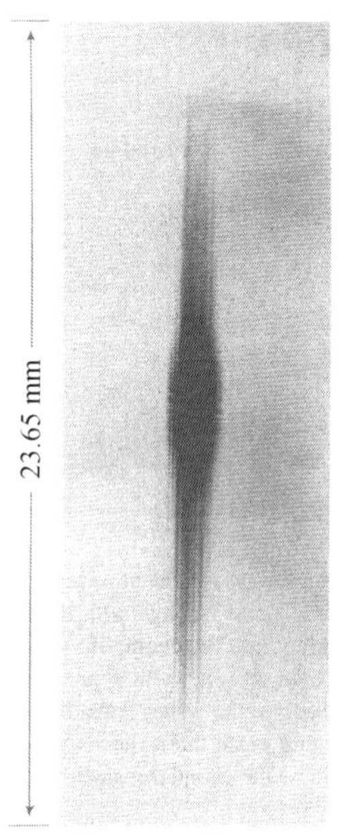

Figure 2.19 Diffraction pattern of a single horizontal slit 0.05mm wide. The photographic film was placed 29.1cm from the slit. The reference marks along the edge of the film were 23.65mm apart. Diffraction makes the vertical image taller than the slit width.

The explanation for the pattern of single-slit diffraction is similar to the explanation for the double-slit pattern. Figure 2.20 shows the geometry for the diffraction of light by a single slit of width b. When the paths from the slit to the screen for light beams passing across the upper and lower edges of the slit differ by an integral multiple of λ, a dark region appears on the screen. This happens because light from the center of the slit is out of phase with light from the edges. (That is, the path difference is one half-wavelength.) Thus, we have a condition for minima in the single-slit pattern,

$$\boxed{m\lambda = b \sin\theta, \quad m = 1, 2, 3, \ldots \quad \text{(single-slit minima).}} \quad (2.7)$$

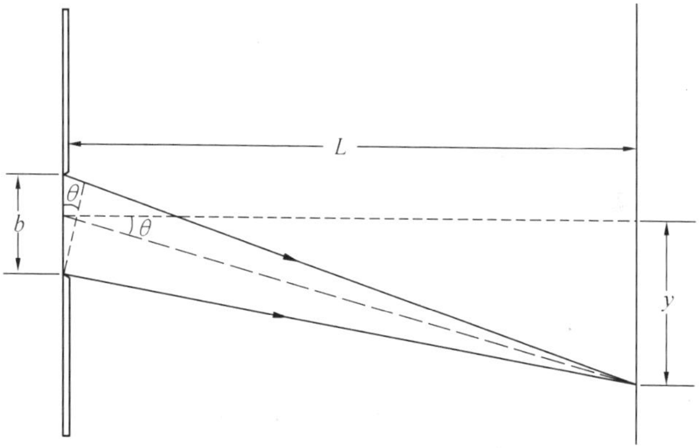

Figure 2.20 Geometry of the single-slit diffraction pattern.

Between each pair of minima is a maximum. The brightest maximum occurs right in the center, and the other maxima get successively dimmer (Figure 2.21). We can calculate the intensities and positions of these maxima, but the process requires techniques beyond the scope of this book.

2.6 Multiple-Slit Diffraction and Gratings

By now you may be wondering about the distinction between diffraction and interference. These phenomena are inseparably linked and are not really different. The pattern that results from the diffraction of light by a single slit could be thought of as the self-interference of light passing through the slit. In double-slit interference, the light is diffracted by each slit. Thus, the patterns describe both interference and diffraction at the same time. However, we generally use the term *diffraction* to describe the

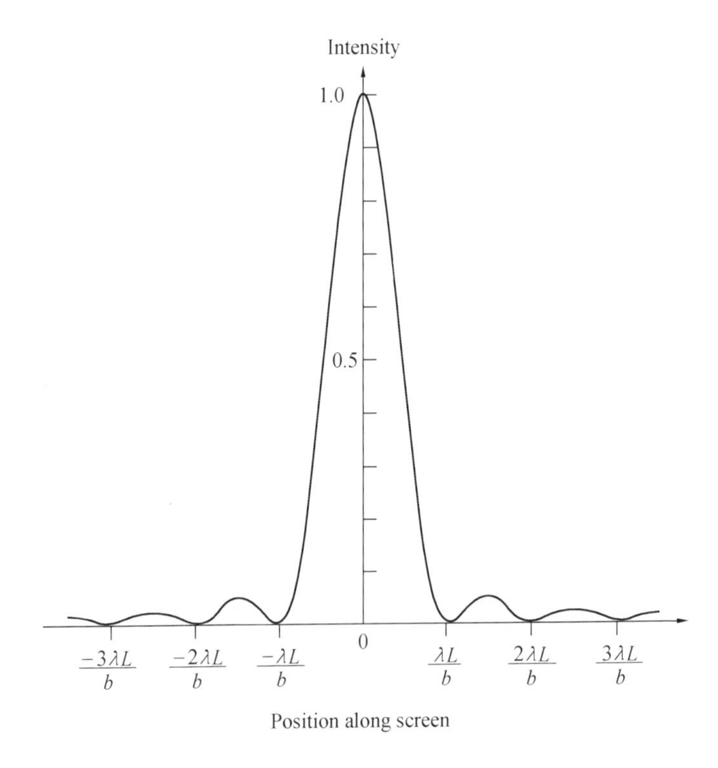

Figure 2.21 Light intensity along the screen for diffraction by a single slit of width b located a distance L from the screen. By far the brightest spot occurs at the center line of the slit.

effect of a wave encountering an obstacle, while we use the term *interference to describe the effects of combining multiple sources or parts of a wave*.

In our initial treatment of the double-slit experiment, we were not concerned with the effect of the finite slit width. However, the slit width does determine the overall extent of the double-slit pattern, in that we find the interference pattern only in the region where light is diffracted by each single slit. Figure 2.23 illustrates the interference pattern of two slits superposed on the intensity pattern of a single slit. For this case, we have chosen the slit width to be small compared with the spacing between the slits. Consequently, since the spreading of the patterns is inversely related to slit spacing and to slit width, the two first minima of the single-slit pattern are spread farther apart than are the double-slit minima. The first minimum of the single-slit pattern occurs at an angle θ given by

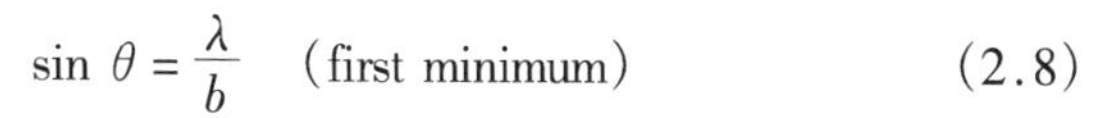

$$\sin\theta = \frac{\lambda}{b} \quad \text{(first minimum)} \tag{2.8}$$

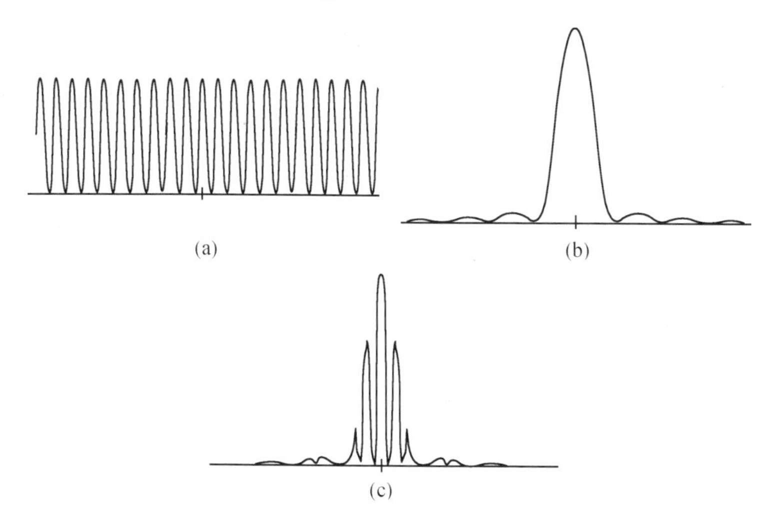

Figure 2.22 (a) Intensity pattern expected from double-slit interference without considering effects due to the width of the slits. (b) Single-slit diffraction intensity pattern. (c) Double-slit interference pattern showing the effects of the single-slit diffraction envelope. This figure is drawn for the slit separation to be three times the slit width.

The double-slit maxima are separated by an angle θ' given by Eq. (2.4a)

$$\sin\theta' = \frac{\lambda}{d}$$

Thus, illuminating two narrow slits separated by a distance d several times their width b produces a broadly spread diffraction pattern with closely spaced interference fringes.

What happens to the resulting pattern if we increase the number of slits? If we use three equally spaced slits instead of two, the pattern looks like [Figure 2.23 (a)]. The intensity peaks still occur at the same positions as in the double-slit pattern, since these are the same positions for which the light from all three slits arrives in phase. However, the peaks are narrower, and subsidiary maxima appear between the principal maxima [Figure 2.23(b)]. For two slits, no light appears at the angle where the two waves are exactly out of phase. But when three waves are present, one wave remains when the other two exactly cancel. Thus, the smaller maxima in the three-slit pattern occur at the positions of the minima in the two-slit pattern.

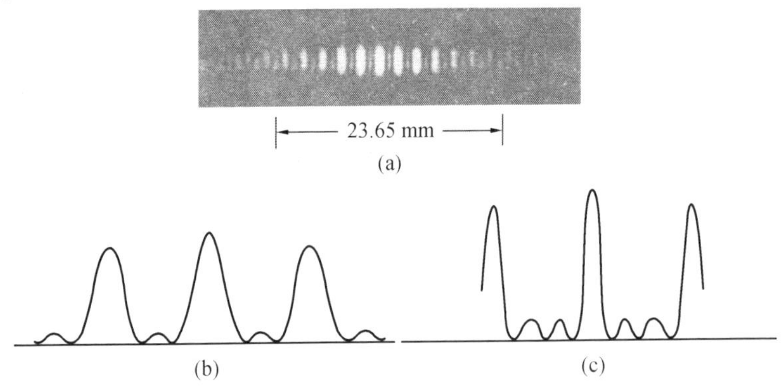

Figure 2.23 (a) Photo of a three-slit diffraction pattern. The conditions were the same as in [Figure 2.11(b)] Sketch of the intensity profile due to three slits. (c) Sketch of the intensity profile due to four slits.

When we add a fourth slit, another subsidiary peak occurs between the principal maxima [Figure 2.23(c)]. The ratio of the intensities of the smaller maxima to that of the principal maxima is even smaller in the four-slit pattern than for the three-slit pattern. In fact, as more and more slits are used, the general behavior in the resulting interference pattern is that the subsidiary peaks are suppressed and the principal peaks become narrower. If a large number of slits are used, the principal maxima can become quite sharp. An array of a large number of paralled, equally spaced slits is called a **diffraction grating.** We can analyze its resulting pattern of light intensity as due to the interference of many slits.

When white light from the sun passes through a diffraction grating, the light forms a spectrum, a rainbowlike pattern of colors. Spectra can also be produced by reflection from a grating. For example, you can see colors in the light reflected from the ridged surface of a black phonograph record. The grooves of the record act as a diffraction grating. Similar color patterns can be seen on a compact disk (CD), these being due to diffraction from the spiraling line of dimples pressed into the disk. Brilliant colors are also produced from embossed plastic film gratings with reflective backings (Figure 2.24).

Figure 2.24 One of the authors (RLC) holds a sphere covered with diffraction foil. White light striking the foil is spread into many colors.

A spectrum is produced because white light consists of a mixture of colors, each with its own wavelength, as we saw in Chapter 1. When white light passes through a grating, each wavelength is diffracted through its own characteristic angle. Thus, the light disperses into a spectrum of component colors. Diffraction gratings are routinely used for spreading light into a spectrum (Section 2.9).

The angle of diffraction of each wavelength of light passed through a grating is given by the diffraction equation,

$$\boxed{m\lambda = d \sin\theta, \quad m = 0, 1, 2, 3, \ldots \quad \text{(grating maximum)}} \quad (2.9)$$

where d is the spacing between the centers of the slits in the grating, and we have assumed the incident light to be normal to the plane of the grating. This equation is the same one we used to describe the maxima of double-slit interference, and it occurs here for exactly the same reasons.

In diffraction, larger wavelengths are deflected through larger angles. Thus, diffraction spectra are formed with violet closer to the normal and red farther away. When the adjacent path lengths differ by two or more wavelengths, corresponding to $m = 2$ or more, a secondary spectrum occurs at an even greater angle of deflection. In general, light falling on a grating is diffracted into several principal maxima or orders simultaneously, in addition to the zeroth-order beam that goes straight through.

2.7 Resolution and the Rayleigh Criterion

An important prediction of the wave theory of light is that the ability of an optical instrument to produce distinct images of objects that are very close together is limited. The *resolving power*, or *resolution* as it is often called, is a measure of this ability to produce sharp images. No matter how carefully a lens is designed and made, there is a limit to its resolution. This limit is determined by the diffraction pattern of the lens.

In Section 2.5, we saw that when parallel light passes through a single slit of width b, the light spreads out by diffraction. The resulting diffraction pattern consists of a bright central region bordered by alternating dark and bright bands. The angle θ between the center of the central maxima and the first minima (Figure 2.20) is obtained from Eq. (2.7)

$$\sin\theta = \frac{\lambda}{b},$$

where λ is the wavelength of the light. For small values of the angle θ, we can approximate $\sin\theta$ by θ so that the equation becomes

$$\theta = \frac{\lambda}{b}$$

As we will see, this angle determines the minimum separation that must exist between objects in order for them to be imaged individually.

Let's start by considering two point sources of light, P and Q, that subtend an angle a at the slit S (see Figure 2.25). The light passing through the slit is imaged on a screen by a converging lens. A ray drawn through the center of the lens from each source point determines the positions on the screen of their images P' and Q'. However, the images are not simply points, but are spread into a larger area as indicated. If the angle α is larger than the angular spread θ of each image due to diffraction, the two sources are imaged distinctly [Figure 2.26(a,b)]. We say the images are well resolved into two images[Figure 2.26(d)]. When $\alpha = \theta$, the variation in intensity across the pattern is discernible to the human eye [Figure 2.26(c)]. For this separation, $\alpha = \theta$, the first minimum of one diffraction image falls on the central maximum of the other image. Lord Rayleigh① chose this separation as the minimum angle that could be resolved. His choice is often referred to as **Rayleigh's criterion.**

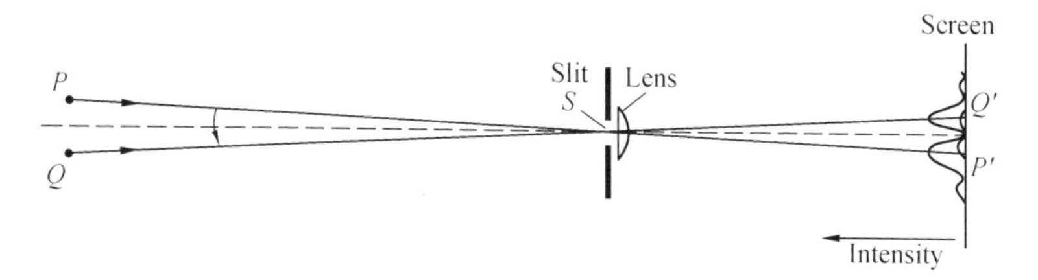

Figure 2.25 The light from two point sources P and Q is imaged on a screen after passing through a slit S. The images P' and Q' are not points, but are spread out by diffraction.

When the diffracting aperture is a circle rather than a slit, the angle of the first minimum in the diffraction pattern is somewhat larger than that given above. The angular position of the first minimum in the diffraction

---

① John William Strutt. Baron Rayleigh(1842 – 1919) is famous for his study of waves. He was awarded the 1904 Nobel Prize in physics for his discovery of argon gas.

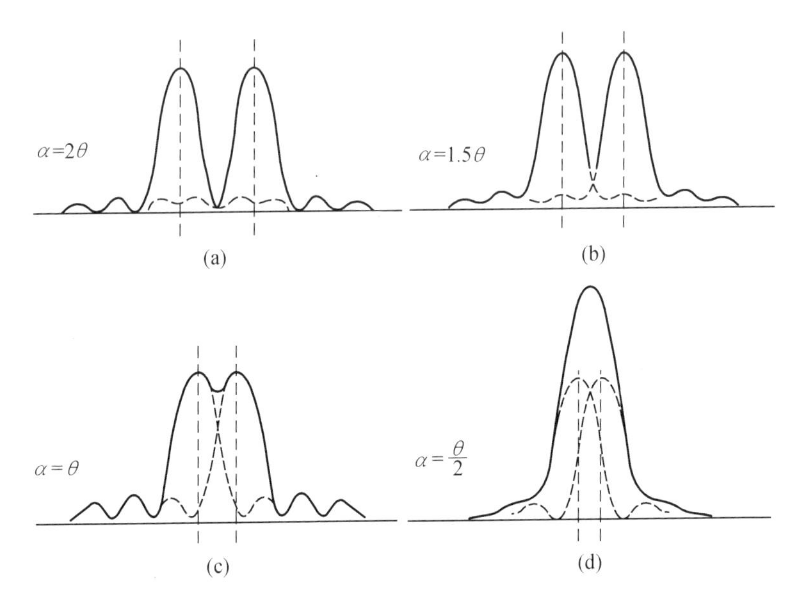

Figure 2.26 The diffracted image of two point sources at different angular separations α. Rayleigh's criterion for just resolving the two images is shown in (c).

pattern of a circular aperture of diameter D occurs at an angle

$$\boxed{\theta_m = 1.22\frac{\lambda}{D}} \tag{2.10}$$

By Rayleigh's criterion, θ_m is the limiting angular separation of two objects that are just resolved. This minimum separation is a theoretical lower limit. In practice, the angular separation of the two objects may have to be even greater in order to achieve distinct images. Since the limiting angle is inversely proportional to the diameter D, we can obtain improved resolution by making D larger or λ smaller.

The diameter of the objective lens (or mirror) sets a limit on the resolution of a telescope. Astronomical telescopes are made with large-diameter objectives for this reason — and for increased light-gathering ability — rather than to obtain high magnification. No matter what the magnification, a telescope's ability to resolve two nearby stars is limited by the diffraction pattern of its objective. In addition, atmospheric turbulence sets a practical limitation on the resolution of ground-based telescopes. For this reason. telescopes are now being placed in earth-orbiting satellites to put them above the image-degrading atmosphere.

The Hubble Space Telescope (Figure 2.27) was launched into earth orbit from the space shuttle in 1990. Traveling above the atmospheric turbulence and absorption, it can detect radiation from wavelengths of 105 mm in the altravioler through the visible to infrared wavelengths of 1 100 nm. The Hubble's 2.4 m diameter mirror was designed to have an angular resolution of about 0. 08 arcsecond at a wavelength of 632. 8 nm. However, shortly after launch it was discovered that although the mirror was polished to an exceptional degree of smoothness, its curvature corresponds to the wrong shape, thus creating spherical aberrations that seriously degrade the images. Because the mirror surface is so good, it was possible to sharpen up the direct images using computer image enhancement techniques. As a result, images have been obtained with resolution of about 0.05 arcsecond (Figure 2.28).

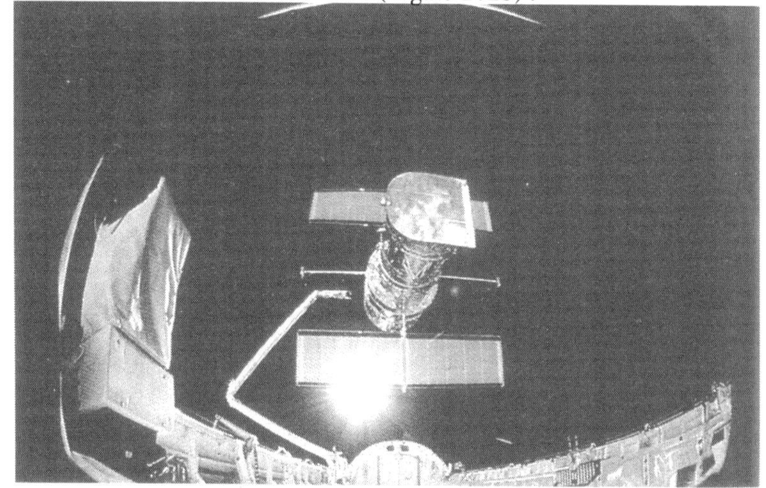

Figure 2. 27 The Hubble Space Telescope as it was launched from the Space Shuttle Discovery on April 25, 1990.

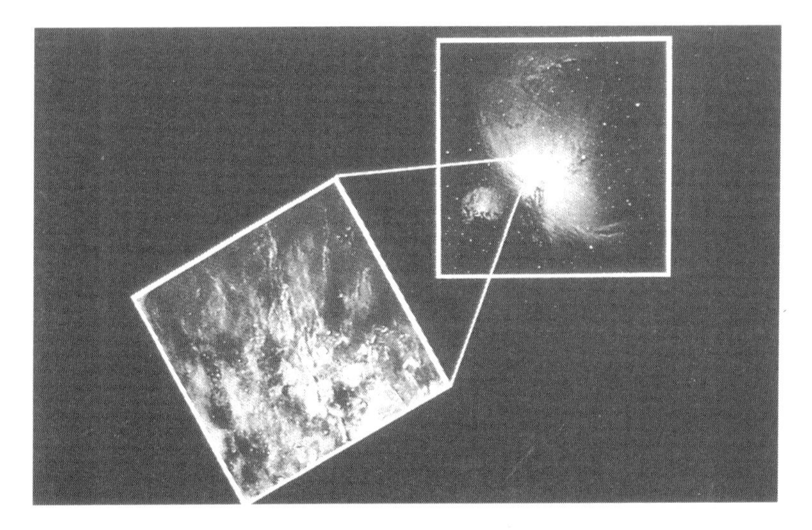

Figure 2.28 Orion Nebula as seen with the groundbased telescope at the Anglo Australian Southern Observatory (upper right) and with the improved resolution of the Hubble Space Telescope's Wide Field Planetary Camera(lower left).

Although the algorithms used on the Hubble pictures significantly sharpened the images, they did not allow the recovery of faint objects. In December 1993, corrective optics were installed to compensate for the spherical aberration in the primary mirror. The improvement in resolution was dramatic. Images taken with the Wide Field Planetary Camera before and after the corrections were made show the difference. Not only are the images sharpened but faint objects can be seen that were previously invisible .

2.8 Dispersion

White light can be spread into its component colors by a glass prism as well as by diffraction. Isaac Newton showed that one prism could separate white light into a spectrum and another prism turned the opposite way could combine that light back into a white beam again. He also showed that a small portion of the spectrum could not be spread into any other colors by passing the light through a second prism. However, the spreading of white light into colors is not just a trick you can do with a prism; it is also responsible for chromatic aberration (see Section 1.8), for the attractiveness of diamonds and cut glass, and even for the beauty of the rainbow.

The production of a continuous spectrum by a prism is due to the variation of wave velocity with wavelength that commonly occurs in transparent media. As stated in Chapter 1, the dependence of the index of refraction upon the wavelength (or color) of light is called *dispersion*. Water, glass, transparent plastics, and quartz are all dispersive materials. Generally, shorter wavelengths travel with slightly smaller wave velocities than do longer wavelengths. This means that the prism's index of refraction is not constant across the visible spectrum, but decreases continuously as the wavelengths increase from violet to red. Since the index of refraction is greater for violet than for red, violet light deflects through a greater angle than does red. (See Figure 1.22) This effect of a prism is just the opposite of the dispersive effect of a diffraction grating, which deflects red light more than violet. The graph in Figure 2.29 shows the typical dispersive behavior of decreasing refractive index with increasing wavelength for a transparent plastic.

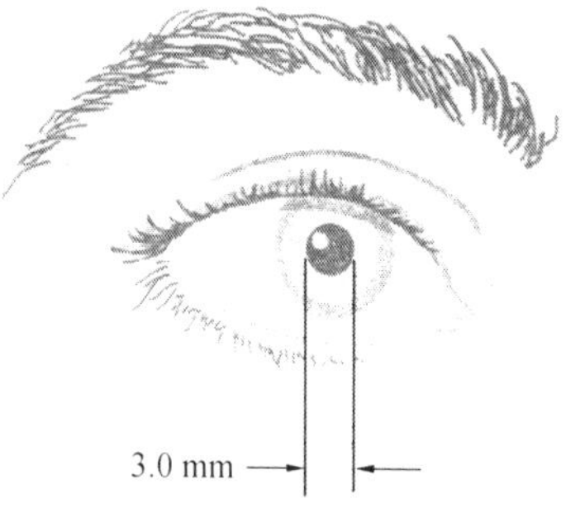

Figure 2.29 In bright light the pupil narrows to 3.0mm diameter.

Diamonds are highly valued as gems because they are rare and beautiful. Their beauty is in part due to their large index of refraction and their large dispersion. Gems are cut and polished to refract and reflect the light incident on them. The brilliant colors you see in an otherwise colorless diamond are due to the dispersion of the light as it is refracted and reflected from within. Slightly rotating a properly faceted diamond causes the colors to change and flash from the different facets as the angle of incidence of the light changes.

The formation of a rainbow by drops of water in the atmosphere is another example of dispersion. In a rainbow, water drops disperse reflected sunlight, revealing the colors of the spectrum across the sky.

To see a rainbow, you must have the sun to your back while you book toward a cloud of water drops. (These drops may be in a natural cloud or in a spray or mist that you could make with a garden hose.) For you to see the primary rainbow (there is also a secondary bow), the angle between the direction from the sun to you and your line of sight to the cloud of mist must be about 42°. Normally, we see rainbows when the sun is low in the sky and its light refracts and reflects from many drops of water in the air. The complete rainbow is a circle and can be seen as such from an airplane. Usually we see only the top portion of the circular arc, as the rest lies below the horizon, where it is not easily seen against the background and where the number of reflecting drops in our line of sight is small. For this same reason, we do not see rainbows when the sun is higher than 42° above the horizon.

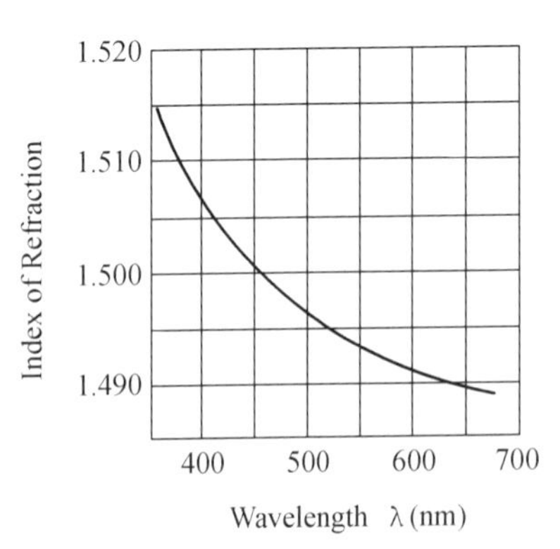

Figure 2.30 Index of refraction as a function of wavelength for polymethylmethacrylate, a transparent plastic used in lenses and other optical components.

Figure 2.31 illustrates the refraction of light by a spherical drop of water, showing those rays responsible for the rainbow. The effect of refraction and reflection bunches the rays so that more leave the drop near the angles indicated than in other directions. As a result, the reflected and refracted light is enhanced at the angles shown. Dispersion by the water drop causes light from the sun to spread out into a spectrum. If there are many drops (Figure 2.32), the eye receives the different colors from drops at different heights. The result is that the top of the rainbow appears red and the inner arc appears violet.

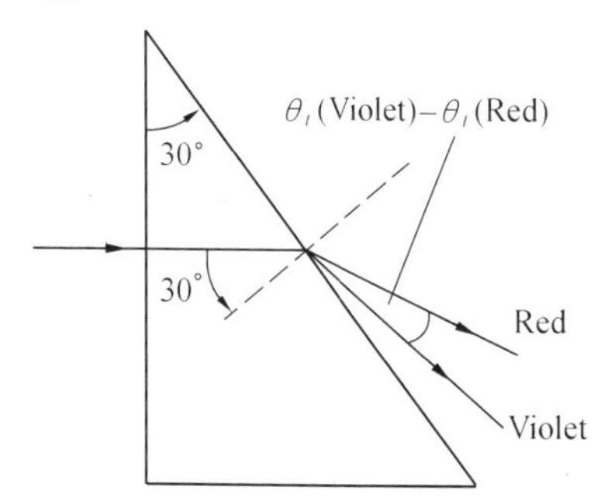

Figure 2.31 Light from a hydrogen source is normally incident on a 30° glass prism. The light is dispersed as it emerges.

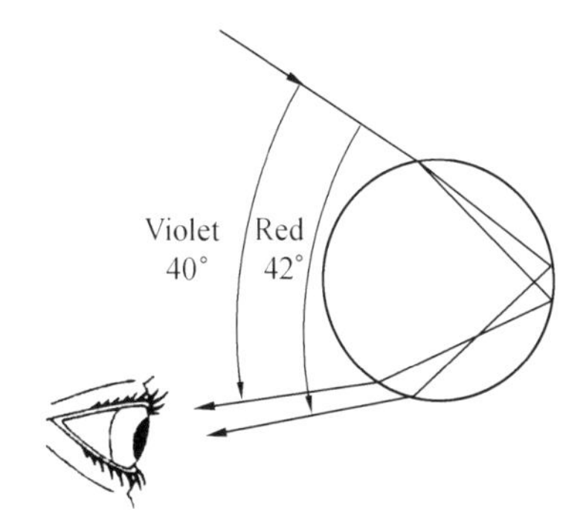

Figure 2.32 Refraction and dispersion of light by a spherical drop of water.

2.9 Spectroscopes and Spectra

If you look through a prism or a diffraction grating at a light source, you will see a band of overlapping images, each in a different color. If the source is an ordinary incandescent lamp with a frosted bulb, the images overlap so that most of the resulting broadened image appears white, with a violet border on one side and a red one on the other. If you place a narrow slit in front of the lamp so that you can see only a thin strip of the source through the spis, the band of images has very little overlap and you see bright spectral colors across the band.

A **spectrometer** is an optical instrument designed to enhance this effect and permit analysis of spectra. A *spectroscope* (Figure 2.33) is a spectrometer that permits direct observation of spectra. Light from the source we wish to analyze passes through a narrow slit and is focused into a parallel beam by a *collimating lens*. (A collimating lens is a converging lens that focuses a diverging light beam into a parallel beam.) After this beam passes through a prism, parallel beams emerge, each at a different angle according to its wavelength. A telescope focuses the parallel beams and allows an observer to see an image of the slit. Most spectroscopes also have a calibrated circular scale. This scale enables the observer to measure the angle of the emerging light for each image, which is then used to determine the wavelengths of the light in the spectrum. Each image of the slit is called a spectral line. For an incandescent source, the lines merge together into a continuum. Other sources, however, give characteristic lines. As we will see in Chapter 27, knowing what wavelengths are emitted by a light source gives us important information about the composition of the source.

Most modern spectrometers are made with a diffraction grating in place of a prism because the dispersion of a grating can be made much greater than that of a prism. Other instruments called *spectrographs*, which

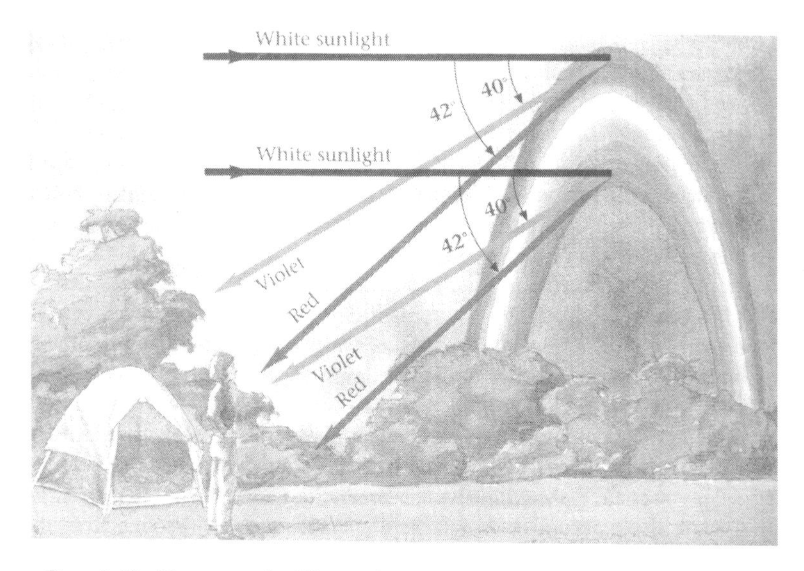

Figure 2.33 The eye sees the different colors of the rainbow in the light reflected from drops at different heights.

are very similar to spectroscopes, permit recording of the spectra photographically. *Spectrophotometers* are devices that convert the resultant light intensity to an electrical signal and then use a chart recorder or computer terminal to display the spectra graphically. Figure 2.34 shows the spectrum taken with a spectrophotometer of light transmitted by a green-colored glass filter. The spectrum is displayed as the percentage of light passed at each wavelength. As you can see, the filter transmits green light(540 nm) but does not pass red or blue light.

In recent years, the public has become more aware of the harm to the eyes caused by ultraviolet radiation. This concern has been heightened by the reduction of the protective layer of ultraviolet-absorbing ozone in the upper atmosphere. This loss of ozone, thought to be due to pollutants such as chlorofluorocarbons, means that the intensity of ultraviolet radiation at ground level is increased. For this reason, many people now have their sunglasses and their regular glasses treated with a protective coating that is

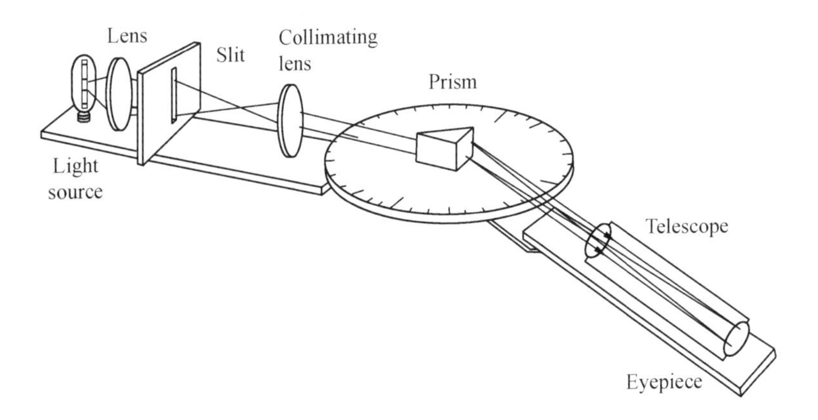

Figure 2.34 A prism spectroscope. The image of the slit formed by the red light is deviated less than the image formed by the violet light. The angle of deviation can be measured and used to calculate the wavelengths of the light.

transparent in the visible range but opaque to the ultraviolet range. The effectiveness of such coatings can be seen in Figure 2.35, which compares the transmission spectra of treated and untreated lenses.

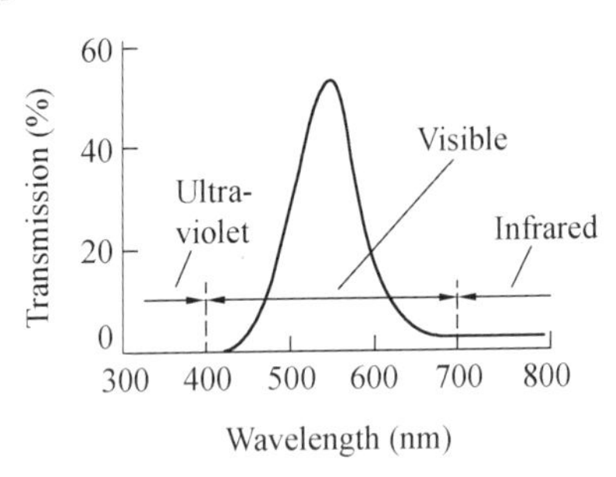

Figure 2.35 Intensity spectrum of the light transmitted by a green glass filter (Hoya # G – 533).

2.10 Polarization

So far in this chapter, we have discussed the optical properties of light that are due to interference and diffraction, showing how these characteristics support a wave theory of light. However, another important property of light, called polarization, is due not just to light being a wave, but to light being a transfer wave. Longitudinal waves, such as sound in air, do not exhibit polarization, but light, as well as other electromagnetic radiation, can be polarized.

The distinguishing characteristic of transverse waves is that their oscillating motion occurs in planes perpendicular to the drection in which the wave itself is moving. Electromagnetic waves, including visible light, are transverse waves, consisting of sinusoidally oscillating electric and magnetic fields perpendicular to each other.

Electromagnetic waves are usually generated by oscillating electric charges, such as the oscillating current in a radio or TV antenna or an electron in an atom. The direction of the oscillation determines the orientation of the wave's electricfield. In most common light sources, such as a candle flame, the sun, or an incandescent lamp, the many oscillating atoms are randomly oriented. Although the electric field of each wave produced by a particular atom lies in a single plane, the overall beam of light contains electric fields oscillating in all planes[Figure 2.36(a)]. However, if the oscillating charges are confined to move in only one plane, as is the oscillating current in an antenna, then the entire electric field of the resulting beam oscillates in only that direction[Figure 2.36(b)]. Such a wave is said to be polarized, and the orientation of the electric field vector is taken as the direction of **polarization**. The complete wave has a sinusoidally oscillating magnetic field coupled to the electric field, but traditionally we refer to lnly the electric field to indicate the direction of polarization.

A useful mechanical analogy for polarized light is the motion of transverse waves along a rope. If you shake a rope up and down, the wave is vertically plane-polarized. Such a wave can travel through a vertical slit [Figure 2.37(a)], but not through a horizontal slit. Similarly, if you shake a rope side to side, the wave is confined to a horizontal plane; it can

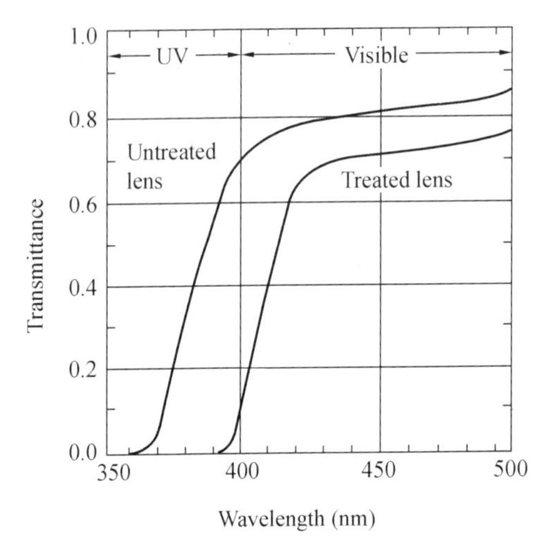

Figure 2.36 Transmission spectra of eyeglass lenses showing the effect of treatment for blocking ultraviolet(UV) radiation.

pass through a horizontal slit, but not a vertical one[Figure 2.37(b)].

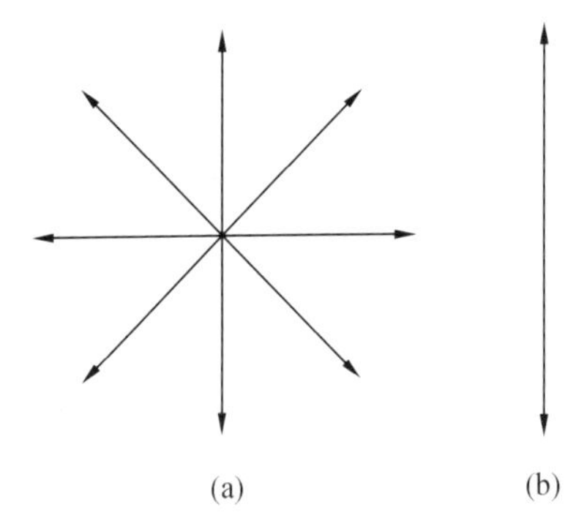

Figure 2.37 (a) Light from an ordinary source contains a mixture of waves oscillating in different directions. (b) A planepolarized light beam contains only one direction of oscillation.

It is possible to produce polarized light by filtering ordinary light through a polarizer, which is a material that transmits only those waves that oscillate in a single plane. Several materials have this property, including the naturally occurring mineral tourmaline. When an ordinary beam of light strikes a tourmaline crystal, part of the light is absorbed and part is transmitted (Figure 2. 38a). If a second tourmaline crystal is placed behind the first and aligned in the same orientation, light is transmitted. If the second crystal is rotated by 90° about the axis of the light beam, no light passes through [Figure 2.38(b)]. We can explain this observation in terms of polarization. The intial light beam is unpolarized. We think of it as containing a mixture of random polarizations. When this beam strikes the first crystal, the tourmaline passes only light polarized along the polarization axis of the crystal. The light emerging from the first crystal is completely polarized along the direction shown. When this light falls on the second tourmaline crystal, all the light is absorbed, because the second crystal has been turned so that its polarization axis is 90° away from that of the first crystal.

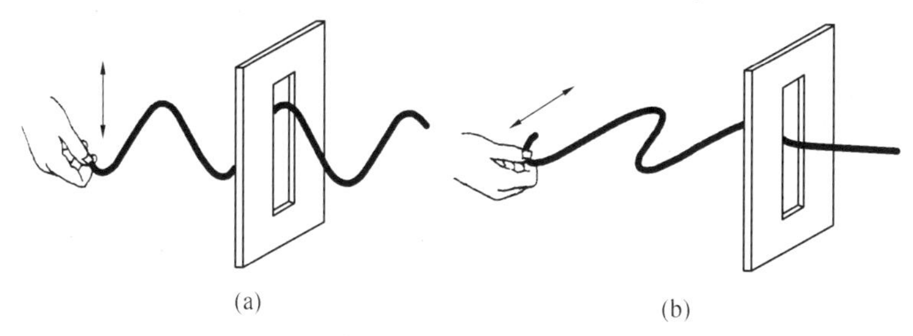

Figure 2.38 (a) Shaking a rope up and down produces a vertically polarized wave that passes through a vertical slit. (b) Shaking a rope from side to side produces a horizontally polarized wave that will not pass through a vertical slit.

Crystal polarizers like tourmaline are not used much now because of the development of inexpensive plastic polarizers called *Polaroids*. Synthetic sheet polarizers were invented in 1928 by Edwin H. Land and improved by him ten years later. When viewed individually, these transparent sheets have a neutral grey appearance. However, when two polarizers overlap with their polarization directions at right angles, the region of overlap looks black (Figure 2.39).

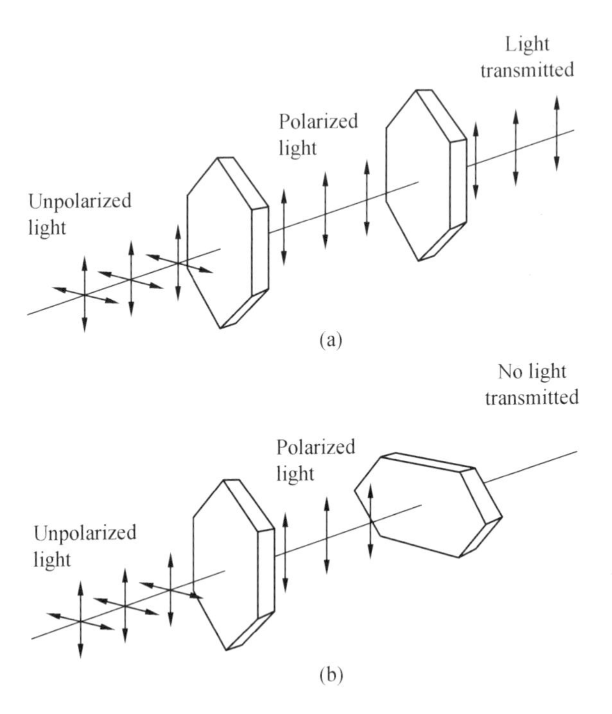

Figure 2.39 Polarization of light by tourmaline crystals. (a) When the polarization axes of the crystals are parallel, planepolarized light passes through. (b) When the axes are perpendicular, no light passes through.

If the angle between the two Polaroids is made less than 90°, the area of overlap is no longer opaque. It gradually gets more and more transparent as the angle decreases to 0°. The intensity of light that passes through two polarizers that are aligned with an angle θ between their polarization directions is

$$\boxed{I = I_{\mathrm{m}} = \cos^2\theta} \tag{2.11}$$

where I_{m} is the maximum amount of light transmitted when the two polarizers are aligned along the same direction of polarization. Equation (2.11) is known as **Malus's law** after its discoverer, who first observed this effect experimentally in 1809.

An ideal polarizer passes 100% of the incident light that is plane-polarized along its polarization direction and 0% of the light that is polarized at 90° to its polarization direction. Since unpolarized light is a mix of all possible polarizations, an ideal polarizer transmits only 50% of the incident unpolarized light intensity.

Malus also discovered that light becomes polarized upon reflection from glass windows and surfaces of water. In 1814, some six years after this discovery, Dayvid Brewster① (1781 ~ 1868) found that at a particular angle of incidence, now called the Brewster angle, polarization of the reflected light is complete, with polarization perpendicular to the plane of incidence (Figure 2.40). At other angles, the reflected light is only partially polarized, with more light polarized in this direction than in any other.

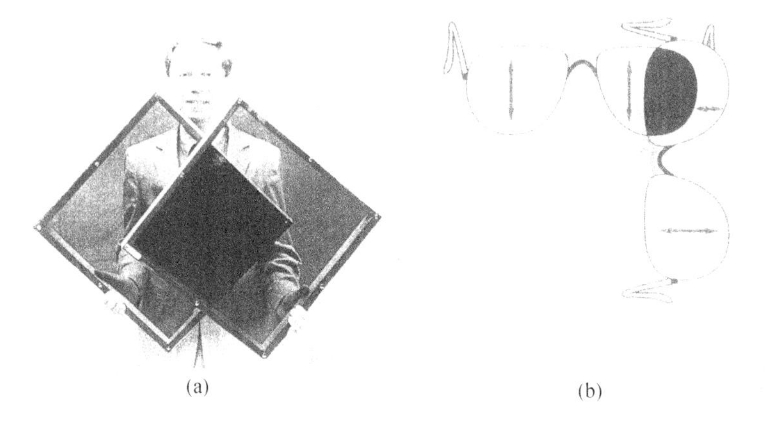

(a) (b)

Figure 2.40 (a) One of the authors (ERJ) holding two large polarizing filters with their polarization directions at a right angle. (b) You can demonstrate the same effect yourself with two pairs of polarizing sunglasses.

---

① windows and surfaces of Brewster is also known for his invention of the kaleidoscope and for his improvements on the stereoscope invented by Sir Charles Wheatstone.

Brewster found that maximum polarization occurs when the reflected ray and the refracted ray are at right angles to each other. By combining Brewster's observation with Snell's law, we can find a relation between a material's index of refraction and its Brewster angle. Maximum polarization occurs for $\theta_r + \theta_t = 90°$, where θ_r is the angle of reflection and θ_t is the angle of transmission (refraction). From Snell's law we can show that

$$n_i \sin \theta_i = n_t \sin \theta_t = n_t \sin(90° - \theta_r) = n_t \cos \theta_r$$

This equation reduces to

$$n_i \sin \theta_i = n_t \cos \theta_i$$

When we use the fact that $\theta_i = \theta_r$. At this condition of maximum polarization, we can denote the angle of incidence as the Brewster angle θ_B and rewrite this equation as

$$\boxed{\tan \theta_B = \frac{n_t}{n_i}} \tag{2.12}$$

This last equation is known as **Brewster's law.** Note that the Brewster angle depends on the index of refraction of the materials on both sides of the reflecting dielectric surface.

Sunglasses made from Polaroid sheet are widely used because of their ability to block the glare of specularly reflected sunlight from water surfaces or from surfaces of other smooth reflecting objects such as automobile windows. The glasses are made with a vertical axis of polarization, because, as can be seen from Figure 2.42(a), glare resulting from the sun is usually horizontally polarized. You can see this by rotating such glasses by 90°. In this new position, the glare light passes through because the transmission axis of the glasses is aligned with the polarzation of the light. Normally, of course, the polarization direction of the glasses is crossed with that of the glare light so that the glare is extinguished.

2.11 Scattering

We conclude this chapter by examining one more optical phenomenon that is best explained by a wave theory of light. This time, we answer an interesting question about our physical world — why is the sky blue?

In a perfectly transparent, homogeneous, and structureless medium, a beam of parallel light proceeds without broadening or lateral spreading. A beam of light in a vacuum behaves this way. But all real materials have some molecular or crystal structure and are not perfectly transparent to visible light. Furthermore, most materials contain impurities; even air has water vapor, dust, smoke, and other minute particles suspended in it. When a beam of light, such as a search light beam, passes through fog or dust-laden air, you can see the beam itself from the side. We say that the light has been scattered; molecules of air and its impurities have absorbed some light from the beam and then reradiate it in other directions, a process we call **scattering**. Most of the light travels in the initial beam direction, but some is scattered to the sides and even backwards.

The details of the scattering process depend on the relative size of the scattering particles compared with the wavelength of light. When the particles that causse the scattering are smaller than the wavelength of the incident light, we have what is called **Rayleigh scattering**. This type of scattering was first explained by Lord Rayleigh, who showed that the scattering was proportional to the fourth power of the frequency of the light.

We can see what this means by examining a simplified version of an experiment for viewing Rayleigh scattering (Figure 2.41). Light from a bright source of white light passes through a glass tank and strikes a screen. The collimating lens makes the beam parallel in the tank; the second lens focuses an image of the end of the tank on the screen. The

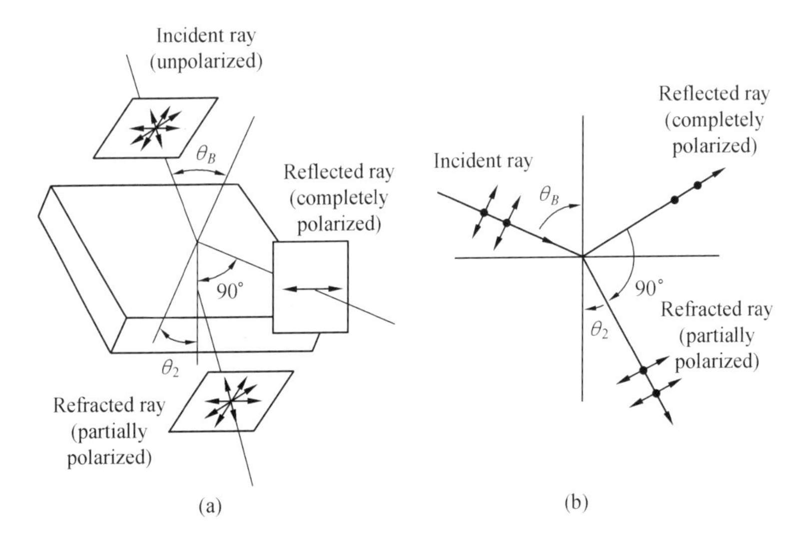

Figure 2.41 (a) Polarization of light by reflection at the Brewster angle. (b) The reflected light is completely polarized in a direction perpendicular to the plane of incidence. The refracted light. which is partially polarized parallel to the plane of incidence, makes an angle of 90° with the reflected ray.

trank is filled with water, into which we have mixed fine microscopic particles. (Such a mixture can be made by adding a few drops of milk to a tank of water.) From the side, the path of the beam is visible and has a bluish tint. Blue light has a higher frequency than red light and, according to Rayleigh's f^4 law, is scattered more. In fact, it is scattered in all directions, as can be observed by walking around the tank. Because the blue end of the spectrum is scattered out of the beam, less blue and relatively more red appears as the beam proceeds. The light emerging from the right-hand end of the tank appears somewhat reddish-orange. This is not due to any change in the scattering, but happens because this end of the tank is illuminated by light that is now deficient in blue.

Light from the sun incident on the earth's atmosphere is scattered most strongly in the violet and blue regions of thespectrum. In fact, the sky would appear violet except that our eyes are not very sensitive to violet. The combination of Rayleigh scattering and the sensitivity of our eyes explains the blue of the daytime sky. When the sun is on the horizon, as at sunrise or sunset, the optical path through the atmosphere is much longer than when the sun is directly overhead, so more blue is scattered out of the direct beam. Thus, both the rising and the setting sun appear red (Figure 2.42). These effects are due to Rayleigh scattering of the molecules of the air itself, which have dimensions of the order of a fraction of a nanometer, and not to suspended dust or vapor in the atmosphere.

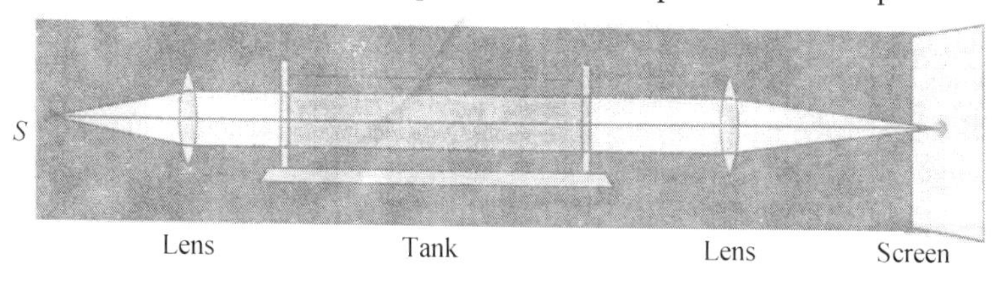

Figure 2.42 An experiment to demonstrate Rayleigh scattering. As the blue light is scattered out of the beam, the transmitted light appears red-orange.

Smoke rising from the end of a lighted cigarette or from around the edges of a pile of burning leaves appears blue because of Rayleigh scattering. On the other hand, exhaled smoke appears white because it contains large water droplets. When the scattering particles are many times larger than the wavelength of light the light is reflected and refracted as in geometrical optics, not absorbed and reradiated, so the scattered light is white. This type of scattering is called Tyndall scattering and is responsible for the appearance of fog, clouds, powders, and ground glass.

Suppose we look through a polarizing filter at the light scattered out the sides of the tank in Figure 2.42, looking at right angles to the incident beam. We find that the intensity of the light grows and diminishes as we rotate the filter. This demonstrates that the scattered light is polarized, an effect that was also predicted by Rayleigh. Because light is a transverse wave, the plane of polarization is perpendicular to the plane containing the incident light beam and the perpendicular line of singht of the viewer (Figure 2.44). When the light is viewed at other angles, the polarzation is not complete.

Figure 2.43 The setting sun looks red because the atmosphere scatters the blue light from the rays that reach our eyes.

If a polarizer is placed in the path of the light entering the tank in Figure 2. 42, the scattered light is a maximum in the direction perpendicular to the direction of polarization and zero along the direction of polarization. Thus, when one is looking horizontally into the tank, the scattered light will appear brightest if the incident beam is polarized vertically and weakest if the beam is polarized horizontally.

If you look at the sky through a polarizing filter in a direction perpendicular to the direction of the sun's rays, you will see the sky change from bright to dark as you rotate the filter. The polarization of the light is not 100%, but may be as much as 70% to 80% when the air is clear. Some insects, especially bees, are able to find their way over long distances by using the direction of polarization as a sort of compass.

Some materials cause the direction of polarization to rotate as the light passes through, an effect known as **optical activity.** As the light moves through the medium, the direction of polarization traces out a helical (screwlike) path. This effect is due to the geometric shape of the molecules in the material. In concentrated solutions of ordinary sugar (dextrose) the pitch of the helix changes with wavelength (color) because of dispersion. Consequently, passing polarized white light through a column of corn syrup creates a multicolored "barber pole" (Figure 2.45).

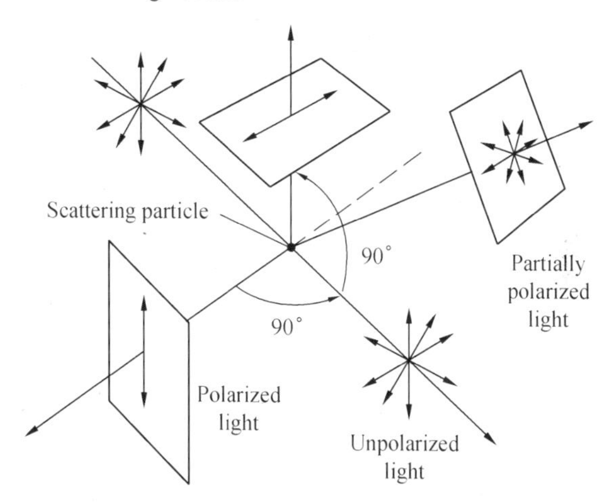

Figure 2.44 Light scattered at right angles to the intial beam direction is polarized

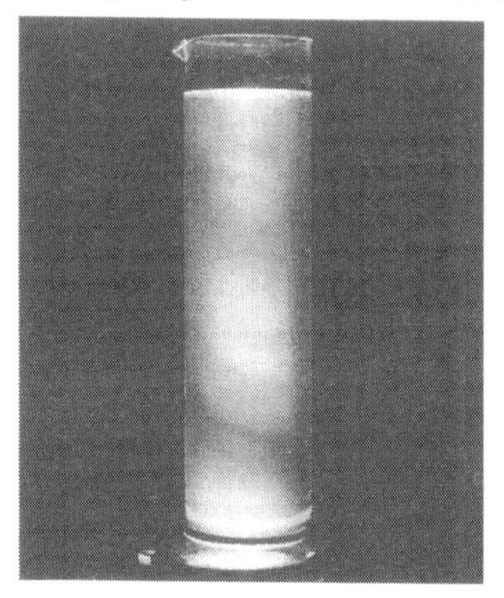

Figure 2.45 A multicolored "barber pole" produced by polarized white light entering a column of corn syrup from below.

Summary

Useful Concepts

● The laws of geometrical optics, such as the law of reflection and Snell's law, can be derived from a wave interpretation of light using Huygens' principle. This principle considers each point on a wavefront to be a point source for expanding wavelets. The resulting wave is the envelope of these wavelets and, in turn, can be considered as a source for finding a still later position of the wave.

● For Young's double-slit experiment, the angular locations of the maxima and minima are given by

$$m\lambda = d \sin \theta,\ m = 0,1,2,3,\ldots \quad \text{(double-slit maxima)},$$

$$\left(m + \frac{1}{2}\right)\lambda = d \sin \theta,\ m = 0,1,2,3,\ldots \quad \text{(double-slit minima)},$$

where d is the separation between the slits.

● When light is normally incident in air on a thin film of index of refraction n, the interference between the front and back surface reflections gives maxima and minima when

$$\left(m + \frac{1}{2}\right)\lambda = 2nt,\ m = 0,1,2,3,\ldots \quad \text{(reflection maxima)},$$

$$m\lambda = 2nt,\ m = 0,1,2,3,\ldots \quad \text{(reflection minima)},$$

where t is the thickness of the film and λ is the wavelength of the light in free space.

● Antireflection coatings can be made by coating a surface with a thin film of thickness t and index of refraction n that is intermediate between the index of the surface and the index of the surrounding medium (usually air) when

$$\left(m + \frac{1}{2}\right)\lambda = 2nt,\ m = 0,1,2,3,\ldots \quad \text{(reflection minima)},$$

● The spreading out of light passing through a small aperture or around a sharp edge is called diffraction. For diffraction by a single slit of width b, the angular locations of the minima are given by

$$m\lambda = b \sin \theta,\ m = 0,1,2,3,\ldots \quad \text{(single-slit minima)}.$$

● For a diffraction grating, the angle of diffraction of each wavelength is given by

$$m\lambda = d \sin \theta,\ m = 0,1,2,3,\ldots \quad \text{(grating maxima)},$$

where d is the spacing between slits.

- For a circular aperture of diameter D, the minimum angle of resolution θ_m given by the Rayleigh criterion is

$$\theta_m = 1.22\,\frac{\lambda}{D}$$

- A polarized light beam contains waves whose electric field vectors are all oriented along the same direction.

- The intensity of light passing through two polarizers is given by Malus's law:

$$I = I_m \cos^2 \theta$$

- The angle of maximum polarization by reflection θ_B is given by Brewster's law:

$$\tan\,\theta_B = \frac{n_t}{n_i}$$

Important Terms

You should be able to write the definition or meaning of each of the following:

Huygens' principle	Rayleigh's criterion
wavefront (or wave surface)	spectrometer
constructive interference	polarization
destructive interference	Malus's law
coherence	Brewster's law
antireflection coating	scattering
diffraction	Rayleigh scattering
diffraction grating	optical activity

New Words and Expressions

wave optics *n*. 波动光学

geometrical optics *n*. 几何光学

sunglasses *n*. 太阳镜;墨镜

electromagnetism *n*. 电磁,电磁学

quantum mechanics *n*. [物]量子力学

cosmological *adj*. 宇宙哲学的,宇宙论的

Huygens' Principle *n*. 惠更斯原理

ripple *n*. 波纹 *v*. 起波纹

wavefront	*n*. 波阵面
wavelet	*n*. 小浪,微波
magnifying glass	*n*. 放大镜,放大器
eyepiece	*n*. 目镜
subtend	*v*. 双向,包在叶腋内
triangle	*n*. [数]三角形
vertex	*n*. 顶点,最高点
angular magnification	*n*. 角度放大率
Snell's law	*n*. 斯涅尔定律,折射定律
Product	*n*. 乘积
double-slit	*n*. 双缝干涉
algebraic	*adj*. 代数的,关于代数学的
vibrate	*v*. (使)振动,(使)摇摆
bob	*n*. 振动,短发,振子锤
wedge-shaped	*adj*. 楔形的,V形的
constructive interference	*n*. (全息)结构干涉,相长干涉
destructive interference	*n*. 相消(性)干扰,破坏性干扰
amplitude	*n*. 振幅,物理学名词
helium-nuon laser	*n*. 氦氖激光器
odd mumber	*n*. 奇数
thin film	*n*. 薄膜
transparent film	*n*. 透射膜
monochromatic	*adj*. [物]单色的,单频的
refractive index	*n*. 折射率
curved bottom chord	*n*. 曲线形下弦杆,弧形下弦杆
incidence	*n*. [物理]入射
antireflection coating	*n*. 减发射敷层;抗反射膜
diffraction	*n*. 衍射
circular symmetry	*n*. 圆对称
dimmer	*n*. 调光器

multiple-slit diffracting	*n*. 多缝衍射
subsidiary	*adj*. 辅助的,补充的
diffraction grating	*n*. 衍射光栅
groove	*n*. (唱片等的)凹槽,惯例 *v*.开槽于
deflection	*n*. 偏斜,偏转,偏差
zeroth order	*n*. 零阶
resolving power	*n*. (光学仪器等)分辨能力
Rayleigh criterion	*n*. 瑞利数据;瑞利准则
diffraction aperture	*n*. 衍射孔径
agon	*n*. [化]氩
angular separation	*n*. 角间距
earth-orbit	*n*. 绕地轨道
arc-second	*n*. 角秒
aberration	*n*. 失常
faint object	*n*. 暗天体
ink jet printer	*n*. 喷墨打印机
light-year	*n*. 光年
electrostatic	*adj*. 静电的,静电学的
electron microscope	*n*. 电子显微镜
dispersion	*n*. 散射
chromatic aberration	*n*. 色差,像差
rainbow	*n*. 彩虹
gem	*n*. 宝石,珍宝,精华,被喜爱的人,美玉
transparent plastic	*n*. 透明塑料
violet	*n*. 紫罗兰 *adj*. 紫罗兰色的
spectroscope	*n*. [物]分光镜
spectra	*n*. 范围,光谱
incandecent lamp	*n*. 白炽灯,白热灯
frosted bulb	*n*. 磨砂灯泡
collimating lens	*n*. 准直透镜

spectrograph	*n*. [物]光谱摄制仪,声谱仪
spectrophotometer	*n*. 分光光度计
ozone	*n*. 新鲜的空气,[化]臭氧
chlorofluorocarbon	*n*. 氯氟
polarization	*n*. [物]偏振(现象),极化(作用)
perpendicular	*adj*. 垂直的,正交的 *n*. 垂线
sinusoidal oscillation	*n*. 正弦振荡
transverse wave	*n*. 横波
tourmaline	*n*. [矿]电气石
srystal	*adj*. 结晶状的 *n*. 水晶,结晶,晶体
polaroid	*n*. 偏振片;人造偏光片
Malus' law	*n*. 马吕斯定律
plane-polarized	*adj*. 平面偏振的
Brewster angle	*n*. 布儒斯特角;极化角;偏振角
kaleidoscope	*n*. 万花筒
Rayleigh scattering	*n*. 散射
dust-laden	*adj*. 多尘的,含尘的
microscopic	*adj*. 用显微镜可见的,精微的
geometrical optics	*n*. 几何光学
optical activity	*n*. [物]旋光性[度]
helical	*adj*. 螺旋状的

NOTES

1. The wavelength inside a material of index of refraction n is smaller than the free-space wavelength because, as we saw in Chapter 15, the velocity of a wave is the product of its wavelength and frequency, $v = f\lambda$. 在介质折射率为 n 的材料内的波长比在自由空间的波长小,这是因为,如第 5 章中讨论的一样,光波的速度是波长和频率的乘积。

2. The emerging waves produce an interference pattern exactly like the one described for two sources. 正在形成的波产生的干涉图与两个

光源的干涉图非常相似。

3. Young recognized that for interference to occur, the light falling on the two slits must be coherent. 杨认识到想要干涉能够产生，那么照射到两个狭缝上的光波必须相同。

4. We can see this effect when a piece of glass with a slightly curved bottom(such as a lens with a large radius of curvature) is placed on a flat glass plate and illuminated from overhead by a point source of light. 当一块具有底面稍微弯曲的玻璃(例如一块具有大曲率半径的玻璃)被放置在一块平面玻璃上，同时在高处用一个点光源照明，我们就能够看到这种效应。

5. One of the most important applications of interference in thin films occurs when we place a thin film in contact with a third medium of still-greater index of refraction. 薄膜干涉最重要的应用之一发生在薄膜与折射率更大的第三介质的接触上。

6. We are familiar with shadows thrown on a wall by an object, such as a hand, that blocks part of a beam of light, and we know that the shadow has approximately the same geometric shape as the objects. 我们对投射到墙上的物体阴影比较熟悉，例如一只手，它能挡住部分光骊，我们也知道阴影跟物体的几何形状是大体相似的。

7. Consequently, since the spreading of the patterns is inversely related to slit spacing and to slit width, the two first minima of the single-slit pattern are spread farther apart than are the double-slit minima. 从而，因为图案的分布与狭缝的度和空间成反比，单缝(干涉)的初级主级小的图案间距比双缝大。

8. However, shortly after launch, it was discovered that although the mirror was polished to an exceptional degree of smoothness, its curvature corresponds to the wrong shape, thus creating spherical aberrations that seriously degrade the images. 然而，在发射后不久，就发现虽然镜子已经被抛光到一个异常光滑的程度，它的曲率对应了错误的形状，这样就产生了严重降低图像质量的球差。

9. After the discovery of the wave properties of electrons, it was soon realized that electrons could easily be accelerated to high velocities at which their wavelengths were a thousand times shorter than the wavelengths of visible light. 在发现了电子的波动属性以后，不久又发现了电子能够很容易地就加速到很高的速率，在这样的高速下，电子产生的波长比可见光的波长小一千倍。

10. The production of a continuous spectrum by a prism is due to the variation of wave velocity with wavelength that commonly occurs in transparent media. 从棱镜得到的连续光谱是由不同波长对应不同的波速引起的，这种现象通常发生在透明介质中。

11. Usually we see only the top portion of the circular arc, as the rest lies below the horizon, where it is not easily seen against the background and where the number of reflecting drops in our line of sight is small. 通常我们只能看到彩虹的顶端部分，因为剩余的部分在地平线之下，并且不容易把它从背景中区分，及在我们视角内的反射雨滴数量很少。

12. This scale enables the observer to measure the angle of the emerging light for each image, which is then used to determine the wavelengths of the light in the spectrum. 这个尺度使观察者能够测量每幅图像所产生光的角度，这个角度随之被用来确定光谱中光的波长。

13. As we discussed in Chapter 4, electromagnetic waves, including visible light are transverse waves, consisting of sinusoidal oscillating electric and magnetic fields perpendicular to each other. 正如我们在第4章中讨论的，电磁波，包括可见光中的横波，是由各自相互正交的正弦电磁场震荡所组成。

14. Although the electric field of each wave produced by a particular atom lies in a single plane, the overall beam of light contains electric fields oscillating in all planes. 虽然每个波的电场都是由位于单个平面的一个特殊原子产生，但全部光束却包含了所有平面上的电场

震荡。

15. These effects are due to Rayleigh scattering of the molecules of the air itself, which have dimensions of the order of a fraction of a nanometer, and not to suspended dust or vapor in the atmosphere. 这些效应是由空气分子本身的瑞利散射引起的,这些空气分子的尺寸比一个 nm 要大,而不是在大气中悬浮的灰尘或蒸汽。

16. When the scattering particles are many times larger than the wavelength of light, the light is reflected and refracted as in geometrical optics, not absorbed and reradiated, so the scattered light is white. 当散射粒子比光波波长大很多倍时,光波遵循几何光学的方式反射和折射,而不是被吸收或二次辐射,因此散射光是白色的。

3

Optical Instruments

3.1 The Eye

The human eye is one of the most familiar optical instruments, yet it is also one of the most complex. The overall mechanism of vision is extremely complicated and even now is not fully understood. Yet we can understand some of the important principles of vision by considering the optical properties of the human eye.

Figure 3.1 is a schematic drawing of a human eye, seen in cross section from above. An image focused on the retina by the optical elements of the eye is the stimulus for nerve signals that are ultimately interpreted by the brain. The principal optical elements are the cornea (a transparent sheath across the front of the eye), the pupil (an opening), the iris (which adjusts the size of the pupil), a fluid called the aqueous humor in front of the lens, the lens, the ciliary muscle (which controls the shape of the lens), the gel-like vitreous humor, and the retina (the light-sensitive layer that lines the back of the eye). However, when we consider the way in which the incident light is refracted by the eye, we cannot treat it as a thin lens because there are more than two refracting boundaries and the object and image are in different optical media. The greatest relative change in the index of refraction takes place between the air ($n \approx 1.000$) and the cornea ($n \approx 1.376$). Consequently. the greatest refraction takes place at that surface also. The eye lens is actually a gradient-index lens with a larger index ($n \approx 1.406$) near the center than near the edges ($n \approx 1.386$). Because the aqueous and vitreous humors that surround the lens have indices of $n \approx 1.336$, the refraction at the lens surfaces is small compared with the refraction at the cornea.

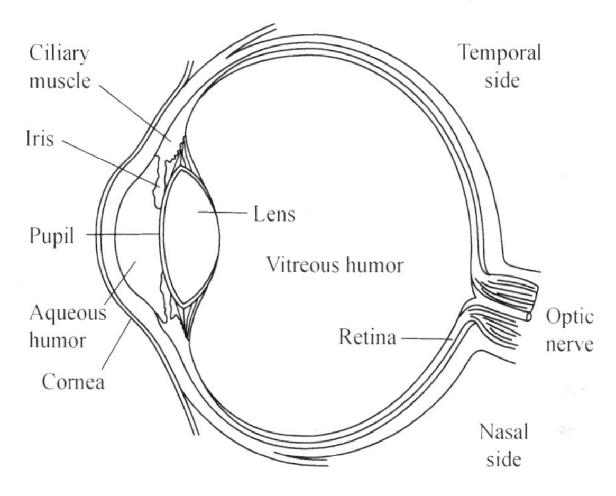

Figure 3.1 Cross section of a human eye viewed from above, showing the main optical elements. The function of each part is described in the text.

The normal relaxed eye presents distant objects in focus on the retina [Figure 3.2(a)]. For close objects to be in focus on the retina, some optical parameters of the eye must change. If a distant object moves toward the eye, the eye changes to keep the image in focus on the retina. This effect is called **accommodation** and is accomplished primarily by contraction of the ciliary muscles, which changes the shape of the lens. When the eye focuses on nearby objects, the lens becomes thicker and the surfaces more curved [Figure 3.2(b)]. Because the lens has a higher index of refraction than the surrounding medium, the net effect is to shorten the focal length of the lens.

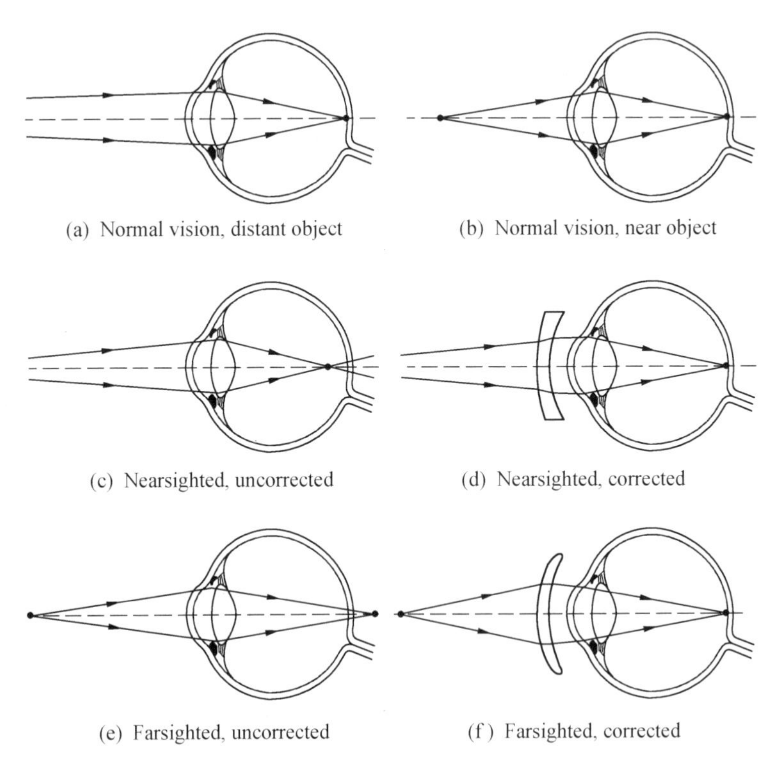

Figure 3.2 (a) The relaxed, normal eye forms an image of distant objects on the retina. (b) For close objects, the lens of the normal eye changes shape to focus the image on the retina. (c) In a nearsighted eye, the image of a distant object forms in front of the retina. (d) A diverging lens corrects for nearsightedness. (e) In a farsighted eye, the image of a nearby object forms beyond the retina. (f) A converging lens corrects for farsightedness.

The ability of the eye to accommodate is limited. We may not move an object any closer than a distance called the near point and still view it clearly. The **near point** is the distance from the unaided eye that produces the largest retinal image without blurring. Objects closer than the near point cannot be brought into focus. The average value for the near point is about 25 cm, although there is considerable individual variation, even for people with so-called normal vision. By convention, the standard near point is taken to be 25 cm. The **far point** is the greatest distance from the unaided eye that produces a distinct image. The far point of the normal eye is at infinity.

Most people do not have ideal vision. The range of visual ability that is considered normal is quite broad. Two common types of vision defects are simply related to the optical properties of the eye and can be readily corrected. They are called nearsightedness (*myopia*) and farsightedness (*hyperopia*).[①] A nearsighted eye is unable to accommodate over the normal range from 25 cm to infinity. Instead, there is a far point beyond which vision is not distinct. The image of a more distant object comes to a focus in front of the retina, so only a blurred image from on the retina [Figure 3.2(c)]. Clear vision of distant objects is restored by placing a diverging lens in front of the eye. This diverging lens forms an image of the distant object within the accommodation range of that particular eye [Figure 3.2(d)]. For a farsighted eye, the near point is farther away than the normal 25 cm[Figure 3.2(e)]. The image of a nearby object comes to focus behind the retina, so again the image is blurred. To see things as close as 25 cm from the eye requires the use of a converging lens, which images the object at least as far away as the actual near point of the particular eye in question [Figure 3.2(f)].

Another common visual problem, **astigmatism**, occurs when the cornea or lens surfaces are not spherical. An astigmatic eye images point objects as lines. Usually the shape of an astigmatic eye can be approximated by the combination of a spherical surface with a cylindrical deformation superposed. Corrections are made for astigmatism by using a compensating cylindrical eyeglass lens.

Opticians generally express the strength of lenses used to correct

---

① Farsightedness commonly occurs in people after middle age because of thickening and lack of flexibility in the lens. For this reason it is also known as *presbyopia*, from the Greek words *presbys* for old and *ops* for eye.

visual defects in terms other than of their focal length. The common unit is the **diopter**. The strength of a lens in diopters is the reciprocal of the focal length expressed in meters. That is.

$$\boxed{D = \frac{1}{f}} \tag{3.1}$$

Where f is the focal length in meters and D is the strength of the lens in diopters. For example, a lens with a focal length of $+0.25$ m has a strength of $+4.0$ diopters. Shorter focal lengths correspond to greater dioptic strengths. The use of lens strength is especially convenient when several thin lenses are used in close proximity. Then the strength of the combination is just the sum of the strengths of the individual lenses. (See Example 1.8). Selecting a lens of proper strength for correcting a visual defect of the kind shown in Figure 3.2 is usually accomplished by trying various combinations of lenses in front of the eye until the clearest vision is obtained. Then a single lens is chosen that has the same strength as the combination of trial lenses.

For people whose eyes are unable to accommodate fully at both ends of the range, lenses are available with two distinct regions of different dioptic strength. Such lenses are known as *bifocals*. They have an upper region of dioptic strength appropriate for distant vision and a lower region designed for close vision. In some cases, *trifocal* lenses are used to provide better vision at intermediate distances. The numerical example below illustrates the use of a corrective lens.

3.2 The Magnifying Glass

Many optical devices are used to produce magnified images of objects. The simplest magnifying instrument of all is the **magnifying glass**, or simple microscope. The magnifying glass is a single converging lens that, when held near the eye, gives an image whose size on the retina is larger than that observed by the unaided eye. By adjusting the distances at which the lens and object are held from the eye, the viewer can obtain maximum magnification without undue eye strain or blurring of the retinal image.

Magnifying glasses are not only used singly to enlarge such things as fine print and small objects, they are also used to enlarge the images formed by other lenses. When they are utilized in this way, they are referred to as eyepieces. The primary function of magnifying glasses and eyepieces is to increase the angular size of the image and to allow viewing with a relaxed eye.

As an object is moved closer, it subtends① a larger angle at the eye, so that the image produced covers a larger part of the retina (Figure 3.3). However, we can not bring the object any closer than the near point and still see it clearly. A magnifying glass allows us to increase the visual angle that an object subtends at the eye — that is, to form a larger image on the retina — without requiring our eye to focus closer than the near point. For example, [Figure 3.4(a)] shows an object of height h placed at the standard near point (25 cm), where it subtends an angle θ. If the angle θ is small, we can write θ in radians as

$$\theta = \frac{h}{25}$$

Where both h and the near point are expressed in centimeters. In fig. 3.4(b), we have placed a magnifying glass (converging lens) next to the eye and moved the object within the focal length of the lens. The positions of the lens and the object have been adjusted so that the enlarged virtual image of the object falls at the eye's near point. Since the image is on the same side of the lens as the original object, the image distance is negative.

---

① Recall that in a triangle, a line *subtends* the angle opposite to it; that is, the line extends across the entire angle. The object in Figure 3.3 forms one side of a triangle whose opposite vertex is at the observer's eye. Thus, the object subtends an angle at the eye.

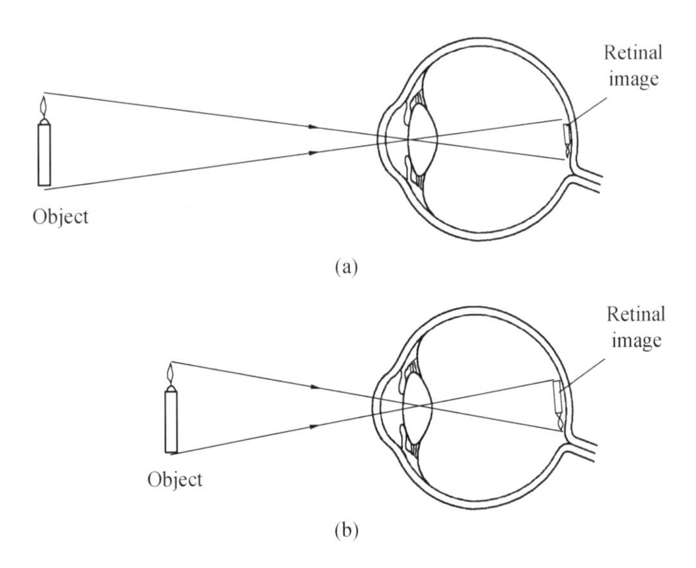

Figure 3.3 (a) The image of an object seen with the unaided eye forms on the retina. (b) As the object comes closer, its image covers a larger part of the retina.

What magnification have we achieved? To find out, we first calculate the object distance, using the thin-lens formula. If distances are measured in centime-ters, we get the result

$$o = \frac{25f}{f + 25}$$

From the figure we see that the angle θ' subtended by the virtual image is approximately

$$\tan\theta' \approx \theta' = \frac{h}{0} = \frac{h(f+25)}{25f}$$

The magnification of interest for a magnifying glass is the angular magnification. In general, the **angular magnification** is defined by

$$\boxed{M \equiv \frac{\theta'}{\theta}} \tag{3.2}$$

which for this case becomes

$$M = \frac{25\ \text{cm}}{f} + 1 \quad (\text{image at near point}) \tag{3.3}$$

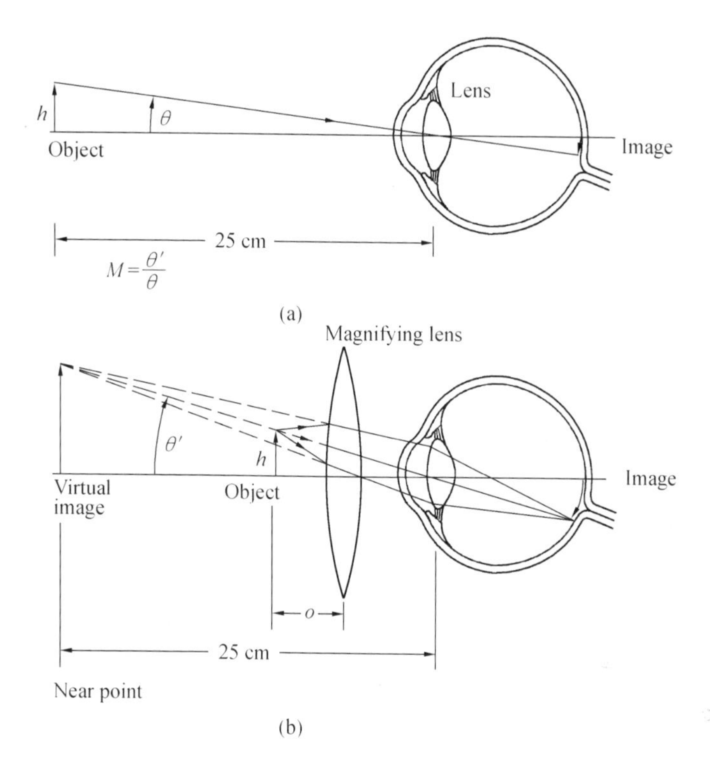

Figure 3.4 (a) An object placed at the near point of an unaided eye. (b) The same object viewed through a magnifier with the virtual image at the near point. The angular magnification M is θ'/θ

Remember that f is measured in centimeters in this equation.

If the object is held at or just inside the focal point of the lens, the image forms very far away, essentially at infinity, rather than at the near point. This corresponds to the most comfortable viewing distance because the eye is relaxed. (See Problem 3.18.) In this case, $\theta' = h/f$, so the magnification is given by

$$\boxed{M = \frac{25\ \text{cm}}{f}} \quad \text{(image at infinity)} \tag{3.4}$$

Manufacturers commonly use Eq. (3.4) to specify the magnification of a magnifying glass. Thus, a magnifying glass of 10cm focal length is marked 2.5×.

3.3 Cameras and Projectors

In its simplest form, a photographic camera is a light-tight box with a lens set into one side and a light-sensitive material (film) placed on the side opposite the lens (Figure 3.5). Normally, light is prevented from entering the camera by means of a shutter, either at the lens position or just in front of the film. Unlike the eye, the camera lens has a fixed focal length. Therefore, you focus a camera by moving the lens closer to or farther from the film, depending on the object's distance from the lens. To take a photograph, you first adjust the lens position so that a real, inverted image of the object is in focus in the plane of the film. Then, when you press a button, the shutter momentarily opens and an image forms on the film. This image is stored in the light-sensitive material for later chemical processing, yielding a reproduction of the scene as its reflected light originally fell on the film.

Modern cameras, such as the 35 mm camera of Figure 1.22, are basically the same as the simple camera. They have well-corrected lenses of the type discussed in Section 1.9. Many have automatic exposure control and automatic focusing, features that are made possible through the use of integrated circuit electronics. Computer-aided design of lenses and computer-aided manufacturing allow us to have inexpensive lenses of high quality that were not available for any price only twenty years ago. Many cameras take interchangeable lenses that allow photographers control over the composition of their pictures (Figure 3.6).

Other cameras, such as digital still cameras, motion picture cameras, and television cameras, use the same basic principle. In a digital camera, the light is imaged on a semiconductor detector rather than photographic film. The information in the image is stored in the same manner as

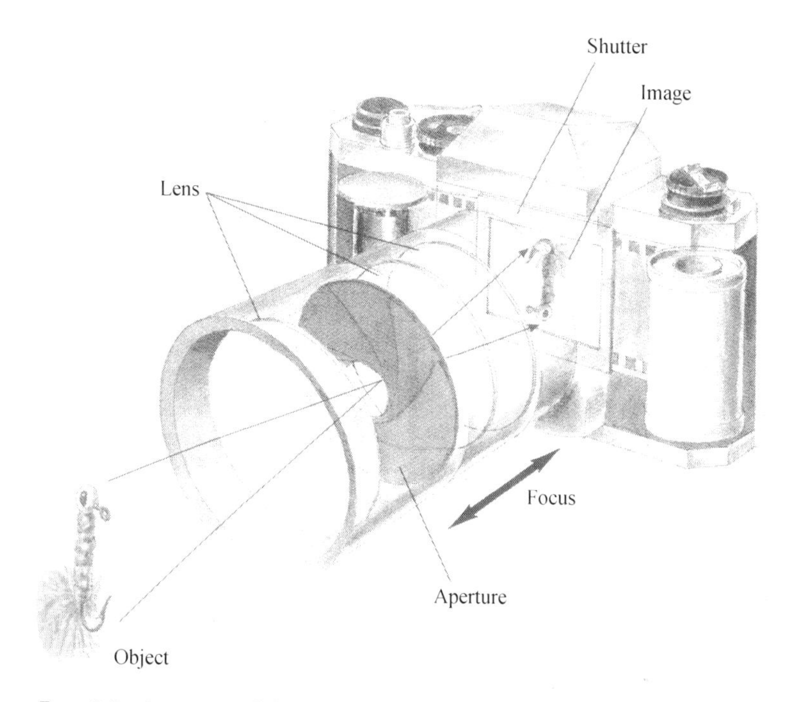

Figure 3.5 A camera is a light-tight box for holding the film and lens in proper position. It also includes a shutter to control the time during which light reaches the film.

Figure 3.6 These three photographs were taken from the same place, using three different lenses. The focal length used were (a)28mm, (b)55mm, and (c)200mm. The longer the focal length of the lens, the greater the angular magnification of the image.

computer data and may be displayed on a computer monitor or printed with a computer printer. A motion picture camera takes a rapid sequence of still pictures, which are eventually projected at the same rate at which they were taken. The apparent smoothness of the motion is due to the fact that the rate at which the pictures are taken and projected, ordinarily 24 frames per second, is higher than the rate at which we can distinguish between individual images. The persistence of vision gives us the appearance of smooth continuous motion. In a television camera, although the image is detected, transmitted, and reproduced by electronic means, the optical principles are the same as for a movie or still camera.

With modern photographic films and with adequate lighting, the shutter need be open only a small fraction of a second to successfully record an image on film. Many cameras have a range of available shutter speeds — that is, lengths of time during which the shutter is open. Shutter speeds of 1/30, 1/60, 1/125, 1/250, 1/500, 1/100, and 1/2 000 second are standard. Note that each of these is approximately half as long (or, in photographic terminology, "twice as fast") as the preceding one. The faster the shutter speed, the faster the object can be moving and still produce a sharp image.

Getting the "correct exposure" in a photograph corresponds to allowing the appropriate amount of light to strike the film.① This amount is different for different types of film. The amount of light that strikes the film is determined not only by how long the shutter stays open, but also by how large the effective lens opening is. Thus, it is analogous to the amount of water flowing from a faucet, which depends on both the length of time the faucet stays open and the cross-sectional area of the opening. The size of the lens opening, or aperture, is often measured by what is called the *f*-value or ***f*-number**. This is defined to be

$$\boxed{f - \text{number} \equiv \frac{\text{focal length of lens}}{\text{diameter of lens}} = \frac{f}{d}} \tag{3.5}$$

① By "amount of light." we mean the energy deposited on the film oy the light.

For example, a lens with a diameter one-half its focal length has an *f-number of 2, which is written f/2*. A variable-diameter opening in the camera, called an iris diaphragm (after the iris in your eye), can be adjusted to decrease the effective lens opening and therefore increase the *f*-number. By being able to adjust both shutter speed and *f*-number. the photographer can give the proper exposure to the film while still having the option of using a particular shutter speed or a particular lens opening.

Table 3.1 lists the standard *f*-number intervals. These numbers are often referred to as *f*-stops or simply stops. Notice that the values of successive *f*-stops differ by a factor of approximately $\sqrt{2}$. Thus, if you change the *f*-number of a lens from $f/4$ to $f/5.6$, you reduce the diameter of the aperture by a factor of $1/\sqrt{2}$. The area of the aperture decreases by the square of this factor. So. if the shutter speed remains constant. the amount of light passing through the lens opening is cut in half . A change in aperture equivalent to going from one *f*-stop to the next successive one is known as a change of one full stop. It increases or decreases the light passed through the lens by a factor of 2.

Table 3.1 Standard Full-Stop *f*-numbers

f-number	(f-number)^2
0.7	0.49
1.0	1
1.4	1.96
2.0	4
2.8	7.84
4	16
5.6	31.4
8	64
11	121
16	256
22	484

A useful consequence of the *f*-number method of classifying lens openings is that for the same shutter speed, lenses of different focal lengths give proper exposure at the same f-numbers. This result is due to two compensating factors. First, the amount of light that passes through the aperture is proportional to its area, and hence to the square of its diameter, d^2. Second, the light per unit area that reaches the film depends inversely on the area of the image. For the usual situation, in which the object distance is large compared with the focal length, area of the image is proportional to f^2. The rate at which a photographic image is formed, or the *speed* of the lens, is then

$$\text{speed of lens} \propto \frac{d^2}{f^2} = \frac{1}{(f\text{-number})^2} \tag{3.6}$$

For most purses, the speed of the lens depends only on the f-number and is independent of the particular focal length.

The film camera has an optical inverse — the slide projector. The basic components of a slide projector are shown in Figure 3.7. The transparent slide with the likeness of the photographed subject, the lens, and the image projected on the screen are the inverse of the camera used to make to slide. By choosing the proper focal-length lens, you can make the image of the slide completely fill the screen at the chosen projector-screen distance. The need to accommodate different sizes of screens and rooms has led to the use of variable-focal-length, or zoom, lenses for home projectors.

The illuminating system of a projector is equally as important as the image-forming system. It usually consists of a lamp and reflector, a condensing lens, and a piece of heat-absorbing glass to protect the slide from the heat of the lamp. The reflector simply directs more light in the direction of the slide. The condenser lens is a strongly converging lens, ensuring that light passing through the edges of the slide passes through

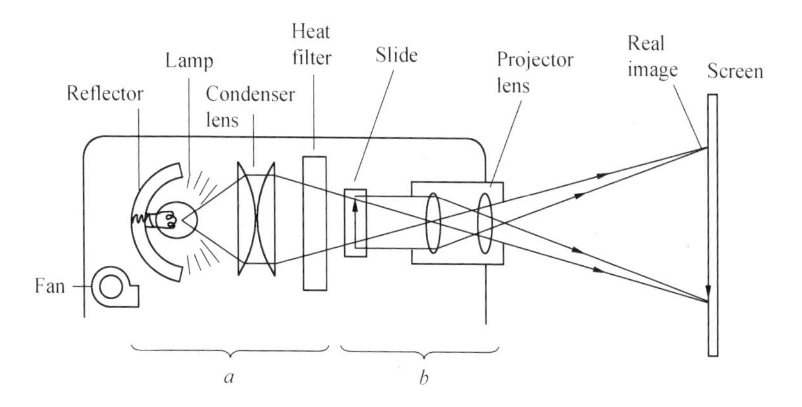

Figure 3.7 The basic optical components of a slide projector. The illuminating system is indicated by bracket *a* and the image-forming system by bracket *b*. Most projectors also include a fan to cool the slide and prevent damage due to overheating.

the projector lens. Thus, the condenser not only increases the amount of light passing through the slide, but also increases the amount of light reaching the screen. If the condenser lens is removed, the entire slide will still be illuminated, but the image from the edges of the slide will not appear on the screen, and only the central part of the picture will be seen.

The optical principles of a movie projector are the same as those of a slide projector. However, a movie projector also contains a mechanical means for moving the film and for interrupting the light when the film is moving. The result is that the viewer sees a series of individual still pictures. As mentioned earlier, the pictures are presented at a rate of 24 per second, and the brain blends the sequence of individual images into a smoothly flowing scene. Ordinarily, a three-bladed shutter is used to block the light when the film is moving. Because the shutter makes one rotation for each frame, it also interrupts the light during the presentation of each individual picture. The shutter interrupts the light 72 times per second, a rate so fast that the eye cannot see the flicker.

3.4 Compound Microscopes

The microscope and the telescope were developed at about the same time in the early 1600s. Early naturalists soon utilized microscopes to make important discoveries. Marcello Malpighi's discovery of capillaries, Anton van Leeuwen hoek's discovery of protozoa, and Robert Hooke's beautiful drawings of magnified cells were important contributions to the understanding of biological processes. These advance would have been impossible without the microscope. The microscope and telescope have perhaps played a greater role than any other scientific instruments in establishing our current understanding of natural laws.

A typical **compound microscope** consists of a tube with a converging lens at each end [Figure 3.8 (a)]. Though in modern microscopes each lens may actually consist of a group of lenses to minimize distortion from lens aberrations, we can understand the optical principles by treating each group as a single lens. The lens close to the object being viewed is called the **objective.** The lens through which one looks is called the **eyepiece** or **ocular.**

[Figure 3.8(b)] is a ray diagram of a compound microscope. The objective is a lens of comparatively short focal length f_0. When an object is placed just outside the focal point of the objective, a real, enlarged image forms at the plane A. If a screen was placed at A, the image of the object would appear there. The eyepiece functions as a magnifier for viewing this image. Thus, the image formed by the objective lens serves as the object for the eyepiece lens. The eyepiece, in turn, produces an enlarged virtual image.

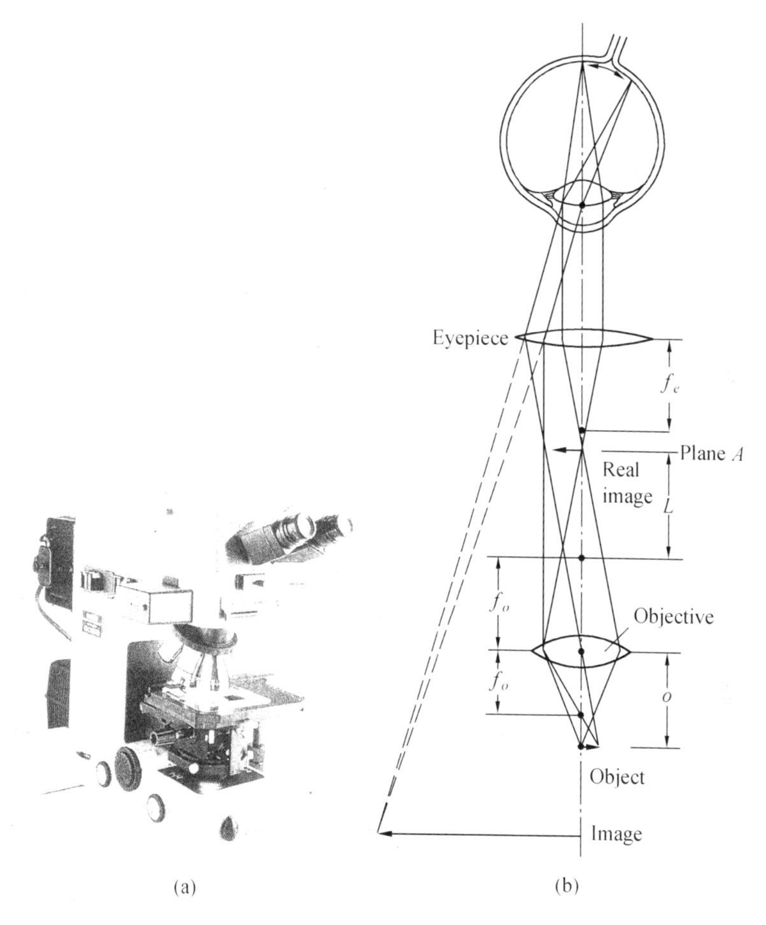

Figure 3.8 (a) A modern microscope. (b) Ray diagram of a compound microscope.

The image formed at plane A by the objective has a *linear magnification* m_o. The Newtonian expression for this magnification (see Problem 1.60) is

$$m_o \approx \frac{L}{f_o}$$

where L is the distance from the focal point to image plane. This image is then viewed through the eyepiece with an *angular magnification*

$$M_e = \frac{25\ \text{cm}}{f_e}$$

The overall magnification M of the microscope is the product of the linear magnification of the objective and the angular magnification of the eyepiece, resulting in a magnification

$$M = m_o M_e = \left(-\frac{L}{f_o}\right)\left(\frac{25\ \text{cm}}{f_e}\right) \tag{3.7}$$

For practical microscopes, f_o is much less than L and fe is less than 25 cm, resulting in a large value for the magnification. The negative sign in Eq. (3.7) indicates that the image is inverted.

Microscope objectives and eyepieces are commonly labeled according to their effective magnifications when used with a standard separation between them. Most (but not all) manufacturers design their microscopes so that the distance L, between the focal point of the objective and that of the eyepiece, is 16.0cm. The magnification of an objective lens can then be expressed as $m_o = 16.0\ \text{cm}/f_o$. If we know the magnifications of the eyepiece and the objective lens, we can use Eq. (3.7) to find the overall magnification. A 10 × eyepiece has an angular magnification of 10 times, and a 10 × objective has a linear magnification of 10 times. Used together, they give an overall magnification of 100 × .

3.5 Telescopes

Though different in purpose, the telescope has a great deal in common with the microscope as an optical instrument. A telescope also consists of a long tube with an objective lens toward the object and an eyepiece lens toward the viewer. Several types of telescopes exist. [Figure 3.9(a)] shows a diagram of a **refracting astronomical**, or **inverting**, **telescope.** In this type of telescope, the objective lens has a relatively long focal length and the object being viewed is faraway compared with this focal length. As a result, rays from the object come in nearly parallel and a real image forms near the focal point of the objective. This image would show up clearly on a screen placed there, just as in the case of the microscope. Not, however, that in this case the image is smaller than the physical size of the object. Again we use an eyepiece to view the inverted image that has been brought to a focus by the objective lens. The angular size of the viewed image is larger than the angular size of the object when viewed without the telescope, and so the initial object appears closer.

We define the magnification M of a telescope as the ratio of the angle θ' subtended by the object when viewed through the telescope to the angle θ subtended when the object is viewed with the unaided eye. The angle subtended at the eye is essentially the same as that subtended at the objective by the object. If we let h be the height of the image (the blue arrow) formed by the objective lens, we can show from Figure 3.9 that $\theta \cong h/f_o$ and $\theta' \cong h/f_e$ so

$$\boxed{M = \frac{\theta'}{\theta} \cong -\frac{f_o}{f_e}} \tag{3.8}$$

If we replace the converging eyepiece with a diverging lens, we get an erect image. Telescopes of this type are called **Galilean telescopes** after Galileo[(Figure 3.9(b)]. If the eyepiece were not present, incoming rays from a distant object would come to focus essentially at the focal point of the objective lens. A real inverted image would be formed on a screen placed at this point. However, when the diverging eyepiece lens is placed within the focal length of the objective, so that the rays striking the eyepiece emerge from it parallel to each other, a virtual image is produced. From the diagram, we see that the magnification is again given by Eq. (3.8). Note that when f_e is negative, as it is here, the magnification is positive, indicating an erect image.

Binoculars are essentially twin refracting telescopes mounted side by side (Figure 3.10). The prisms, which give most binoculars their characteristic shape, are used to erect the image that would otherwise be

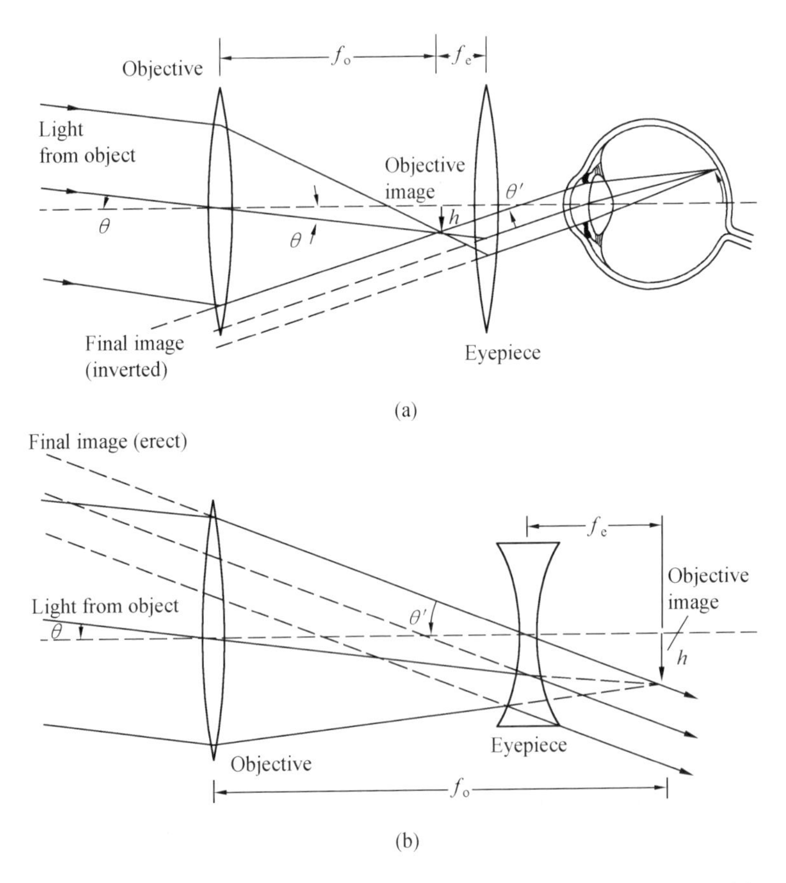

Figure 3.9 (a) Principle of operation of a refracting astronomical (or inverting) telescope. (b) Ray diagram of a Galilean telescope. The diverging eyepiece produces an erect, virtual image.

inverted. (The light is totally internally reflected in the prisms.) Because the index of refraction of the prisms is greater than that of air and because they fold the light path, the tubes are shorter than the tubes of a simple telescope of the same magnification. Opera glasses differ from binoculars in that they consist of a pair of side-by-side Galilean telecopes. Because their images are already erect, no prisms are needed.

Telescopes may also be constructed with mirrors for the objective.

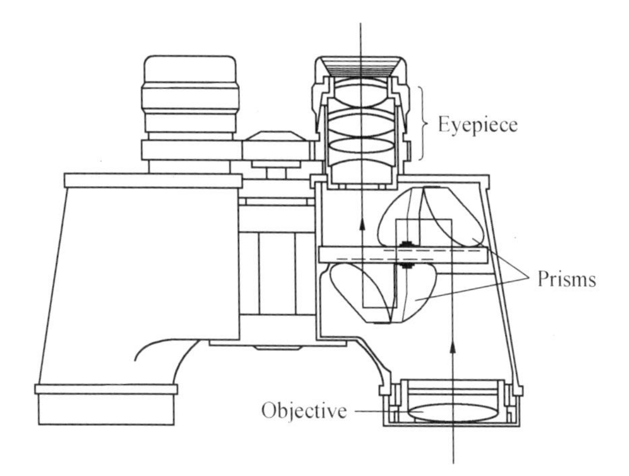

Figure 3.10 Prisms are used to erect the image in a binocular.

Isaac Newton constructed such a telescope in 1668 in order to avoid the problems of chromatic aberration found in lenses. The **Newtonian telescope** has a converging (concave) mirror as its first element (Figure 3.11). A flat mirror mounted diagonally in the middle of the tube reflects the light so that it comes to a focus just outside the tube. An eyepiece is used to view the resulting image. Other types of reflecting telescopes differ mainly in where the image is brought to focus. Modern telescopes used in astronomy are reflecting telescopes.

3.6 Other Lenses

One of the marvels of modern technology is the zoom lens, widely used in photography and television. The zoom lens can be changed quickly from a wide-angle lens (short focal length) to a telephoto lens (long focal length). A true **zoom lens** maintains focus throughout the entire zoom range at any focusing distance, provided you have focused sharply on an object. A **varifocal lens** must be refocused whenever you change its focal length.

Modern zoom lenses used with television cameras are available with

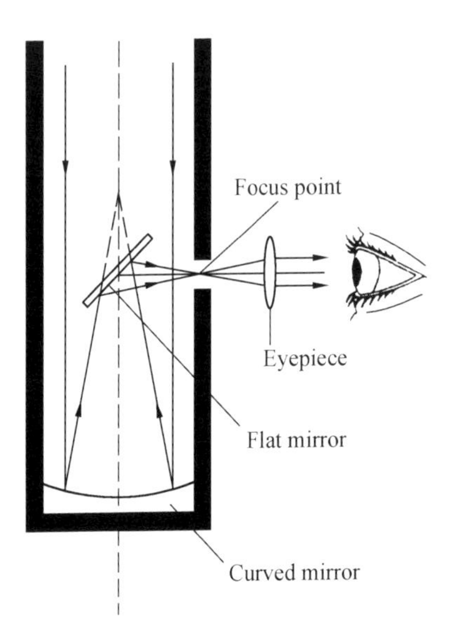

Figure 3.11 Ray diagram of a Newtonian telescope, which uses a mirror instead of a lens to focus light.

focallength changes of as much as 20 to 1. Lenses used on 35mm lens to the 7.5:1 ratio of a 28 – 210mm lens. Thus, a single lens can be adjusted to obtain the focal length needed to provide the desired composition and magnification (Figure 3.6).

To understand the operation of a varifocal lens, consider the behavior of two thin lenses of equal and opposite focal lengths as the separation between them changes (Figure 3.12). Suppose the lenses have focal lengths of + 12 cm and – 12 cm. When the lenses are separated by 4cm, light incident on the converging lens parallel to its axis is ultimately brought to focus 24cm from the diverging lens. If the separation of the lenses is increased to 8cm, the light converges only 6cm from the diverging lens, resulting in a wider-angle field of view.

Commercial zoom lenses are much more complicated than a simple combination of two thin lenses. A typical zoom lens has 12 or more

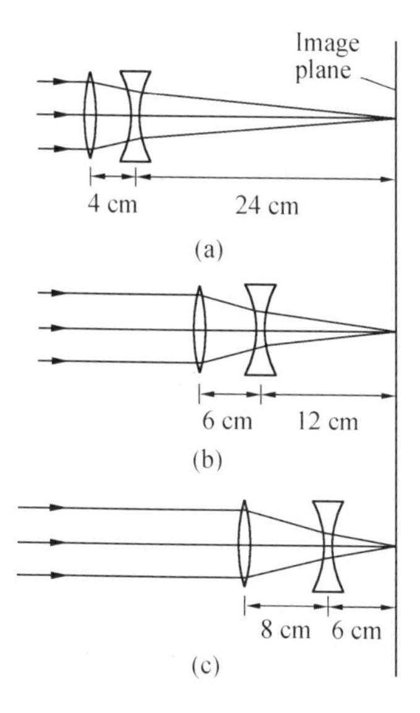

Figure 3.12 A varifocal lens made from two components of equal but opposite focal length (12cm). The distance between the diverging lens and the effective focal point depends on the separation of the lenses: (a)4cm, (b)6cm, and (c)8cm.

separate glass lenses arranged in four groups (Figure 3.13). Two or more of these groups move relative to the others in order to vary the effective focal length, all the while maintaining a constant image position so that focus is maintained. The numerous lenses in each group are needed to minimize aberrations and produce a lens that is not restricted to paraxial rays.

Because a zoom lens has so many elements, it is extremely important that each surface be coated with an antireflection layer. (See Section 2.4.) These coatings not only increase the amount of light passing through the system, but also increase the contrast and reduce the flare that results from multiple reflections from the surfaces. Some reduction in the number of elements is achieved by using aspheric lenses. In the future, the incorporation of gradient-index lenses may further reduce the number of elements, making zoom lenses lighter and cheaper.

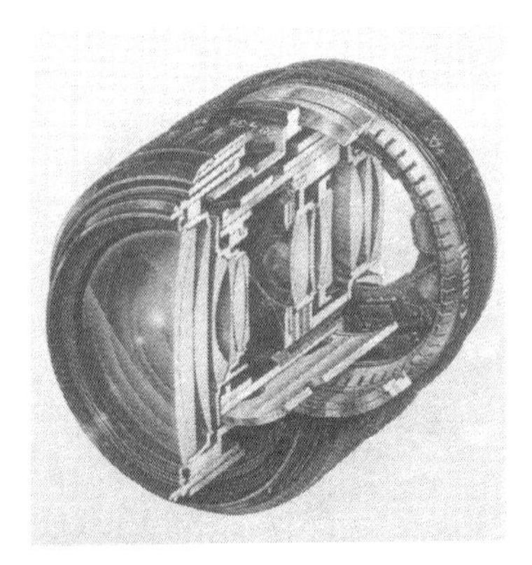

Figure 3.13 A typical zoom lens for a 35 mm camera has 12 or more individual lenses arranged in four groups. The Canon 28 ~ 105mm lens has 15 elements.

Another special lens was developed by Augustin Jean Fresnel early in the nineteenth century to solve the problem of how to make large lenses for light houses. Fresnel recognized that refraction occurs only at the lens surfaces. By designing a lens that eliminated some of the material between the surfaces, he was able to produce a thin lens that was equivalent to a much thicker one. Thus a **Fresnel lens** has concentric rings with the surface contour of the equivalent ring of a thick lens but with each successive ring stepped back to eliminate the unnecessary material between the front and back surfaces (Figure 3.14).

Fresnel lenses do not produce high quality images, but they can be made very thin and light. Originally used in lighthouses (Figure 3.15), they are now used in other ways, such as condenser lenses in overhead projectors, as light collectors for solar cells, and as flat pocket magnifiers (Figure 3.16).

One of the more recent advances in lens technology is the development of the **gradient-index lens** (GRIN), in which the index of refraction decreases as a function of the radius. Light rays incident near the edge of the lens are not refracted as much as those near the center of the lens. Consequently, a gradientindex lens can be used to correct for

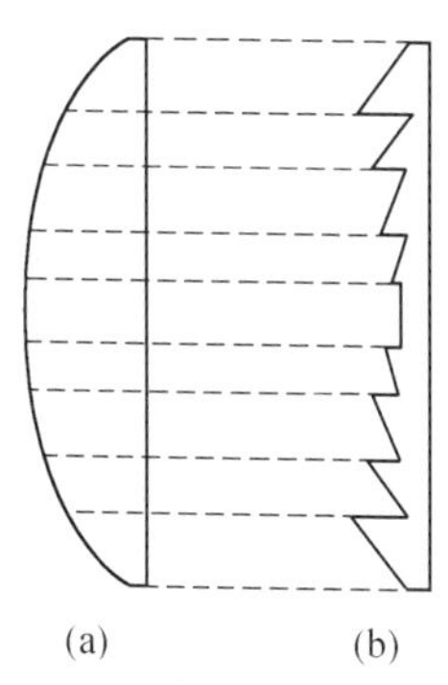

Figure 3.14 Cross sections of (a) an ordinary plano-convex lens and (b) the corresponding Fresnel lens. Note that the surface contours are the same for the corresponding rings.

spherical aberration (Figure 3.17). Lenses of this type may soon be used in cameras, but at present they are limited to small diameter rods with parallel, flat faces. The rod axis is the optical axis of the lens. Because of their nonuniform index of refraction, they can be used for coupling light sources to optical fibers, even though both faces are flat.

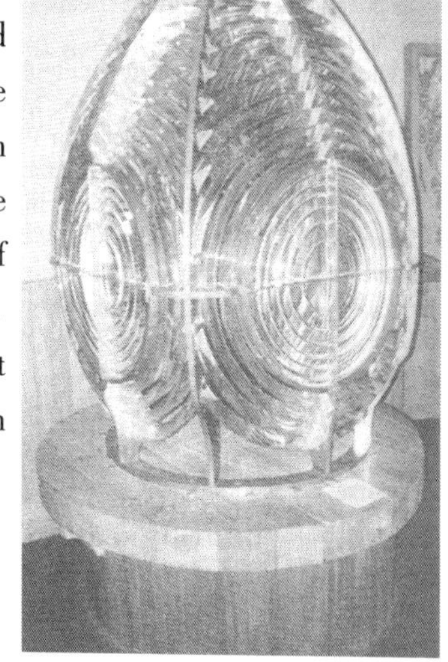

Figure 3.15 A fresnel lighthouse lens

Summary

Useful Concepts

● The strength of a lens in diopters is the reciprocal of the focal length when the focal length is expressed in meters.

$$D = \frac{1}{f}$$

● A converging lens may be used as a magnifying glass. The important magnification is the angular magnification,

Figure 3. 16 Fresnel lenses are often used as magnifiers

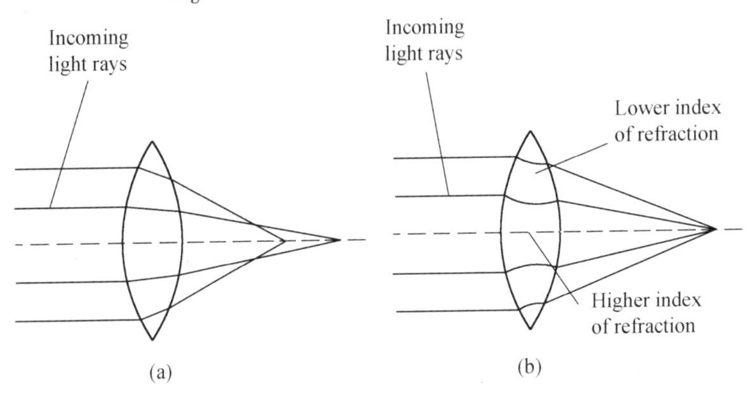

Figure 3.17 (a) An ordinary lens bends light more at the edges than near the center. (b) A gradient-index lens brings all the light to the same focus.

$$M \equiv \frac{\theta'}{\theta}$$

- If the lens is held close to the eye, the angular magnification for the relaxed eye is

$$M = \frac{25\ \text{cm}}{f}$$

where 25 cm is used for the standard near point and corresponds to having the image at infinity. This equation is normally used to specify the magnification of a magnifying glass.

- The f-number of a lens is given by

$$f\text{-number} \equiv \frac{\text{focal length}}{\text{diameter}}$$

- The overall magnification of a compound microscope is the product of m_o, the linear magnification of the objective lens given by $m_o = \frac{L}{f_o}$, and M_e, the angular magnification of the eyepiece.

- The magnification of a telescope is the ratio of the focal length of the objective lens to the focal length of the eyepiece

$$M = -\frac{f_o}{f_e}$$

- Zoom and varifocal lenses have multiple elements that move relative to one another to allow the effective focal length of the combination to be changed.

Important Terms

You should be able to write the definition or meaning of each of the following

accommodation	eyepiece
near point	ocular
far point	refracting astronomical
astigmatism	telescope
diopter	Galilean telescope
magnifying glass	Newtonian telescope
angular magnification	zoom lens
f-number	varifocal lens
compound microscope	Fresnel lens
objective lens	gradient-index lens

New Words and Expressions

optical instrument *n*. 光学仪器

microscope *n*. 显微镜

diameter	*n*. 直径
telescope	*n*. 望远镜
infrared light	*n*. 红外光
aberration	*n*. 失常
acoustic wave	*n*. 声波
ultrasonic	*adj*. 超声的,超音速的 *n*. 超声波
cornea	*n*. [医]角膜
iris	*n*. [解]虹膜
pupil	*n*. 瞳孔
ciliary muscle	*n*. 睫状肌
gel	*n*. 凝胶体 *v*. 成冻胶
vitreous humor	*n*. [解](眼睛的)玻璃状液
retina	*n*. [解]视网膜
refraction	*n*. 折光,折射
aqueous humor	*n*. (眼球的)水状体
optic nerve	*n*. 视神经
nasal	*adj*. 鼻的,鼻音的,护鼻的 *n*. 鼻音
temporal	*adj*. [解]颞的
myopia	*n*. 近视
hyperopia	*n*. [医]远视
presbyopia	*n*. 老花眼,远视眼
astigmatism	*n*. [医]散光
spherical	*adj*. 球的,球形的
cylindrical	*adj*. [计]圆柱的
diopter	*n*. [物]屈光度
reciprocal	*adj*. 相应的,倒数的,彼此相反的 *n*. 倒数,互相起作用的事物
bifocal	*adj*. 双焦点的
trifocal	*adj*. 三焦距的 *n*. 三焦距透镜
magnifying glass	*n*. 放大镜,放大器

eyepiece	*n*. 目镜
subtend	*v*. 双向,包在叶腋内
triangle	*n*. [数]三角形
vertex	*n*. 顶点,最高点
angular magnification	*n*. 角度放大率
retina	*n*. [解]视网膜
visual angle	*n*. 视角,视界
near point	*n*. 近点
projector	*n*. 放映机
still camera	*n*. 静物摄影机
shutter	*n*. 快门
aperture	*n*. (照相机,望远镜等的)光圈,孔径
terminology	*n*. 术语学
faucet	*n*. 龙头,旋塞,(连接管子的)插口
iris diaphragm	*n*. [摄]虹彩光圈,可变光圈
slide projector	*n*. 幻灯片放映机,幻灯机
condenser lens	*n*. 聚光透镜
television camera	*n*. 电视摄像机
still camera	*n*. 照相机
ocular	*adj*. 眼睛的,视觉的 *n*. 目镜,眼睛
object lens	*n*. [物]物镜
astronomical telescope	*n*. 天文望远镜
Galileo telescope	*n*. 伽利略望远镜
binoculars	*n*. 双眼望远镜
prism	*n*. [物]棱镜,[数]棱柱
opera glasses	*n*. (视剧用的)小型双眼望远镜
Newton telescope	*n*. 牛顿望远镜
zoom lens	*n*. 变焦透镜
wide-angle lens	*n*. 广角镜头
telephoto lens	*n*. 远摄镜头,长焦镜头

varifocal	*adj*. [摄]变焦距
fresnel	*n*. 菲(涅耳)(频率单位,百亿赫兹)
plano-convex	*adj*. 平凸的
antireflection layer	*n*. 吸声层,抗反射膜
aspheric	*adj*. [物]非球面的
objective lens	*n*. 物镜

NOTES

1. However, when we consider the way in which the incident light is refracted by the eye, we cannot treat it as a thin lens because there are more than two refracting boundaries and the object and image are in different optical media. 然而,当我们考虑入射光被眼睛折射的方式时,我们不能够视它为一个薄透镜,因为折射界面超过两个,而且物体和像是在不同的光学介质中。

2. To see things as close as 25 cm from the eye requires the use of a converging lens, which images the object at least as far away as the actual near point of the particular eye in question. 为了能用眼睛看清近在25 cm处物体,需要用一个会聚透镜,它能使目标物体至少成像在每个人眼的实际近点距离上,该人眼实际近点距离尚在论证中。

3. For people whose eyes are unable to accommodate fully at both ends of the range, lenses are available with two distinct regions of different dioptic strength. 对那些眼睛不能完全调节两个末稍范围的人,透镜有助于区分两个不同的屈光度强度区域。

4. The magnifying glass is a single converging lens that, when held near the eye, gives an image whose size on the retina is larger than that observed by the unaided eye. 放大镜是一个单个的会聚透镜,当它放在眼睛近处时,它在视网膜上得到的像尺寸大于直接用肉眼观察到的像尺寸。

5. A magnifying glass allows us to increase the visual angle that an object subtends at the eye — that is, to form a larger image on the retina — without requiring our eye to focus closer than the near point. 放

大镜使我们可以增大眼睛的视角——也就是说，在视网膜上形成一个更大的图像——不需要我们的眼睛离视觉进点更为靠近。

6. Computer-aided design of lenses and computer-aided manufacturing allow us to have inexpensive lenses of high quality that were not available for any price only twenty years ago. 半自动的计算机透镜设计和全自动的计算机制造技术使我们拥有廉价而又高质量的透镜，而这些价格在二十年前是不可能得到的。

7. The apparent smoothness of the motion is due to the fact that the rate at which the pictures are taken and projected, ordinarily 24 frames per second, is higher than the rate at which we can distinguish between individual images. 外表上的平滑运动是基于这些图片被采用和放映时的速率，通常一般是每秒 24 桢，这高于我们能区分单个图片的速率。

8. In a television camera, although the image is detected, transmitted, and reproduced by electronic means, the optical principles are the same as for a movie or still camera. 在电视摄像机内，虽然影像用电子手段检测、传输和再生，而光学原理对电影或者照相机同样适用。

9. By being able to adjust both shutter speed and f-number, the photographer can give the proper exposure to the film while still having the option of using a particular shutter speed or a particular lens opening. 通过调整快门速度和 F 数，摄影师能够给出胶卷的正确曝光，但是他们仍然能够选择用特殊快门速度和特殊透镜。

10. If the condenser lens is removed, the entire slide will still be illuminated, but the image from the edges of the slide will not appear on the screen, and only the central part of the picture will be seen. 如果聚光透镜被移动了，整个幻灯片仍然能够被照亮，但是幻灯片的边缘图像不能在屏幕上显示，仅有图片的中间部分被照亮。

11. Though in modern microscopes each lens may actually consist of a group of lenses to minimize distortion from lens aberrations, we can

understand the optical principles by treating each group as a single lens. 虽然为了尽可能减小透镜的像差,现代显微镜的每个镜片实际上都是由一组透镜构成,我们仍然能将每组镜片看做是单个镜片来考察其光学原理。

12. Microscope objectives and eyepieces are commonly labeled according to their effective magnifications when used with a standard separation between them. 在显微镜中做标准区分时,它的物镜和目镜通常都按照它们的有效放大率来做标记。

13. As we will see in the next chapter, the minimum distance between two object points that can be resolved in an image depends on the wav-length of the illumination and on the diameter of the lens. 正如我们将在下一章节中看到的一样,一个图像中两个物点的最小可分辨距离依赖于照明发出的波长和透镜的直径。

14. We define the magnification M of a telescope as the ratio oof the angle θ' subtended by the object when viewed through the telescope to the angle θ subtended when the object is viewed with the unaided eye. 我们定义望远镜的放大率 M 为通过望远镜观察物体的张角 θ' 和肉眼直接观察物体的张角 θ 之比。

15. however, when the diverging eyepiece lens is placed within the focal length of the objective, so that the rays striking the eyepiece emerge from it parallel to each other, a virtual image is produced. 然而,当发散目镜被放置在物距范围内,以致射到目镜上的光线各自平行,这就产生了虚像。

16. Modern zoom lenses used with television cameras are available with focal-length changes of as much as 20 to 1. Lenses used on 35mm cameras come in a range of zoom ratios from the limited 2:1 ratio of a 35 ~ 70 mm lens to the 7.5:1 ratio of a 28 ~ 210 mm lens. 现代的应用于电视摄像机放大镜头焦距变化范围在 20 ~ 1 之内都是可以得到的。在 35 毫米摄像机上应用的镜头的放大率范围从 35 ~ 70 毫米镜头的 2:1 到 28 ~ 210 毫米镜头的 7.5:1。

17. The magnitude of the effort required for construction of the Hale raised doubts that significantly larger telescopes would ever be built due to the technological difficulties. 构建 Hale 望远镜所需要的巨大努力使人们怀疑具有深远意义的大型望远镜由于制造技术难度的可行性。

18. These coatings not only increase the amount of light passing through the system, but also increase the contrast and reduce the flare that results from multiple reflections from the surfaces. 这些覆盖物不仅可以增加透过系统的光量,而且还可以增加对比度,减少由于表面的多重反射而产生的闪光。

19. Thus, a Fresnel lens has concentric rings with the surface contour of the equivalent ring of a thick lens but with each successive ring stepped back to eliminate the unnecessary material between the front and back surfaces. 这样,菲涅尔透镜由一系列同心环组成,这些环的表面轮廓与厚透镜相应的环等效。但是,每一阶的环前后表面中间的不是必须的那部分材料被去除了。

4

Electromagnetic Theory

4.1 Introduction

In this chapter we derive some of the basic results concerning the propagation of plane single-frequency, electromagnetic waves in homogeneous isotropic media, as well as in anisotropic crystal media. Starting with Maxwell's equations we obtain expressions for the dissipation, storage, and transport of energy resulting from the propagation of waves in material media. We consider in some detail the phenomenon of birefringence, in which the phase velocity of a plane wave in a crystal depends on its direction of polarization. The two allowed modes of propagation in uniaxial crystals—the "ordinary" and "extraordinary" rays—are discussed using the formalism of the index ellipsoid.

We also derive the Fresnel-Kirchhoff diffraction integral. This integral, the key for work in coherent and Fourier optics, will be used extensively throughout this book.

4.2 Complex-Function Formalism

In problems that involve sinusoidally varying time functions, we can save a great deal of manipulation and space by using the complex-function formalism. As an example consider the function

$$a(t) = |A| \cos(\omega t + \phi_a) \tag{4.2.1}$$

where ω is the circular (radian) frequency① and ϕ_a is the phase. Defining the complex amplitude of $a(t)$ by

$$A = |A| e^{i\phi_a} \tag{4.2.2}$$

we can rewrite (4.2.1) as

$$a(t) = \mathrm{Re}[A e^{i\omega t}] \tag{4.2.3}$$

We will often represent $a(t)$ by

$$a(t) = A e^{i\omega t} \tag{4.2.4}$$

instead of by (4.2.1) or (4.2.3). This, of course, is not strictly correct so that when this happens *it is always understood* that what is meant by (4.2.4) is the real part of $A\exp(\mathrm{iwt})$. In most situations the replacement of (4.2.3) by the complex form (4.2.4) poses no problems. The exceptions are cases that involve the product (or powers) of sinusoidal functions. In these cases we must use the real form of the function (4.2.3). To illustrate the case where the distinction between the real and complex form is not necessary, consider the problem of taking the derivative of $a(t)$. Using (4.2.1) we obtain

$$\frac{da(t)}{dt} = \frac{d}{dt}[|A| \cos(\omega t + \phi_a)] = -\omega |A| \sin(\omega t + \phi_a) \tag{4.2.5}$$

If we use instead the complex form (4.2.4), we get

$$\frac{da(t)}{dt} = \frac{d}{dt}(A e^{i\omega t}) = i\omega A e^{i\omega t}$$

Taking, as agreed, the real part of the last expression and using (4.2.2), we obtain (4.2.5).

As an example of a case in which we have to use the real form of the

① The radian frequency ω is to be distinguished from the real frequency $v = \omega/2\pi$

function, consider the product of two sinusoidal functions $a(t)$ and $b(t)$, where

$$a(t) = |A| \cos(\omega t + \phi_a) = \frac{|A|}{2}[e^{i(\omega t + \phi_a)} + e^{-i(\omega t + \phi_a)}] = \mathrm{Re}[Ae^{i\omega t}] \tag{4.2.6}$$

and

$$b(t) = |B| \cos(\omega t + \phi_b) = \frac{|B|}{2}[e^{i(\omega t + \phi_b)} + e^{-i(\omega t + \phi_b)}] = \mathrm{Re}[Be^{i\omega t}] \tag{4.2.7}$$

with $A = |A| \exp(i\phi_a)$ and $B = |B| \exp(i\phi_b)$. Using the real functions, we get

$$a(t)b(t) = \frac{|A||B|}{2}[\cos(2\omega t + \phi_a + \phi_b) + \cos(\phi_a - \phi_b)] \tag{4.2.8}$$

Were we to evaluate the product $a(t)b(t)$ using the complex form of the functions, we would get

$$a(t)b(t) = ABe^{i2\omega t} = |A||B| e^{i(2\omega t + \phi_a + \phi_b)} \tag{4.2.9}$$

Comparing the last result to (4.2.8) shows that the time-independent (dc) term $\frac{1}{2}|A||B|\cos(\phi_a - \phi_b)$ is missing, and thus the use of the complex form led to an error.

Time-Averaging of Sinusoidal Products[2]

Another problem often encountered is that of finding the time average of the product of two sinusoidal functions of the same frequency

$$\overline{a(t)b(t)} = \frac{1}{T}\int_0^T |A| \cos(\omega t - \phi_a) |B| \cos(\omega t + \phi_b) dt \tag{4.2.10}$$

where $a(t)$ and $b(t)$ are given by (4.2.6) and (4.2.7) and the horizontal bar denoter time-averaging. $T = 2\pi/\omega$ is the period of the oscillation. Since the integrand in (4.2.10) is periodic in T, the averaging can be performed over a time T. Using (4.2.8) we obtain directly

$$\overline{a(t)b(t)} = \frac{|A||B|}{2}\cos(\phi_a - \phi_b) \qquad (4.2.11)$$

This last result can be written in terms of the complex amplitudes A and B, defined immediately following (4.2.7), as

$$\overline{a(t)b(t)} = \frac{1}{2}\mathrm{Re}(AB^*) \qquad (4.2.12)$$

This important result will find frequent use throughout the book.

4.3 Considerations of Energy and Power in Electromagnetic Fields

In this section we derive the formal expressions for the power transport, power dissipation, and energy storage that accompany the propagation of electromagnetic radiation in material media. The starting point is Maxwell's curl equations (in MKS units)

$$\nabla \times \boldsymbol{h} = \boldsymbol{i} + \frac{\partial \boldsymbol{d}}{\partial t} \qquad (4.3.1)$$

$$\nabla \times \boldsymbol{e} = -\frac{\partial \boldsymbol{d}}{\partial t} \qquad (4.3.2)$$

and the constitutive equations relating the polarization of the medium to the displacement vectors

$$\boldsymbol{d} = \varepsilon_0 \boldsymbol{e} + \boldsymbol{p} \qquad (4.3.3)$$

$$\boldsymbol{b} = \mu_0(\boldsymbol{h} + \boldsymbol{m}) \qquad (4.3.4)$$

where $\boldsymbol{i}$ is the current density (amperes per square meter); $\boldsymbol{e}(\boldsymbol{r},t)$ and $\boldsymbol{h}(\boldsymbol{r},t)$ are the electric and magnetic field vectors, respectively; $\boldsymbol{d}(\boldsymbol{r},t)$ and $\boldsymbol{b}(\boldsymbol{r},t)$ are the electric and magnetic displacement vectors; $\boldsymbol{p}(\boldsymbol{r},t)$ and $\boldsymbol{m}(\boldsymbol{r},t)$ are the electric and magnetic polarizations (dipole moment per unit volume) of the medium; and ε_0 and μ_0 are the electric and magnetic permeabilities of vacuum, respectively. We adopt the convention of using lowercase letters to denote the time-varying functions, reserving capital letters for the amplitudes of the sinusoidal time functions. For a detailed discussion of Maxwell's equations, the reader is referred to any standard text on electromagnetic theory such as Reference [1].

Using (4.3.3) and (4.3.4) in (4.3.1) and (4.3.2) leads to

$$\nabla \times \boldsymbol{h} = \boldsymbol{i} + \frac{\partial}{\partial t}(\varepsilon_0 \boldsymbol{e} + \boldsymbol{p}) \tag{4.3.5}$$

$$\nabla \times \boldsymbol{e} = -\frac{\partial}{\partial t}\mu_0(\boldsymbol{h} + \boldsymbol{m}) \tag{4.3.6}$$

Taking the scalar (dot) product of (4.3.5) and **e** gives

$$\boldsymbol{e} \times \nabla \times \boldsymbol{h} = \boldsymbol{e} \cdot \boldsymbol{i} + \frac{\varepsilon_0}{2}\frac{\partial}{\partial t}(\boldsymbol{e} \cdot \boldsymbol{e}) + \boldsymbol{e} \cdot \frac{\partial \boldsymbol{p}}{\partial t} \tag{4.3.7}$$

where we used the relation

$$\frac{1}{2}\frac{\partial}{\partial t}(\boldsymbol{e} \cdot \boldsymbol{e}) = \boldsymbol{e} \cdot \frac{\partial \boldsymbol{e}}{\partial t}$$

Next we take the scalar product of (4.3.6) and $\boldsymbol{h}$:

$$\boldsymbol{h} \times \nabla \times \boldsymbol{e} = -\frac{\mu_0}{2}\frac{\partial}{\partial t}(\boldsymbol{h} \cdot \boldsymbol{h}) - \mu_0 \boldsymbol{h} \cdot \frac{\partial \boldsymbol{m}}{\partial t} \tag{4.3.8}$$

Subtraction (4.3.8) from (4.3.7) and using the vector identity

$$\nabla \cdot (\boldsymbol{A} \times \boldsymbol{B}) = \boldsymbol{B} \cdot \nabla \times \boldsymbol{A} - \boldsymbol{A} \cdot \nabla \times \boldsymbol{B} \tag{4.3.9}$$

results in

$$-\nabla \cdot (\boldsymbol{e} \times \boldsymbol{h}) = \boldsymbol{e} \cdot \boldsymbol{i} + \frac{\partial}{\partial t}\left(\frac{\varepsilon_0}{2}\boldsymbol{e} \cdot \boldsymbol{e} + \frac{\boldsymbol{\mu}_0}{2}\boldsymbol{h} \cdot \boldsymbol{h}\right) + \boldsymbol{e} \cdot \frac{\partial \boldsymbol{p}}{\partial t} + \mu_0 \boldsymbol{h} \cdot \frac{\partial \boldsymbol{m}}{\partial t} \tag{4.3.10}$$

We integrate the last equation over an arbitrary volume V and use the Gauss theorem [1]

$$\int_V (\nabla \cdot \boldsymbol{A}) \mathrm{d}v = \int_S \boldsymbol{A} \cdot \boldsymbol{n} \mathrm{d}a \tag{4.3.10a}$$

where **A** is any vector function, **n** is the unit vector normal to the surface S enclosing V, and $\mathrm{d}v$ and $\mathrm{d}a$ are the differential volume and surface elements, respectively. The result is

$$-\int_V \nabla \cdot (\boldsymbol{e} \times \boldsymbol{h}) \mathrm{d}v = -\int_S (\boldsymbol{e} \times \boldsymbol{h}) \cdot \boldsymbol{n} \mathrm{d}a =$$

$$\int_V \left[\boldsymbol{e} \cdot \boldsymbol{i} + \frac{\partial}{\partial t}\left(\frac{\varepsilon_0}{2}\boldsymbol{e} \cdot \boldsymbol{e}\right) + \frac{\partial}{\partial t}\left(\frac{\mu_0}{2}\boldsymbol{h} \cdot \boldsymbol{h}\right) + \boldsymbol{e} \cdot \frac{\partial \boldsymbol{p}}{\partial t} + \mu_0 \boldsymbol{h} \cdot \frac{\partial \mathbf{m}}{\partial t} \right] \mathrm{d}v \tag{4.3.11}$$

According to the conventional interpretation of electromagnetic theory, the left side of (4.3.11), that is

$$-\int_S (\boldsymbol{e} \times \boldsymbol{h}) \times \boldsymbol{n} \mathrm{d}a$$

gives the total power flowing *into* the volume bounded by S. The first term on the right side is the power expended by the field on the moving charges; the sum of the second and third terms corresponds to the rate of increase of the vacuum electromagnetic stored energy $\mathscr{E}_{\text{vac}}$ where

$$\mathscr{E}_{\text{vac}} = \int_V \left[\frac{\varepsilon_0}{2}\boldsymbol{e} \cdot \boldsymbol{e} + \frac{\mu_0}{2}\boldsymbol{h} \cdot \boldsymbol{h} \right] \mathrm{d}v \tag{4.3.12}$$

Of special interest in this book is the next-to-last term

$$\boldsymbol{e} \cdot \frac{\partial \boldsymbol{p}}{\partial t} \tag{4.3.13}$$

which represents the power per unit volume expended by the field *on* the electric dipoles. This power goes into an increase in the potential energy stored by the dipoles as well as into supplying the dissipation that may accompany the change is $\boldsymbol{p}$. We will return to it again in Chapter 5, where we treat the interaction of radiation and atomic systems.

Dipolar Dissipation in Harmonic Fields

According to the discussion in the preceding paragraph, the average power per unit volume expended by the field on the medium electric polarization is

$$\overline{\frac{\text{Power}}{\text{Volume}}} = \overline{\boldsymbol{e} \cdot \frac{\partial \boldsymbol{p}}{\partial t}} \tag{4.3.14}$$

where the horizontal bar denotes time-averaging. Let us assume for the sake of simplicity that $\boldsymbol{e}(t)$ and $\boldsymbol{p}(t)$ are parallel to each other and take their time dependence to be

$$e(t) = \mathrm{Re}[E\mathrm{e}^{\mathrm{i}\omega t}] \tag{4.3.15}$$

$$p(t) = \mathrm{Re}[P\mathrm{e}^{\mathrm{i}\omega t}] \tag{4.3.16}$$

where E and P are the complex amplitudes. The electric susceptibility χ_e of the medium is defined by

$$P = \varepsilon_0 \chi_e E \tag{4.3.17}$$

and is thus a complex number, in general a function of the frequency ω. Substituting (4.3.15) and (4.3.16) in (4.3.14) and using (4.3.17) gives

$$\overline{\frac{\text{Power}}{\text{Volume}}} = \overline{\mathrm{Re}[E\mathrm{e}^{\mathrm{i}\omega t}]\mathrm{Re}[i\omega P\mathrm{e}^{\mathrm{i}\omega t}]} = \frac{1}{2}\mathrm{Re}[i\omega\varepsilon_0\chi_e EE^*] = \frac{\omega}{2}\varepsilon_0 \mid E \mid^2 \mathrm{Re}(i\chi_e) \tag{4.3.18}$$

where in going from the first to the second equality we used (4.2.12). Since χ_e is complex, we can write it in terms of its real and imaginary parts as

$$\chi_e = \chi_e' - \mathrm{i}\chi_e'' \tag{4.3.19}$$

which, when used in (4.3.17), gives

$$\overline{\frac{\text{Power}}{\text{Volume}}} = \frac{\omega\varepsilon_0\chi_e''}{2} \mid E \mid^2 \tag{4.3.20}$$

which is the desired result.

We leave it as an exercise (Problem 4.3) to show that in anisotropic media in which the complex field components are related by

$$P_i = \varepsilon_0 \sum_j \chi_{ij} E_j \tag{4.3.21}$$

the application of (4.3.14) yields

$$\overline{\frac{\text{Power}}{\text{Volume}}} = \frac{\omega}{2}\varepsilon_0 \sum_{i,j} \mathrm{Re}(i\chi_{ij}) E_i^* E_j \tag{4.3.22}$$

The study of power exchange between electrons (bound or free) and electromagnetic fields is central to this book. It is thus instructive to redrive Equation (4.3.18), obtained here formally from Maxwell's equations, using another, and possibly more familiar, point of view.

Consider the case of a single localized electric dipole $\boldsymbol{\mu}$. In this case, the power flow from *the dipole to the field* is obtained by replacing $\boldsymbol{p}$ by $\boldsymbol{\mu}$ in (4.3.13).

$$\underset{\text{dipole}\to\text{field}}{\text{Power}} = -\boldsymbol{e}\cdot\frac{\partial\boldsymbol{\mu}}{\partial t} \tag{4.3.23}$$

This simplest oscillating dipole imaginable is arguably that of an electron whose position is given by

$$x = x_0\cos(\omega t - \phi_e) \tag{4.3.24}$$

which is subject to an electric field

$$e_x = E_0\cos\omega t \tag{4.3.25}$$

The dipole moment of the oscillating electron (whose charge is $-e$) is

$$\mu_x = -ex = -ex_0\cos(\omega t + \phi_e)$$

which leads to

$$\underset{\text{elect.}\to\text{field}}{\text{Power}} = e\,e_x\frac{\partial x}{\partial t} = e\,e_x(t)v(t) = -Fv \tag{4.3.26}$$

where $F = -e\,e_x$ is the force on the electron, while $v = \dfrac{\partial x}{\partial t}$ is its velocity. We have thus shown that the dipolar result for power exchange, Equation (4.3.23) or equivalently (4.3.13) is equivalent to Equation (4.3.27), well familiar from classical dynamics. It is now easy to understand why the power flow from a field to the electron (polarization) depends on their relative phase. The case $\phi_e = -\dfrac{\pi}{2}$, for example, is one where

$$\underset{\text{elect.}\to\text{field}}{\text{Power}} = \omega e x_0 E_0\cos^2\omega t \tag{4.3.27}$$

The electron *is subject to a braking force* at all times and continually loses power to the field. If $\phi_e = \dfrac{\pi}{2}$, the reverse is true; the electron is always accelerated and the power flow is given by (4.3.28) with a minus sign.

The reader is encouraged to make plots of the power flow vs. t during one optical cycle for, say, $\phi_e = 0, \pi$, the power flow reverses sign four times per (optical) period, thus averaging out to zero.

4.4 Wave Propagation in Isotropic Media

Here we consider the propagation of electromagnetic plane waves in homogeneous and isotropic media so that ε and μ are scalar constants. Vacuum is, of course, the best example of such a "medium." Liquids and glasses are material media that, to a first approximation, can be treated as homogeneous and isotropic.① We choose the direction of propagation as z and, taking the plane wave to be uniform in the x-y plane, put $\partial/\partial x = \partial/\partial y = 0$ in (4.3.1) and (4.3.2). Assuming a lossless ($\sigma = 0$) medium, (4.3.1) and (4.3.2) become

$$\nabla \times \boldsymbol{e} = -\mu \frac{\partial \boldsymbol{h}}{\partial t} \tag{4.4.1}$$

$$\nabla \times \boldsymbol{h} = \varepsilon \frac{\partial \boldsymbol{e}}{\partial t} \tag{4.4.2}$$

$$\frac{\partial e_y}{\partial z} = \mu \frac{\partial h_x}{\partial t} \tag{4.4.3}$$

$$\frac{\partial h_y}{\partial z} = -\varepsilon \frac{\partial e_x}{\partial t} \tag{4.4.4}$$

$$\frac{\partial e_x}{\partial z} = -\mu \frac{\partial h_y}{\partial t} \tag{4.4.5}$$

$$\frac{\partial h_x}{\partial z} = \varepsilon \frac{\partial e_y}{\partial t} \tag{4.4.6}$$

$$0 = \mu \frac{\partial h_z}{\partial t} \tag{4.4.7}$$

① The individual molecules making up the liquid or glass are, of course, anisotropic. This anisotropy, however, is averaged out because of the very large number of molecules with random orientations present inside a volume $\sim \lambda^3$.

$$0 = \varepsilon \frac{\partial e_z}{\partial t} \tag{4.4.8}$$

From (4.4.7) and (4.4.8) it follows that the time dependent parts of h_z and e_z are both zero; therefore, a uniform plane wave in a homogeneous isotropic medium can have no longitudinal field components. We can obtain a self-consistent set of equations from (4.4.3) through (4.4.8) by taking e_y and h_x (or e_x and h_y) to be zero.① In this case the last set of equations reduces to Equations (4.4.4) and (4.4.5). Taking the derivative of (4.4.5) with respect to z and using (4.4.4), we obtain

$$\frac{\partial^2 e_x}{\partial z^2} = \mu\varepsilon \frac{\partial^2 e_x}{\partial t^2} \tag{4.4.9}$$

A reversal of the procedure will yield a similar equation for h_y. Since our main interest is in harmonic (sinusoidal) time variation, we postulate a solution in the form of

$$e_x^{\pm} = E_x^{\pm} e^{i(\omega t \mp kz)} \tag{4.4.10}$$

where $E_x^{\pm}$ exp $(\mp ikz)$ are the complex field amplitudes at z. Before substituting (4.4.10) into the wave equation (4.4.9), we may consider the nature of the two functions $e_x^{\pm}$. Taking first e_x^{+}: if an observer were to travel in such a way as to always exercise the same field value, he would have to satisfy the condition

$$\omega t - kz = \text{constant}$$

where the constant is arbitrary and determines the field value "seen" by the observer. By differentiation of the last result, it follows that the observer must travel in the $+z$ direction with a velocity

$$c = \frac{\mathrm{d}z}{\mathrm{d}t} = \frac{\omega}{k} \tag{4.4.11}$$

This is the *phase velocity* of the wave. If the wave were frozen in time,

① More fundamentally it can be easily shown from (4.4.1) and (4.4.2) (see Problem 1.4) that, for uniform plane harmonic waves, **e** and **h** are normal to each other as well as to the direction of propagation. Thus, **x** and **y** can simply be chosen to coincide with the directions of **e** and **h**.

the sepration between two neighboring field peaks—that is, the wavelength—would be

$$\lambda = \frac{2\pi}{k} = 2\pi \frac{c}{\omega} \tag{4.4.12}$$

The e_x^- solution differs only in the sign of k, and thus, according to (4.4.11), it corresponds to a wave traveling with a phase velocity c in the $-z$ direction.

The value of c can be obtained by substituting the assumed solution (4.4.10) into (4.4.9), which results in

$$c = \frac{\omega}{k} = \frac{1}{\sqrt{\mu\varepsilon}} \tag{4.4.13}$$

or

$$k = \omega\sqrt{\mu\varepsilon}$$

The phase velocity in vacuum is

$$c_0 = \frac{1}{\sqrt{\mu_0\varepsilon_0}} = 3 \times 10^8 \text{ m/s}$$

whereas in material media it has the value

$$c = \frac{c_0}{n}$$

where $n = \sqrt{\varepsilon/\varepsilon_0}$ is the *index of refraction*.

Turning our attention next to the magnetic field h_v, we can express it in a manner similar to (4.4.10), in the form of

$$h_x^{\pm} = H_x^{\pm} e^{i(\omega t \mp kz)} \tag{4.4.14}$$

Substitution of this equation into (4.4.4) and using (4.4.10) gives

$$-ikH_y^+ e^{i(\omega t - kz)} = -i\omega\varepsilon E_x^+ e^{i(\omega t - kz)}$$

Therefore, from (4.4.13),

$$H_y^+ = \frac{E_x^+}{\eta} \quad \eta = \sqrt{\frac{\mu}{\varepsilon}} \tag{4.4.15}$$

In vacuum $\eta_0 = \sqrt{\mu_0/\varepsilon_0} \simeq 377$ ohms. Repeating the same steps with H_y^- and E_x^- gives

$$H_y^- = -\frac{E_x^-}{\eta} \qquad (4.4.16)$$

so that for negative ($-z$) traveling waves the relative phase of the electric and magnetic fields is reversed with respect to the wave traveling in the $+z$ direction. Since the wave equation (4.4.9) is a linear differential equation, we can take the solution for the harmonic case as a linear superposition of e_x^+ and e_x^-

$$e_x(z,t) = E_x^+ \mathrm{e}^{\mathrm{i}(\omega t - kz)} - E_x^- \mathrm{e}^{\mathrm{i}(\omega t + kz)} \qquad (4.4.17)$$

and, similarly,

$$h_y(z,t) = \frac{1}{\eta}[E_x^+ e^{i(\omega t - kz)} - E_x^- e^{i(\omega t + kz)}]$$

where E_x^+ and E_x^- are arbitrary complex constants.

Power Flow in Harmonic Fields

The average power per unit area—that is, the intensity (W/m^2)—carried in the direction of propagation by a uniform plane wave is given by (4.3.11) as

$$|I| = |\overline{\boldsymbol{e} \times \boldsymbol{h}}| \qquad (4.4.18)$$

where the horizontal bar denotes time averaging. Since $\mathbf{e} \parallel x$ and $\mathbf{h} \parallel y$, we can obtain from (4.4.18) for the power flow in the z direction

$$I = \overline{e_x h_y}$$

Taking advantage of the harmonic nature of e_x and h_y, we use (4.4.17) and (4.2.12) to obtain

$$I = \frac{1}{2}\mathrm{Re}\{[E_x H_y^*]\} = \frac{1}{2\eta}\mathrm{Re}[E_x^+ \mathrm{e}^{-\mathrm{i}kz} + E_x^- \mathrm{e}^{\mathrm{i}kz}][(E_x^+)^* \mathrm{e}^{\mathrm{i}kz} - (E_x^-)^* \mathrm{e}^{-\mathrm{i}kz}]\} = \frac{|E_x^+|^2}{2\eta} - \frac{|E_x^-|^2}{2\eta} \qquad (4.4.19)$$

The first term on the right side of (4.4.19) gives the intensity associated with the positive ($+z$) traveling wave, whereas the second term represents the negative traveling wave, with the minus sign accounting for the opposite direction of power flow.

An important relation that will be used in a number of later chapters relates the intensity of the pane wave to the stored electromagnetic energy density. We start by considering the second and fourth terms on the right of (4.3.11)

$$\frac{\partial}{\partial t}\left(\frac{\varepsilon_0}{2}\boldsymbol{e}\cdot\boldsymbol{e}\right)+\boldsymbol{e}\cdot\frac{\partial \boldsymbol{p}}{\partial t}$$

Using the relations

$$\boldsymbol{p} = \varepsilon_0\chi_e\boldsymbol{e}$$
$$\varepsilon = \varepsilon_0(1+\chi_e) \tag{4.4.20}$$

we obtain

$$\frac{\partial}{\partial t}\left(\frac{\varepsilon_0}{2}\boldsymbol{e}\cdot\boldsymbol{e}\right)+\boldsymbol{e}\cdot\frac{\partial \boldsymbol{p}}{\partial t} = \frac{\partial}{\partial t}\left(\frac{\varepsilon}{2}\boldsymbol{e}\cdot\boldsymbol{e}\right) \tag{4.4.21}$$

Since we assumed the medium to be lossles, the last term must represent the rate of change of electric energy density stored in the vacuum as well as in the electric dipoles; that is

$$\frac{\mathscr{E}_{\text{electric}}}{\text{Volume}} = \frac{\varepsilon}{2}\boldsymbol{e}\cdot\boldsymbol{e} \tag{4.4.22}$$

The magnetic energy density is derived in a similar fashion using the relations

$$\boldsymbol{m} = \chi_m\boldsymbol{h}$$
$$\mu = \mu_0(1+\chi_m)$$

resulting in

$$\frac{\mathscr{E}_{\text{magnetic}}}{\text{Volume}} = \frac{\mu}{2}\boldsymbol{h}\cdot\boldsymbol{h} \tag{4.4.23}$$

Considering only the positive traveling wave in (4.4.17), we obtain from (4.4.22) and (4.4.23)

$$\frac{\overline{\mathscr{E}}}{\text{Volume}} = \frac{\overline{\mathscr{E}_{\text{magnetic}}+\mathscr{E}_{\text{electric}}}}{\text{Volume}} = \left(\frac{\varepsilon}{2}\right)\overline{(e_x^+)^2}+\left(\frac{\mu}{2}\right)\overline{(h_y^+)^2} =$$
$$\frac{\varepsilon}{4}|E_x^+|^2+\frac{\mu}{4}|H_y^+|^2 = \frac{\varepsilon}{4}|E_x^+|^2+\frac{\mu}{4}\frac{|E_x^+|^2}{\eta^2} = \frac{1}{2}\varepsilon|E_x^+|^2 \tag{4.4.24}$$

where the second equality is based on (4.2.12), and the third and fourth use (4.4.15). Comparing (4.4.24) to (4.4.19), we get

$$\frac{I}{\overline{\mathscr{E}}/\text{Volume}} = \frac{1}{\sqrt{\mu\varepsilon}} = c \tag{4.4.25}$$

where $\overline{\mathscr{E}} = \overline{\mathscr{E}_{\text{magnetic}}} + \overline{\mathscr{E}_{\text{electric}}}$ is the electromagnetic field energy and c is the phase velocity of light in the medium. In terms of the electric field we get, putting $|E_x^+| \equiv E$,

$$I = \frac{ce\,|E|^2}{2} = \frac{|E|^2}{2\eta}$$

$$\eta = \sqrt{\frac{\mu}{\varepsilon}} = 377 \text{ ohms in free space} \tag{4.4.26}$$

4.5 Wave Propagation in Crystals-the Index Ellipsoid

In the discussion of electromagnetic wave propagation up to this point, we have assumed that the medium was isotropic. This causes the induced polarization to be parallel to the electric field and to be related to it by a (scalar) factor that is independent of the direction along which the field is applied. This situation does not apply in the case of dielectric crystals. Since the crystal is made up of a regular periodic array of atoms (or ions), we may expect that the induced polarization will depend in its magnitude and direction, on the direction of the applied field. Instead of the simple relation (4.4.20) linking **p** and **e**, we have

$$\begin{aligned} P_x &= \varepsilon_0(\chi_{11}E_x + \chi_{12}E_y + \chi_{13}E_z) \\ P_y &= \varepsilon_0(\chi_{21}E_x + \chi_{22}E_y + \chi_{23}E_z) \\ P_z &= \varepsilon_0(\chi_{31}E_x + \chi_{32}E_y + \chi_{33}E_z) \end{aligned} \tag{4.5.1}$$

where the capital letters denote the complex amplitudes of the corresponding time-harmonic quantities. The 3×3 array of the χ_{ij} coefficients is called the electric susceptibility tensor. The magnitude of the χ_{ij} coefficients depends, of course, on the choice of the x, y, and z

axes relative to that of the crystal structure. It is always possible to choose x, y, and z in such a way that the off-diagonal elements vanish, leaving

$$\begin{aligned} P_x &= \varepsilon_0 \chi_{11} E_x \\ P_y &= \varepsilon_0 \chi_{22} E_y \\ P_z &= \varepsilon_0 \chi_{33} E_z \end{aligned} \tag{4.5.2}$$

These directions are called the *principal dielectric axes of the crystal*. In this book we will use only the principal coordinate system. We can, instead of using (4.5.2), describe the dielectric response of the crystal by means of the electric permeability tensor ε_{ij}, defined by

$$\begin{aligned} D_x &= \varepsilon_{11} E_x \\ D_y &= \varepsilon_{22} E_y \\ D_z &= \varepsilon_{33} E_z \end{aligned} \tag{4.5.3}$$

From (4.5.2) and the relation

$$\boldsymbol{D} = \varepsilon_0 \boldsymbol{E} + \boldsymbol{P}$$

we have

$$\begin{aligned} \varepsilon_{11} &= \varepsilon_0(1 + \chi_{11}) \\ \varepsilon_{22} &= \varepsilon_0(1 + \chi_{22}) \\ \varepsilon_{33} &= \varepsilon_0(1 + \chi_{33}) \end{aligned} \tag{4.5.4}$$

Birefringence

One of the most important consequences of the dielectric anisotropy of crystals is the phenomenon of birefringence in which the phase velocity of an optical beam propagating in the crystal depends on the direction of polarization of its $\boldsymbol{e}$ vector. Before treating this problem mathematically, we may pause and ponder its physical origin. In an isotropic medium the induced polarization is independent of the field direction so that $\chi_{11} = \chi_{22} = \chi_{33}$, and, using (4.5.4), $\varepsilon_{11} = \varepsilon_{22} = \varepsilon_{33} = \varepsilon$. Since $c = (\mu\varepsilon)^{-1/2}$, the phase velocity is independent of the direction of polarization. In an anisotropic medium the situation is different. Consider, for example, a wave propagating along z. If its electric field is parallel to x, it will induce, according to (4.5.2), only P_x and will consequently "see" an electric permeability ε_{11}. Its phase velocity will thus be $c_x = (\mu\varepsilon_{11})^{-1/2}$. If, on the other hand, the wave is polarized parallel to y, it will propagate with a phase velocity $c_y = (\mu\varepsilon_{22})^{-1/2}$.

Birefringence has some interesting consequences. Consider, as an example, a wave propagating along the crystal z direction and having at some plane, say $z = 0$, a linearly polarized field with equal components along x and y. Since $k_x \neq k_y$, as the wave propagates into the crystal the x and y components get out of phase and the wave becomes elliptically polarized. This phenomenon is discussed in detail and forms the basis of the electro-optic modulation of light.

Returning to the example of a wave propagation along the crystal z direction, let us assume, as in Section 4.3, that the only nonvanishing field components are e_x and h_y. Maxwell's curl equations (4.4.5) and (4.4.4) reduce, in a self-consistent manner, to

$$\begin{aligned} \frac{\partial e_x}{\partial z} &= -\mu \frac{\partial h_y}{\partial t} \\ \frac{\partial h_y}{\partial z} &= -\varepsilon_{11} \frac{\partial e_x}{\partial t} \end{aligned} \tag{4.5.5}$$

Taking the derivative of the first of Equations (4.5.5) with respect to z and then substituting the second equation for $\partial h_y / \partial z$ gives

$$\frac{\partial^2 e_x}{\partial z^2} = \mu\varepsilon_{11} \frac{\partial^2 e_x}{\partial t^2} \tag{4.5.6}$$

If we postulate, as in (4.4.10), a solution in the form

$$e_x = E_x e^{i(\omega t - k_x z)} \tag{4.5.7}$$

then Equation (4.5.6) becomes

$$k_x^2 E_x = \omega^2 \mu\varepsilon_{11} E_x$$

Therefore, the propagation constant of a wave polarized along x and traveling along z is

$$k_x = \omega\sqrt{\mu\varepsilon_{11}} \tag{4.5.8}$$

Repeating the derivation but with a wave polarized along the y axis, instead of the x axis, yields $k_y = \omega\sqrt{\mu\varepsilon_{22}}$.

Index Ellipsoid

As shown above, in a crystal the phase velocity of a wave propagating along a given direction depends on the direction of its polarization. For propagation along z, as an example, we found that Maxwell's equations admitted two solutions: one with its linear polarization along x and the second along y. If we consider the propagation along some arbitrary direction in the crystal, the problem becomes more difficult. We have to determine the directions of polarization of the two allowed waves, as well as their phase velocities. This is done most conveniently using the so-called index ellipsoid

$$\frac{x^2}{\varepsilon_{11}/\varepsilon_0} + \frac{y^2}{\varepsilon_{22}/\varepsilon_0} + \frac{z^2}{\varepsilon_{33}/\varepsilon_0} = 1 \qquad (4.5.9)$$

This is the equation of a generalized ellipsoid with major axes parallel to x, y, and z whose respective lengths are $2\sqrt{\varepsilon_{11}/\varepsilon_0}$, $2\sqrt{\varepsilon_{22}/\varepsilon_0}$, and $2\sqrt{\varepsilon_{33}/\varepsilon_0}$. The procedure for finding the polarization directions and the corresponding phase velocities for a *given* direction of propagation is as follows: Determine the ellipse formed by the intersection of a plane through the origin and normal to the direction of propagation and the index ellipsoid (4.5.9). The directions of the major and minor axes of this ellipse are those of the two allowed polarizations,① and the lengths of these axes are $2n_1$ and $2n_2$, where n_1 and n_2 are the indices of refraction of the two allowed solutions. The two waves propagate thus, with phase velocities c_o/n_1 and c_o/n_2, respectively, where $c_0 = (\mu_0\varepsilon_0)^{-1/2}$ is the phase velocity in vacuum. A formal proof of this procedure is given in

---

① These are actually the directions of the **D**, not of the **E**, vector. In a crystal these two are separated, in general, by a small angle; see References [2] and [3].

References [2 ~ 4].

To illustrate the use of the index ellipsoid, consider the case of a uniaxial crystal (that is, a crystal with a single axis of threefold, fourfold, or sixfold symmetry). Taking the direction of this axis as z, symmetry considerations dictate that $\varepsilon_{11} = \varepsilon_{22}$.① *Defining the principal indices of refraction* n_o and n_e by

$$n_o^2 \equiv \frac{\varepsilon_{11}}{\varepsilon_o} = \frac{\varepsilon_{22}}{\varepsilon_o} \qquad n_e^2 \equiv \frac{\varepsilon_{33}}{\varepsilon_o} \tag{4.5.10}$$

the equation of the index ellipsoid (4.5.9) becomes

$$\frac{x^2}{n_o^2} + \frac{y^2}{n_o^2} + \frac{z^2}{n_e^2} = 1 \tag{4.5.11}$$

This is an ellipsoid of revolution with the circular symmetry axis parallel to z. The z major axis of the ellipsoid is of length $2n_e$, whereas that of the x and y axes is $2n_o$. The procedure of using the index ellipsoid is illustrated by Figure 4.1.

The direction of propagation is along **s** and is at an angle θ to the (optic) z axis. Because of the circular symmetry of (4.5.11) about z, we can choose, without any loss of generality, the y axis to coincide with the projection of **s** on the x-y plane. The intersection ellipse of the plane normal to **s** with the ellipsoid is shaded in the figure. The two allowed polarization directions are parallel to the axes of the ellipse and thus correspond to the line segments OA and OB. They are consequently perpendicular to **s** as well as to each other. The two waves polarized along these directions have, respectively, indices of refraction given by $n_e(\theta) = |OA|$ and $n_o(\theta) = |OB|$. The first of these two waves, which is polarized along OA, is called the *extraordinary wave*. Its direction of polarization varies with θ following the intersection point A. Its index of refraction is given by the length of OA. It can be determined using Figure

---

① See, for example, J. F. Nye, *Physical Properties of Crystals*. New York: Oxford University Press, 1957.

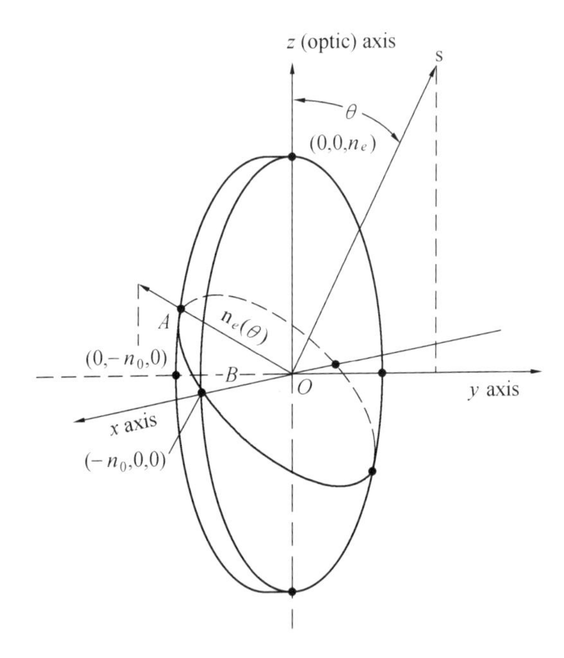

Figure 4.1 Construction for finding indices of refraction and allowed polarization for a given direction of propagation **s**. The figure shown is for a uniaxial crystal with $n_x = n_y = n_o$.

4.2, which shows the intersection of the index ellipsoid with the y-z plane.

Using the relations

$$n_e^2(\theta) = z^2 + y^2$$

$$\frac{z}{n_e^2(\theta)} = \sin\theta$$

and the equation of the ellipse

$$\frac{y^2}{n_o^2} + \frac{z^2}{n_e^2} = 1$$

we obtain

$$\frac{1}{n_e^2(\theta)} = \frac{\cos^2\theta}{n_o^2} + \frac{\sin^2\theta}{n_e^2} \tag{4.5.12}$$

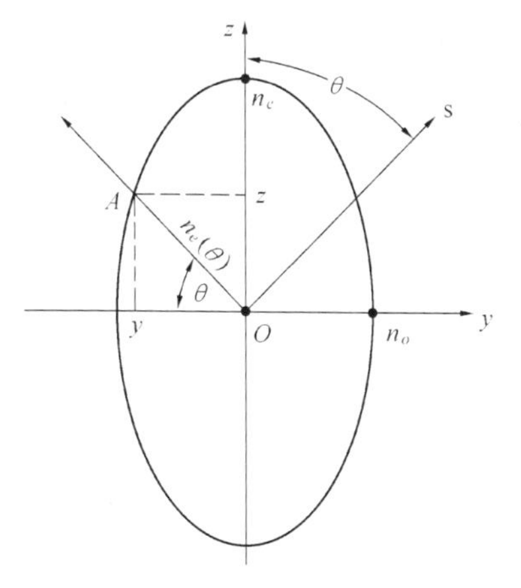

Figure 4.2 Intersection of the index ellipsoid with the *z*-*y* plane. $|OA| = n_e(\theta)$ is the index of refraction of the extraordinary wave propagating in the direction **s**.

Thus, for $\theta = 0°$, $n_e(0°) = n_o$, and for $\theta = 90°$, $n_e(90°) = n_e$.

The ordinary wave remains, according to Figure 4.1, polarized along the same direction *OB* independent of θ. It has an index of refraction n_o. The amount of birefringence $n_e(\theta) - n_o$ thus varies from zero for $\theta = 0°$ (that is, propagation along the optic axis) to $n_e - n_o$ for $\theta = 90°$.

Normal (index) Surfaces

Consider the surface in which the distance of a given point from the origin is equal to the index of refraction of a wave propagating along this direction. This surface. not to be confused with the index ellipsoid, is called the normal surface. It is constructed using the index ellipsoid (Figure 4.1). The normal surface of the extraordinary ray is constructed by measuring along each direction $\mathbf{s}(\theta, \phi)$ the corresponding index $n_e(\theta, \phi)$, which is the distance *OA* in Figure 4.2. For a uniaxial crystal, this results in an ellipsoid of revolution about the *z* axis as illustrated by the outer line in Figure 4.4. For the ordinary ray we plot the distance $OB = n_o$ (which is independent of θ, ϕ), resulting in the inner sphere of Figure 4.4.

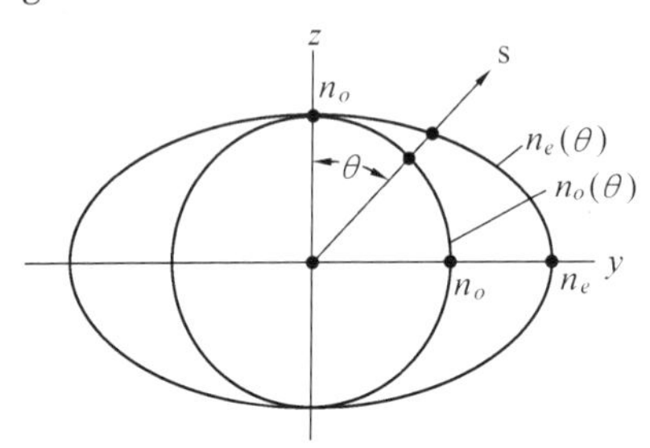

Figure 4.3 Intersection of *s*-*z* plane with normal surfaces of a positive uniaxial crystal ($n_e > n_o$).

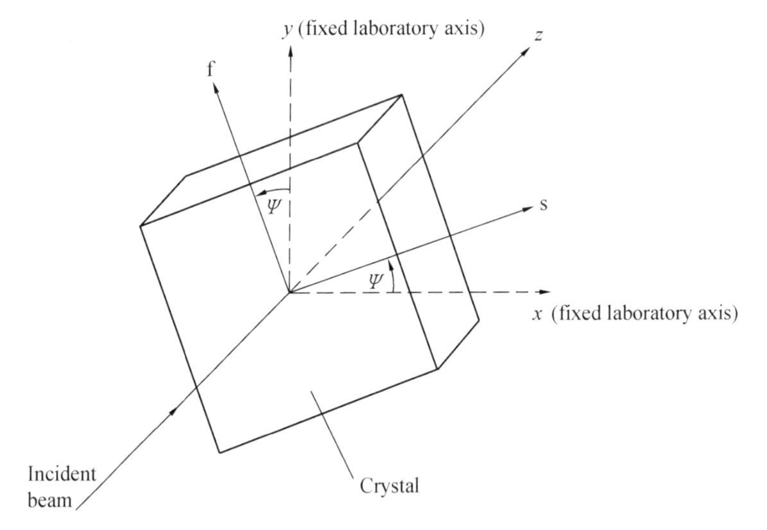

Figure 4.4 A retardation plate rotated at an angle ψ about the z axis. f("fast") and s ("show") are the two principal dielectric axes of the crystal for light propagating along z (see Section 4.4). The x and y axes are fixed in the laboratory frame.

4.6 Jones Calculus and Its Application to Propagation in Optical Systems with Birefringent Crystals

Many sophisticated optical systems, such as electro-optic modulation (to be discussed in Chapter 9) involve the passage of light through a train of polatzers and birefringent (retardation) plates. The effect of each individual element ether polarizer or retardation plate, on the polarization state of the transmitted light can be described by simple means. However, when an optical system consists of many such elements, each oriented at a different azimuthal angle, the calculation of the overall transmission becomes complicated and is greatly facilitated by a systematic approach. The Jones calculus, invented in 1940 by R. C. Jones [5], is a powerful matrix method in which the state of polarization is represented by a two-component actor, while each optical element is represented by a 2×2 matrix. The overall transfer matrix for the whole system is obtained by multiplying all the individual element matrices, and the polarization state of the transmitted light is computed by multiplying the vector representing the input beam by the overall matrix. We will first develop the mathematical formulation of the Jones matrix method and then apply it to some cases of practical interest.

We have shown in the previous section that a unidirectional light propagation in a birefringent crystal generally consists of a linear superposition of two orthogonally polarized waves—the eigenwaves. These eigenwaves, for a given direction of propagation, have well-defined phase velocities and directions of polarization. The birefringent crystals may be either uniaxial ($n_x = n_y, n_z$) or biaxial ($n_x \neq n_y \neq n_z$). However, the most commonly used materials, such as calcite and quartz, are uniaxial. In a uniaxial crystal, these eigenwaves are the so-called *ordinary* and *extraordinary* waves, whose properties were derived in Section 4.5. The directions of polarization for these eigenwaves are mutually orthogonal and are called the *slow* and *fast* axes of the crystal for the given direction of propagation. Retardation plates are usually cut in such a way that the c axis lies in the plane of the plate surfaces. Thus the propagation direction of normally incident light is perpendicular to the c axis.

Retardation plates (also called wave plates) are polarization-state converters, or transformers. The polarization state of a light beam can be converted to any other polarization state by using a suitable retardation plate. In formulating the Jones matrix method, we assume that there is no reflection of light from either surface of the plate and the light is totally transmitted through the plate surfaces. In practice, there is some reflection, though most plates are coated with "antireflection" coatings to greatly reduce such reflection. Referring to Figure 4.4, we consider a light beam that is incident normally on a retardation plate along the z axis with a polarization state described by the Jones column vector

$$\boldsymbol{V} = \begin{pmatrix} V_x \\ V_y \end{pmatrix} \tag{4.6.1}$$

where V_x and V_y are two complex numbers representing the complex field amplitudes along x and y. The x, y and z axes are *fixed* laboratory axes. To determine how the light propagates in the retardation plate, we need to resolve it into a linear combination of the fast and slow eigenwaves of the crystal. This is done by the coordinate transformation

$$\begin{pmatrix} V_s \\ V_f \end{pmatrix} = \begin{pmatrix} \cos\psi & \sin\psi \\ -\sin\psi & \cos\psi \end{pmatrix} \begin{pmatrix} V_x \\ V_y \end{pmatrix} \equiv R(\psi) \begin{pmatrix} V_x \\ V_y \end{pmatrix} \tag{4.6.2}$$

V_s is the slow component of the polarization vector $\mathbf{V}$, whereas V_f is the fast component. The slow and fast axes are fixed in the crystal. The angle between the fast axis and the y direction is ψ. These two components are eigenwaves of the retardation plate and will propagate with their own phase velocities and polarizations as discussed in Section 4.5. Because of the difference in phase velocity, the two components undergo a different phase delay in passage through the crystal. This retardation changes the polarization state of the emerging beam.

Let n_s and n_f be the refractive indices of the slow and fast eigenwaves, respectively. The polarization state of the emerging beam in the crystal coordinate system is thus given by

$$\begin{pmatrix} V'_s \\ V'_f \end{pmatrix} = \begin{pmatrix} \exp\left(-in_s \frac{\omega}{c} l\right) & 0 \\ 0 & \exp\left(-in_f \frac{\omega}{c} l\right) \end{pmatrix} \begin{pmatrix} V_s \\ V_f \end{pmatrix} \tag{4.6.3}$$

Where l is the thickness of the plate and ω is the radian frequency of the light beam. The phase retardation is defined as the difference of the phase delays (exponents) in (4.6.3)

$$\Gamma = (n_s - n_f) \frac{\omega l}{c} \tag{4.6.4}$$

Notice that the phase retardation Γ is a measure of the relative change in phase, not the absolute change. The birefringence of a typical crystal retardation plate is small, that is, $|n_s - n_f| \ll n_s, n_f$. Consequently, the absolute change in phase caused by the plate may be hundreds of times greater than the phase retardation. Let ϕ be the mean absolute phase change

$$\phi = \frac{1}{2}(n_s + n_f) \frac{\omega l}{c} \tag{4.6.5}$$

Then Equation (4.6.3) can be written in terms of ϕ and Γ as

$$\begin{pmatrix} V'_s \\ V'_f \end{pmatrix} = e^{-i\phi} \begin{pmatrix} e^{-i\frac{\Gamma}{2}} & 0 \\ 0 & e^{i\frac{\Gamma}{2}} \end{pmatrix} \begin{pmatrix} V_s \\ V_f \end{pmatrix} \tag{4.6.6}$$

The Jones vector of the polarization state of the emerging beam in the xy coordinate system is given by transforming back from the crystal to the laboratory coordinate system

$$\begin{pmatrix} V'_x \\ V'_y \end{pmatrix} = \begin{pmatrix} \cos\psi & -\sin\psi \\ \sin\psi & \cos\psi \end{pmatrix} \begin{pmatrix} V'_s \\ V'_f \end{pmatrix} \tag{4.6.7}$$

By combining Equations (4.6.2), (4.6.6), and (4.6.7), we can write the transformation due to the retardation plate as

$$\begin{pmatrix} V'_x \\ V'_y \end{pmatrix} = R(-\psi) W_0 R(\psi) \begin{pmatrix} V_x \\ V_y \end{pmatrix} \tag{4.6.8}$$

where $R(\psi)$ is the rotation matrix of (4.6.2) and W_0 is the Jones matrix of (4.6.6) for the retardation plate. These are given, respectively, by

$$R(\psi) = \begin{pmatrix} \cos\psi & \sin\psi \\ -\sin\psi & \cos\psi \end{pmatrix} \tag{4.6.9}$$

and

$$W_0 = e^{-i\phi} \begin{pmatrix} e^{-i\Gamma/2} & 0 \\ 0 & e^{i\Gamma/2} \end{pmatrix} \tag{4.6.10}$$

The phase factor $e^{-i\phi}$ can usually be left out.① A retardation plate, characterized by its phase retardation Γ and its azimuth angle ψ, is represented by the product of three matrices

$$W(\psi, \Gamma) \equiv W = R(-\psi) W_0 R(\psi) =$$

$$\left| \begin{matrix} e^{-i(\Gamma/2)} \cos^2\psi + e^{i(\Gamma/2)} \sin^2\psi & -i\sin\frac{\Gamma}{2}\sin(2\psi) \\ -i\sin\frac{\Gamma}{2}\sin(2\psi) & e^{-i(\Gamma/2)} \sin^2\psi + e^{i(\Gamma/2)} \cos\psi \end{matrix} \right| \tag{4.6.11}$$

Note that the Jones matrix of a wave plate is a unitary matrix, that is,

$$W^+ W = 1$$

where the dagger$^+$ signifies the Hermitian conjugate ($W^*_{ij} = (W^+)_{ji}$). The passage of a polarized light beam through a wave plate is described mathematically as a unitary transformation. Many physical properties are invariant under unitary trans-formations; these include the orthogonal relation between the Jones vectors and the magnitude of the Jones vectors. Thus, if the polarization states of two beams are mutually orthogonal, they will remain orthogonal after passing through an arbitrary wave plate.

The Jones matrix of an ideal, lossless homogeneous and linear, thin

---

① The overall phase factor $\exp(-i\phi)$ is only important when the output field $\mathbf{V}'$ is combined coherently with another field.

plate polarizer oriented with its transmission axis parallel to the laboratory x axis is

$$P_0 = e^{-i\theta}\begin{pmatrix} 1 & 0 \\ 0 & 0 \end{pmatrix} \tag{4.6.12}$$

where θ is the absolute phase accumulated due to the finite optical thickness of the polarizer. The Jones matrix of a polarizer rotated by an angle ψ from the x anis about z is given by

$$P = R(-\psi)P_0R(\psi) \tag{4.6.13}$$

Thus, if we neglect the (in this case unimportant) absolute phase θ, the Jones matrix representations of the polarizers oriented so as to transmit light with electric field vectors parallel to the x and y laboratory axes, respectively, are given by

$$P_x = \begin{pmatrix} 1 & 0 \\ 0 & 0 \end{pmatrix} \quad \text{and} \quad P_y = \begin{pmatrix} 0 & 0 \\ 0 & 1 \end{pmatrix} \tag{4.6.14}$$

To find the effect of an arbitrary train of retardation plates and polarizers on the polarization state of polarized light, we multiply the Jones vector of the incident beam by the ordered product of the matrices of the various elements.

Example: A Half-Wave Retardation Plate

A half-wave plate has a phase retardation of $\Gamma = \pi$. According to Equation (4.6.4), an x-cut[①] (or-cut) uniaxial crystal will act as a half-wave plate, provided the thickness is $l = \lambda/2(n_e - n_o)$. We will determine the effect of a half-wave plate on the polarization state of a transmitted light beam. The azimuth angle of the wave plate is taken as 45° and the incident beam as vertically (y) polarized. The Jones vector for the incident beam can be written as

① A crystal plate is called *x-cut* if its facets are perpendicular to the principal x axis.

$$V = \begin{pmatrix} 0 \\ 1 \end{pmatrix} \tag{4.6.15}$$

and the Jones matrix for the half-wave plate is obtained by using Equation (4.6.11) with $\Gamma = \pi$, $\psi = \pi/4$

$$W = \frac{1}{\sqrt{2}}\begin{pmatrix} 1 & -1 \\ 1 & 1 \end{pmatrix}\begin{pmatrix} -i & 0 \\ 0 & i \end{pmatrix}\frac{1}{\sqrt{2}}\begin{pmatrix} 1 & 1 \\ -1 & 1 \end{pmatrix} = \begin{pmatrix} 0 & -i \\ -i & 0 \end{pmatrix} \tag{4.6.16}$$

The Jones vector for the emerging beam is obtained by multiplying Equations (4.6.16) and (4.6.15); the result is

$$V = \begin{pmatrix} -i \\ 0 \end{pmatrix} = -i\begin{pmatrix} 1 \\ 0 \end{pmatrix} \tag{4.6.17}$$

which corresponds to horizontally (x) polarized light. The effect of the half-wave plate is thus to rotate the input polarization by 90°. It can be shown that for a general azimuth angle ψ, the half-wave plate will rotate the polarization by an angle 2ψ (see Problem 1.7a). In other words, linearly polarized light remains linearly polarized, except that the plane of polarization is rotated by an angle of 2ψ.

When the incident light is circularly polarized, a half-wave plate will convert right-hand circularly polarized light into left-hand circularly polarized light and vice versa, regardless of the azimuth angle. The proof is left as an exercise (see Problem 4.7). Figure 4.5 illustrates the effect of a half-wave plate.

Example: A Quarter-Wave Plate

A quarter-wave plate has a phase retardation of $\Gamma = \pi/2$. If the plate is made of an x-cut (or y-cut) uniaxially anistropic crystal, the thickness is $l = \lambda/4(n_e - n_o)$ (or odd multiplies thereof). Suppose again that the azimuth angle of the plate is $\psi = 45°$ and the incident beam is vertically polarized. The Jones vector for the incident beam is given by Equation (4.6.15). The Jones matrix for this quarter-wave plate is

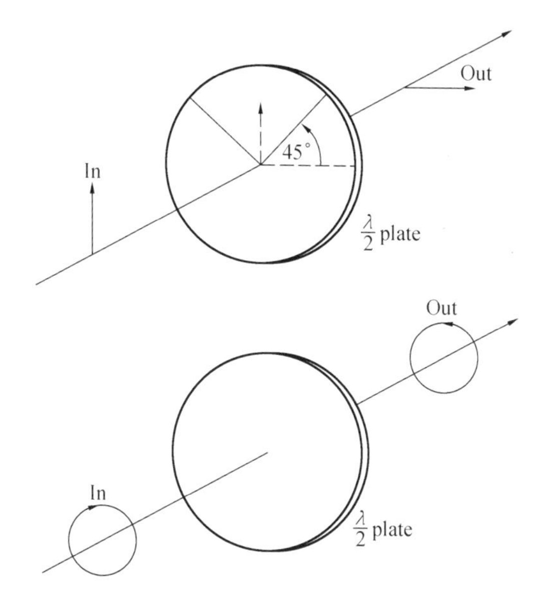

Figure 4.5 The effect to a half-wave plate on the polarization state of a beam.

$$W = \frac{1}{\sqrt{2}}\begin{pmatrix} 1 & -1 \\ 1 & 1 \end{pmatrix}\begin{pmatrix} e^{-i\pi/4} & 0 \\ 0 & e^{i\pi/4} \end{pmatrix}\frac{1}{\sqrt{2}}\begin{pmatrix} 1 & 1 \\ -1 & 1 \end{pmatrix} = \frac{1}{\sqrt{2}}\begin{pmatrix} 1 & -i \\ -i & 1 \end{pmatrix} \tag{4.6.18}$$

The Jones vector of the emerging beam is obtained by multiplying Equations (4.6.18) and (4.6.15) and is given by

$$\boldsymbol{V}' = -\frac{i}{\sqrt{2}}\begin{pmatrix} 1 \\ i \end{pmatrix} \tag{4.6.19}$$

To an observer facing the z direction (direction of propagation), this is a clockwise circularly polarized light. The effect of a 45°-oriented quarter-wave plate is thus to convert vertically polarized light into circularly polarized light. If the incident beam is horizontally polarized, the emerging beam will be circularly polarized in a counterclockwise sense. The effect of this quarter-wave plate is illustrated in Figure 4.6.

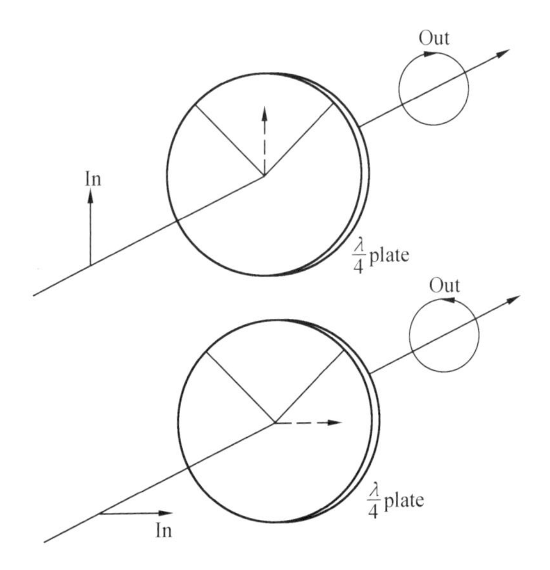

Figure 4.6 The effect of a quarter-wave plate on the polarization state of a linearly polarized input wave.

Intensity Transmission

Up to this point our development of the Joes calculus was concerned with the polarization state of the light beam. In many cases, we need to determine the transmitted intensity. The combination of retardation plates and polarizers is often used to control or modulate the transmitted optical intensity. Because the phase retardation of each wave plate is wavelength-dependent, the polarization state of the emerging beam and its intensity (when polarizers are present) depend on the wavelength of the light. Let us represent the field as a Jones vector

$$\boldsymbol{V} = \begin{pmatrix} V_x \\ V_y \end{pmatrix} \tag{4.6.20}$$

The intensity is taken using (4.2.12) and (4.4.24) as proportional to

$$I = \boldsymbol{V} \cdot \boldsymbol{V}^* = | V_x |^2 + | V_y |^2 \tag{4.6.21}$$

If the output beam is given by

$$V' = \begin{pmatrix} V_x' \\ V_y' \end{pmatrix} \tag{4.6.22}$$

the transmissivity of the optical system is calculated as

$$\frac{|V_x'|^2 + |V_y'|^2}{|V_x|^2 + |V_y|^2} \tag{4.6.23}$$

Example: A Birefringent Plate Sandwiched between Parallel Polarizers

Referring to Figure 4.7, we consider a birefringent plate sandwiched between a pair of parallel polarizers. The plate is oriented so that the slow and fast axes are at 45° with respect to the polarizer. Let the birefringence be $n_e - n_o$ and the plate thickness be d. The phase retardation is then given by

$$\Gamma = 2\pi(n_e - n_o)\frac{d}{\lambda} \tag{4.6.24}$$

and the corresponding Jones matrix is, according to Equation (4.6.11), with $\psi = 45°$

$$W = \begin{pmatrix} \cos\frac{1}{2}\Gamma & -i\sin\frac{1}{2}\Gamma \\ -i\sin\frac{1}{2}\Gamma & \cos\frac{1}{2}\Gamma \end{pmatrix} \tag{4.6.25}$$

The incident beam, after it passes through the front polarizer, is polarized parallel to y and can be represented by

$$V = \begin{pmatrix} 0 \\ 1 \end{pmatrix} \tag{4.6.26}$$

we shall take, arbitrarily, the intensity corresponding to (4.6.26) as unity. The Jones vector representation of the electric field vector of the transmitted beam is obtained as follows

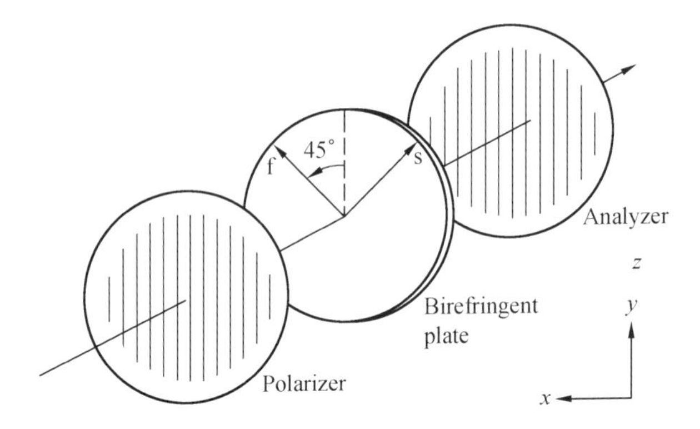

Figure 4.7 A birefringent plate sandwiched between a pair of parallel polarizers.

$$\boldsymbol{V}' = \begin{pmatrix} 0 & 0 \\ 0 & 1 \end{pmatrix} \begin{pmatrix} \cos\frac{1}{2}\Gamma & -i\sin\frac{1}{2}\Gamma \\ -i\sin\frac{1}{2}\Gamma & \cos\frac{1}{2}\Gamma \end{pmatrix} \begin{pmatrix} 0 \\ 1 \end{pmatrix} = \begin{pmatrix} 0 \\ \cos\frac{1}{2}\Gamma \end{pmatrix} \tag{4.6.27}$$

The transmitted beam is y polarized with an intensity given by

$$I = \cos^2\frac{1}{2}\Gamma = \cos^2\left[\frac{\pi(n_e - n_o)d}{\lambda}\right] \tag{4.6.28}$$

It can be seen from Equation (4.6.28) that the transmitted intensity is a sinusoidal function of the wave number (λ^{-1}) and peaks at $\lambda = (n_e - n_o)d$, $(n_e - n_o)d/2$, $(n_e - n_o)d/3, \ldots$. The wave-number separation between transmission maxima increases with decreasing plate thickness.

Example: A Birefringent Plate Sandwiched between a Pair of Crossed Polarizers

If we rotate the analyzer shown in Figure 4.7 by 90°, then the input and output polarizers are crossed. The transmitted beam for this case is obtained as follows:

$$\boldsymbol{V}' = \begin{pmatrix} 1 & 0 \\ 0 & 0 \end{pmatrix} \begin{pmatrix} \cos\frac{1}{2}\Gamma & -i\sin\frac{1}{2}\Gamma \\ -\sin\frac{1}{2}\Gamma & \cos\frac{1}{2}\Gamma \end{pmatrix} \begin{pmatrix} 0 \\ 1 \end{pmatrix} = -i \begin{pmatrix} \sin\frac{\Gamma}{2} \\ 0 \end{pmatrix} \tag{4.6.29}$$

The transmitted beam is horizontally (x) polarized with an intensity relative to the input value given by

$$\frac{I_{\text{out}}}{I_{\text{in}}} = \sin^2\frac{1}{2}\Gamma = \sin^2\left[\frac{\pi(n_e - n_o)d}{\lambda}\right] \tag{4.6.30}$$

This is again a sinusoidal function of the wave number λ^{-1}. The transmission spectrum consists of a series of maxima at $\lambda = 2(n_e - n_o)d$, $2(n_e - n_o)d/3$.... There wavelengths correspond to phase retardations of $\pi, 3\pi, 5\pi, \ldots$, that is, when the wave plate becomes a "half-wave" plate or odd integral multiples of a half-wave plate.

Circular Polarization Representation

Up to this point we represented the state of the propagating field as a vector **V** [Equation (4.6.1)] with components V_x and V_y.

$$\boldsymbol{V} = \begin{pmatrix} V_x \\ V_y \end{pmatrix} \tag{4.6.31}$$

The orthogonal unit vectors (*basis* vector set) in this representation are

$$\boldsymbol{V}_x = \begin{pmatrix} 1 \\ 0 \end{pmatrix} \quad \boldsymbol{V}_y = \begin{pmatrix} 0 \\ 1 \end{pmatrix} \tag{4.6.32}$$

The above choice is most convenient when dealing with birefringent crystals, since the propagating eigenmodes in this case are linearly and orthogonally polarized. It is often more convenient to express the field in terms of "basis" vectors that are circular polarized [6]. This is the case, for example, when we propagate through a magnetic medium. We define a wave of unit amplitude seen rotating in the CCW sense by an observer gazing along the $+z$ axis as $\begin{Bmatrix} 1 \\ 0 \end{Bmatrix}$, while $\begin{Bmatrix} 0 \\ 1 \end{Bmatrix}$ denotes a CW rotating wave. As in the case of the linearly polarized basis vectors $\begin{pmatrix}1\\0\end{pmatrix}$ and $\begin{pmatrix}0\\1\end{pmatrix}$, $\begin{Bmatrix}1\\0\end{Bmatrix}$ and $\begin{Bmatrix}0\\1\end{Bmatrix}$ constitute a complete set that can be used to describe a transverse field of arbitrary polarization. Let $\boldsymbol{V}$ be some such field. We can write

$$\boldsymbol{V} = V_x\begin{pmatrix}1\\0\end{pmatrix} + V_y\begin{pmatrix}0\\1\end{pmatrix} \equiv \begin{pmatrix}V_x\\V_y\end{pmatrix} \tag{4.6.33}$$

or alternatively

$$\boldsymbol{V} = V_+\begin{Bmatrix}1\\0\end{Bmatrix} + V_-\begin{pmatrix}0\\1\end{pmatrix} \equiv \begin{Bmatrix}V_+\\V_-\end{Bmatrix} \tag{4.6.34}$$

The $\begin{pmatrix}V_x\\V_y\end{pmatrix}$ and $\begin{pmatrix}V_+\\V_-\end{pmatrix}$ representations of a given vector can be derived from each other by a 2×2 matrix[①]

$$\begin{Bmatrix}V_+\\V_-\end{Bmatrix} = \frac{1}{2}\begin{vmatrix}1 & i\\1 & -i\end{vmatrix}\begin{pmatrix}V_x\\V_y\end{pmatrix} \equiv T\begin{pmatrix}V_x\\V_y\end{pmatrix} \tag{4.6.35}$$

$$\begin{pmatrix}V_x\\V_y\end{pmatrix} = \begin{vmatrix}1 & 1\\-i & i\end{vmatrix}\begin{Bmatrix}V_+\\V_-\end{Bmatrix} \equiv S\begin{Bmatrix}V_+\\V_-\end{Bmatrix} \tag{4.6.36}$$

so that $T = S^{-1}$. As an example, consider a (unit) field polarized along x. Its rectangular representation is $\begin{pmatrix}1\\0\end{pmatrix}$, while its rotating representation is

$$\begin{Bmatrix}V_+\\V_-\end{Bmatrix} \equiv \begin{vmatrix}1 & i\\1 & -i\end{vmatrix}\begin{pmatrix}1\\0\end{pmatrix} = \begin{Bmatrix}1\\1\end{Bmatrix} \tag{4.6.37}$$

i. e., equal and in-phase admixture of the two counter-rotating eigenmodes. Conversely, a clockwise, circularly polarized unit wave

---

① The form of T implies that at $t = 0$ the rotating waves $\begin{Bmatrix}1\\0\end{Bmatrix}$ and $\begin{Bmatrix}0\\1\end{Bmatrix}$ are parallel to the x axis.

$\begin{Bmatrix} 0 \\ 1 \end{Bmatrix}$, for example, is expressed in the rectangular representation by

$$\begin{pmatrix} V_x \\ V_y \end{pmatrix} = \begin{vmatrix} 1 & 1 \\ -i & i \end{vmatrix} \begin{Bmatrix} 0 \\ 1 \end{Bmatrix} = \begin{pmatrix} 1 \\ i \end{pmatrix} \tag{4.6.38}$$

Faraday Rotation

In certain optical materials containing magnetic atoms or ions, the natural modes of propagation are the two counter-rotating, circularly-polarized (CP) waves described above. The z direction is usually that of an applied magnetic field or that of the spontaneous magnetization As in the case of a birefringent crystal, the two CP modes propagate with different phase velocities or, equivalently, have different indices of refraction. This difference is due to the fact that the individual atomic magnetic moments precess in a unique sense about the zaxis and thus interact differently (have slightly displaced resonances) with the two CP waves. Using the notation of (4.6.34), we can describe the propagation of a wave with arbitrary transverse polarization by first resolving it, at $z = 0$, into its components $\begin{Bmatrix} V_+(0) \\ 0 \end{Bmatrix}$ and $\begin{Bmatrix} 0 \\ V_-(0) \end{Bmatrix}$ and propagating each component with its appropriate phase delay through the magnetic medium

$$\begin{Bmatrix} V_+(z) \\ V_-(z) \end{Bmatrix} = \begin{Bmatrix} V_+(0) \\ 0 \end{Bmatrix} e^{-i(\omega/c)n+z} + \begin{Bmatrix} 0 \\ V_-(0) \end{Bmatrix} e^{-i(\omega/c)n-z} =$$

$$e^{-(i/2)(\theta_+ + \theta_-)} \begin{vmatrix} e^{(i/2)(\theta_- - \theta_+)} & 0 \\ 0 & e^{-(i/2)(\theta_- - \theta_+)} \end{vmatrix} \begin{Bmatrix} V_+(0) \\ V_-(0) \end{Bmatrix} \tag{4.6.39}$$

where $\theta_z \equiv (\omega/c)n \pm z$ is the phase delay for the $(+)$ or $(-)$ circularly polarized wave. Ignoring the prefactor $\exp[-(i/2)(\theta_+ + \theta_-)]$ (it is only relative phase delays that are of interest here) we rewrite (4.6.39) as

$$\begin{Bmatrix} V_+(z) \\ V_-(z) \end{Bmatrix} = \begin{vmatrix} e^{i\theta_F(z)} & 0 \\ 0 & e^{-i\theta_F(z)} \end{vmatrix} \begin{Bmatrix} V_+(0) \\ V_-(0) \end{Bmatrix} \tag{4.6.40}$$

$$\theta_F(z) \equiv \frac{1}{2}(\theta_- - \theta_+) = \frac{\omega}{2c}(n_- - n_+)z \equiv$$

Faraday rotation angle (4.6.41)

The reason for calling θ_F the *Faraday rotation angles* becomes clear if we consider the effect of a magnetic medium on an incident wave that is described in the rectangular component representation

$$\begin{pmatrix} V_x(z) \\ V_y(z) \end{pmatrix} = T^{-1} \begin{vmatrix} e^{i\theta_F(z)} & 0 \\ 0 & e^{-i\theta_F(z)} \end{vmatrix} T \begin{pmatrix} V_x(0) \\ V_y(0) \end{pmatrix} =$$

$$\begin{vmatrix} \cos\theta_F & -\sin\theta_F \\ \sin\theta_F & \cos\theta_F \end{vmatrix} \begin{pmatrix} V_x(0) \\ V_y(0) \end{pmatrix} = \tag{4.6.42}$$

$$R(-\theta_F) \begin{pmatrix} V_x(0) \\ V_y(0) \end{pmatrix} \tag{4.6.43}$$

where $R(-\theta_F)$ is, according to (4.6.2), the matrix representing a *rotation* by $-\theta_F$ about the z axis. The output field is thus rotated by $-\theta_F$ with respect to the input field.

There exists a basic difference between propagation in a magnetic medium and in a dielectric birefringent medium. Consider the latter case first. An x'-polarized eigenwave, for instance, propagating along the z direction in a birefringent crystal has a phase velocity c/n_x, where x' is a principal dielectric axis. The same applies to the wave propagating in the reverse direction. The medium is *reciprocal*. In a magnetic medium the story is quite different. Let a linearly polarized wave traveling from left to right a distance L (in the $+z$ direction) undergo a (Faraday) rotation of its plane of polarization of $+\theta$ (the sign signifies the sense of the rotation about the *direction of propagation*). A wave traveling in the $-z$ direction in the crystal will experience a rotation of $-\theta(L)$ *about the new direction* $(-z)$ *of propagation*. This is because the *magnetic field or, equivalently, the magnetic polarization* now points in the opposite direction relative to the direction of propagation. (The wave can differentiate between $+z$ and $-z$—something that it cannot do in a birefringent crystal). The medium is termed nonreciprocal. The net effect of a round trip through the medium of length L is that the plane of polarization of the beam returning to the starting, $z = 0$ plane, is rotated by $2\theta_F(L)$. This Faraday rotation is used to make optical isolators to block off back-reflected radiation. The basic configuration of a Faraday isolator is illustrated in Figure 4.8 (a) and (b). A linearly polarized incident wave is rotated by 45° in passage through the Faraday medium and then passed fully by the output polarizer. A reflected wave is rotated an additional 45° in the return trip and is thus blocked off by the input

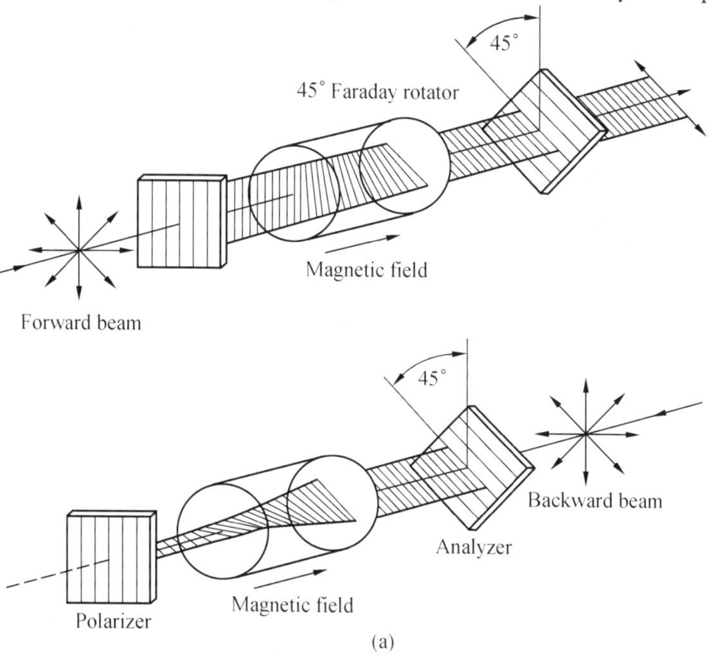

Figure 4.8 (a) A Farday isolator comprised of two polarizers trtated by 45° relative to each other on either side of a magnetic medium with $\theta_F = 45°$

polarizer. Faraday isolators now form an integral part of most optical communication systems employing semiconductor diode lasers since such lasers are extremely sensitive to even small amounts of reflected light that cause instabilities in their power and frequency characteristics.

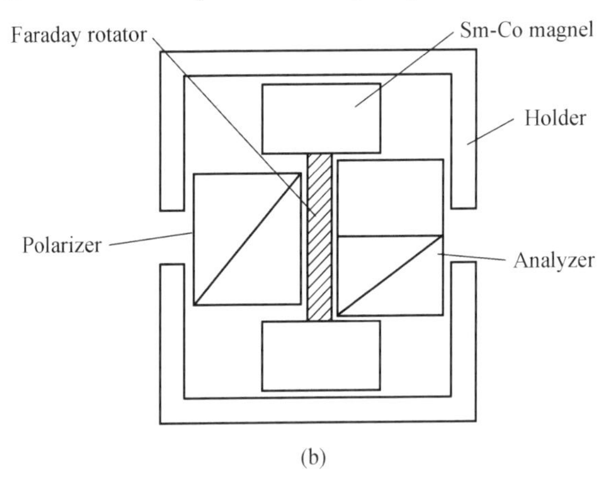

Figure 4.8 (*continued*) (b) A cross-sectional view of a practical commercial isolator. (Courtesy of Namiki Precision Jewel Company.)

4.7 Diffraction of Electromagnetic Waves

In this section, we will derive a most important result, the Fresnel-Kirchhoff Diffraction Integral to describe how an electromagnetic field propagates between any two planes, say, the planes $z = 0$ and $z = L$ of an isotropic medium, as shown in Figure 4.9. The result , Equation (4.7.13), is the starting point to many of the important developments in coherent optics and image processing [9] and is used in this book to treat optical resonators (Section 4.9) and image processing by four-wave mixing.

We will depart from the conventional, Green function, derivation [10] and employ a "linear system" approach that is formally identical to that used to analyze electrical and mechanical systems. As a bonus we will find the mathematics and resulting formulae *identical* to those we will employ in Chapter 3 to describe how narrow, information-bearing, optical pulses propagate in fibers. This analogy will prove both interesting and useful.

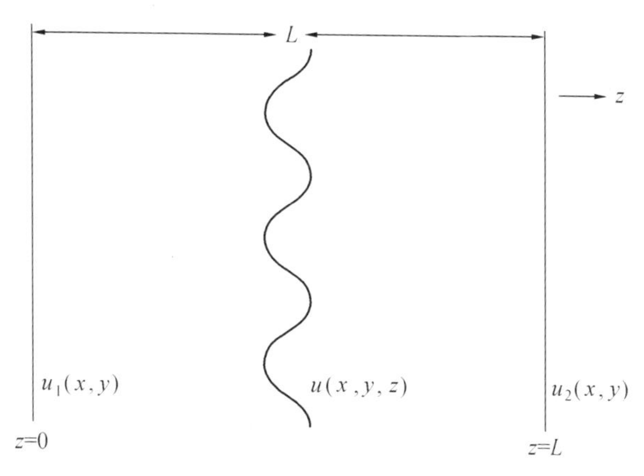

Figure 4.9 Propagation of an optical wave from $z=0$ to $z=L$.

In this section we will make ample use of the Fourier transform (FT) relations

$$f(x,y) = \iint \bar{f}(k_x,k_y)\exp[i(k_x x + k_y y)]\mathrm{d}k_x\mathrm{d}k_y$$
$$\bar{f}(k_x,k_y) = \frac{1}{4\pi^2}\iint f(x,y)\exp[i(k_x x + k_y y)]\mathrm{d}x\mathrm{d}y \tag{4.7.1}$$

between a function f(x, y) and its $FT\bar{f}(k_x, k_y)$. Another important result that follows directly from (4.7.1) is that of the convolution integral

$$\iint \bar{f}_1(k_x,k_y)\bar{f}_2(k_x,k_y)\exp[i(k_x x + k_y y)]\mathrm{d}k_x\mathrm{d}k_y =$$
$$\frac{1}{4\pi^2}\iint f_1(x',y')f_2(x-x',y-y')\mathrm{d}x'\mathrm{d}y' \equiv$$
$$\frac{1}{4\pi^2}f_1 * f_2 \tag{4.7.2}$$

where the * symbol represents the convolution integral.

We consider an electromagnetic field of (radian) frequency ω and a scalar complex amplitude $u(x, y, z)$

$$E(x, y, z, t) = u(x, y, z)\exp(i\omega t) \tag{4.7.3}$$

where E is some Cartesian coordinate of the vector field. The complex amplitude $u(x, y, z)$ obeys

$$\nabla^2 u + k^2 u = 0 \quad k^2 = \omega^2 \mu\varepsilon \tag{4.7.4}$$

which equation results when we substitute (4.7.3) in the equation $\nabla^2 u = \mu\varepsilon \dfrac{\partial^2 u}{\partial t^2}$ which is the three-dimensional extension of Equation (4.4.9) to the case of an isotropic medium.

The scenario considered next is one where we are given an "input" optical field whose complex amplitude at $z = 0$ is

$$u_1(x, y) \equiv u(x, y, 0) \tag{4.7.5}$$

Our task is to find the "output" field u_2 at $z = L$:

$$u_2(x, y) \equiv u(x, y, L) \tag{4.7.6}$$

We have already shown that a simple solution of (4.7.4) is of the form $u = e^{\pm ikz}$, which represents a plane wave propagating along $\mp z$. It can be verified, by direct substitution, that a plane wave propagating along any arbitrary direction, say that of $\boldsymbol{k}$ can be taken as

$$u(x, y, z) = u(\boldsymbol{k})\exp(-i\boldsymbol{k} \cdot \boldsymbol{r}), \; k = \omega\sqrt{\mu\varepsilon} = \frac{2\pi n}{\lambda} \tag{4.7.7}$$

since this expression also satisfies (4.7.4). It follows that an arbitrary superposition of plane waves, each of the form of (4.7.7), propagating along all conceivable directions, i.e.,

$$u(x, y, z) = \iint F(k_x, k_y)\exp[i(k_x x + k_y y) - i\sqrt{k^2 - k_x^2 - k_y^2} z]\mathrm{d}k_x \mathrm{d}y_y \tag{4.7.8}$$

where F is an arbitrary function, also satisfies Equation (4.7.4). The integrand of (4.7.8) represents the amplitude of a plane wave propagating along the direction

$$\boldsymbol{k} = -\hat{x}k_x - \hat{y}k_y + \hat{z}\sqrt{k^2 - k_x^2 - k_y^2} \qquad (4.7.9)$$

so that integration over k_x and k_y takes in waves propagating along all possible directions. The form of k_z in (4.7.9) is dictated by the requirement that $u(x, y, z)$ satisfy the wave equation (4.7.4) or, physically, to ensure that each plane wave component in (4.7.8) has the same wavelength $\lambda/n = 2\pi/k$ as appropriate to a wave propagating in an isotropic medium.

It follows from (4.7.8) that

$$u(x, y, 0) \equiv u_1(x, y) = \iint F(k_x, k_y)\exp[i(k_x x + k_y y)]\mathrm{d}k_x \mathrm{d}k_y \qquad (4.7.10)$$

Equation (4.7.10) is of the form of the Fourier integral transform (FT) of (4.7.1) $F(k_x, k_y)$ is thus the Fourier transform of the input field $u_1(x, y)$.

$$F(k_x, k_y) = \frac{1}{4\pi^2}\iint u_1(x, y)\exp[-i(k_x x + k_y y)]\mathrm{d}k_x \mathrm{d}k_y \equiv \bar{u}_1(k_x, k_y) \qquad (4.7.11)$$

We can thus rewrite (4.7.8) as

$$u(x, y, z) = \iint \bar{u}_1(k_x, k_y)\exp[i(k_x x + k_y y - \sqrt{k^2 - k_x^2 - k_y^2})z]\mathrm{d}k_x \mathrm{d}k_y \qquad (4.7.12)$$

Equation (4.7.12) constitutes a powerful algorithm for the propagation of a monochromatic wave. It states that given an "input" field with an arbitrary complex amplitude $u_1(x, y)$ at some plane, which without loss of generality we take as $z = 0$, we can write down the field amplitude at any other plane z by first deriving the Fourier integral transform $\bar{u}_1(k_x, k_y)$ of $u_1(x, y)$ and then using it in the integral of Equation (4.7.12).

Our main preoccupation in this book is with beamlike optical waves. By this we mean optical beams whose plane wave components propagate either along the z axis of at small angles to it. Mathematically, this is equivalent to stating that in Equation (4.7.12) $\bar{u}(k_x, k_y)$ is appreciable only in a region where k_x, $k_y \ll k$. THis is the, so-called, paraxial condition. When it applies, we can approximate $\sqrt{k^2 - k_x^2 - k_y^2} \approx k\left(1 - \frac{k^2x + k_y^2}{2k^2}\right)$ and re-express (4.7.8), taking $z = L$, as

$$u_2(x,y) \equiv u(x,y,L) = \exp(-ikL)\iint \left\{\overbrace{\bar{u}_1(k_x,k_y)}^{①}\overbrace{\exp\left(i\frac{k_x^2 + k_y^2}{2k}L\right)}^{②}\right] \cdot \exp[i(k_xx + k_yy)]\mathrm{d}k_x\mathrm{d}k_y \tag{4.7.13}$$

We recognize the integral of (4.7.13) as the inverse Fourier transform (IFT) of the product of the two functions, designated as 1 and 2. Using the convolution theorem (4.7.2), we write the integral as the convolution of the two functions

$$u_1(x,y) = \mathrm{IFT}\{\bar{u}_1(k_x,k_y)\} \tag{4.7.14}$$

$$p(x,y) = \mathrm{IFT}\left\{\exp\left(i\frac{k_x^2 + k_y^2}{2k}L\right)\right\} = \frac{2\pi ik}{L}\exp\left(-ik\frac{x^2 + y^2}{2L}\right) \tag{4.7.15}$$

which results in

$$u_2(x,y) = \frac{1}{4\pi^2}\iint u_1(x',y')p(x-x',y-y')\mathrm{d}x'\mathrm{d}y' \equiv \frac{1}{4\pi^2}u_1 * p \tag{4.7.16}$$

$$\frac{i}{\lambda L}\exp(-ikL)\iint u_1(x',y')\exp\left\{\frac{-ik}{2L}[(x-x')^2 + (y-y')^2]\right\} \times \mathrm{d}x'\mathrm{d}y' \tag{4.7.17}$$

Equation (4.7.17) is the celebrated Fresnel-Kirchhoff diffraction integral, which we will use on a number of occasions throughout the book.

Another useful relation results when we compare Equation (4.7.13) to the first of Equations (4.7.1). It follows directly that

$$\bar{u}_2(k_x, k_y) = \bar{u}_1(k_x, k_y)\exp\left(i\frac{k_x^2 + k_y^2}{2k}L\right) \quad (4.7.18)$$

where we left out the constant delay factor exp $(-ikL)$ since it does not depend on k_x, k_y. It can always be restored, on the rare occasions when needed, by multiplying the right side through by exp $(-ikL)$.

The effect of propagation of (4.7.18) a distance L by a monochromatic beam in a homogeneous and isotropic medium can thus be represented by multiplying the Fourier transform $\bar{u}(k_x, k_y)$ of the input amplitude $u_1(x, y)$ by the "transfer function,"

$$T(k_x, k_y) = \exp\left(i\frac{k_x^2 + k_y^2}{2k}L\right) \quad (4.7.19)$$

This point of view, which embodies the spirit of system theory [8], is completely equivalent to the spatial relationship (4.7.17). It is often more convenient to use one or the other of these relations depending on the problem at hand.

Problems

4.1 Consider the problem of finding the time average

$$\overline{a^2(t)} = \frac{1}{T}\int_0^T a^2(t)\mathrm{d}t$$

of

$$a(t) = |A_1|\cos(\omega_1 t + \phi_1) + |A_2|\cos(\omega_2 t + \phi_2) = \mathrm{Re}[V_a(t)]$$

where

$$V_a(t) = A_1 e^{i\omega_1 t} + A_2 e^{i\omega_2 t}$$

and $A_{1,2} = |A_{1,2}|e^{i\phi_{1,2}}$. $V_a(t)$ is called the *analytical signal* of $a(t)$. Assume that $(\omega_1 - \omega_2) \ll \omega_1$ and integrate over a time T, which is long compared to the period $2\pi/\omega_{1,2}$ but short compared to the beat period $2\pi/$

$(\omega_1 - \omega_2)$.① Show that

$$a^2(t) = \frac{1}{2}[V_a(t)V_a^*(t)]$$

4.2 Show how we can sue the analytic functions as defined by Problem 4.1 to find the time average

$$\overline{a(t)b(t)} = \frac{1}{T}\int_0^T a(t)b(t)\mathrm{d}t$$

where $a(t)$ is the same as in Problem 4.1, and the analytic function of $b(t)$ is

$$Vb(t) = [A_3 e^{i\omega_3 t} + A_4 e^{i\omega_4 t}]$$

so that $b(t) = Re[V_b(t)]$. Assume that the difference between any two of the frequencies ω_1, ω_2, ω_3, and ω_4 is small compared to the frequencies themselves. (Answer: $\overline{a(t)b(t)} = \frac{1}{2} Re[V_a(t) V_b^*(t)]$.)

4.3 Derive Equation (4.3.22).

4.4 Starting with Maxwell's curl equations [(4.3.1),(4.3.2)] and taking $\mathbf{i} = 0$, show that in the case of a harmonic (sinusoidal) uniform plane wave, the field vectors **e** and **h** are normal to each other as well as to the direction of propagation. [*Hint*: Assume the wave to have the form $e^{i(\omega t - \boldsymbol{k}\cdot\boldsymbol{r})}$ and show by actual differentiation that we can formally replace the operator ∇ in Maxwell's equations by $-i\boldsymbol{k}$].

4.5 Derive Equation (4.4.19).

4.6 A linearly polarized electromagnetic wave is incident normally at $z = 0$ on the x-y face of a crystal so that it propagates along its z axis. The crystal electric permeability tensor referred to x, y, and z is diagonal with elements ε_{11}, ε_{22} and ε_{11}. If the wave is polarized initially so that it has equal components along x and y, what is the state of its polarization

① When this condition is fulfilled, a(t) consists of a sinusoidal function with a "slowly" varying amplitude and is often called a *quasi-sinusoid*.

at the plane z, where

$$(k_x - k_y)z = \frac{\pi}{2}$$

Plot the position of the electric field vector in this plane at times $t = 0$, $\pi/6\omega$, $\pi/3\omega$, $\pi/2\omega$, $2\pi/3\omega$, $5\pi/6\omega$.

4.7 *Half-wave plate*. A half-wave plate has a phase retardation of $\Gamma = \pi$. Assume that the plate is oriented so that the azimuth angle (i.e., the angle between the x axis and the slow axis of the plate) is ψ.

a. Find the polarization state of the transmitted beam, assuming that the incident beam is linearly polarized in the y direction.

b. Show that a half-wave plate will convert right-hand circularly polarized light into left-hand circularly polarized light, and vice versa, regardless of the azimuth angle of the plate.

c. Lithium tantalate ($LiTaO_3$) is a uniaxial crystal with $n_o = 2.1391$ and $n_e = 2.1432$ at $\lambda = 1\ \mu m$. Find the half-wave-plate thickness at this wavelength, assuming the plate is cut in such a way that the surfaces are perpendicular to the x axis of the principal coordinate (i.e., x-cut).

4.8 *Quarter-wave plate*. A quarter-wave plate has a phase retardation of $\Gamma = \pi/2$. Assume that the plate is oriented in a direction with azimuth angle ψ.

a. Find the polarization state of the transmitted beam, assuming that the incident beam is polarized in the y direction.

b. If the polarization state resulting from (a) is represented by a complex number on the complex plane, show that the locus of these points as ψ varies from 0 to $\frac{1}{2}\pi$ is a branch of a hyperbola. Obtain the equation of the hyperbola.

c. Quartz ($\alpha = SiO_2$) is a uniaxial crystal with $n_o = 1.53283$ and $n_e = 1.541\ 52$ at $\lambda = 1.159\ 2\ \mu m$. Find the thickness of an x-cut quartz quarter-wave plate at this wavelength.

4.9 A matrix $\boldsymbol{A}$ is called unitary if

$$\boldsymbol{A}^{+}\boldsymbol{A} = \boldsymbol{A}\boldsymbol{A}^{+} = 1$$

where **1** is the unity matrix and the Hermitian conjugate A + of matrix $\boldsymbol{A}$ is defined by $(A+)_{ij} = A_{ji}^{*}$. *Show that if* $\boldsymbol{A}$ is unitary

$$\sum_{j} A_{ji}^{*} A_{jk} = \delta_{ik}$$

This property will be needed in Problem 1.10d.

4.10 Polarization transformation by a wave plate. A wave plate is characterized by its phase retardation Γ and azimuth angle ψ.

a. Find the polarization state of the emerging beam, assuming that the incident beam is polarized in the x direction.

b. Use a complex number to represent the resulting polarization state obtained in (a).

c. The polarization state of the incident x-polarized beam is represented by a point at the origin of the complex plane. Show that the transformed polarization state can be anywhere on the complex plane, provided Γ can be varied from 0 to 2π and ψ can be varied from 0 to $\frac{1}{2}\pi$. Physically, this means that any polarization state can be produced from linearly polarized light, provided a proper wave plate is available.

d. Show that the Jones matrix **W** of a wave plate is unitary, that is

$$W^{+} \cdot W = 1, (W^{+})_{ij} \equiv W_{ji}^{*}$$

where the dagger indicates Hermitian conjugation [see Equation (4.6.11)].

e. Let $\boldsymbol{V}_1'$ and $\boldsymbol{V}_2'$ be the transformed Jones vectors of $\boldsymbol{V}_1$ and $\boldsymbol{V}_2$, respectively. Show that if $\boldsymbol{V}_1$ and $\boldsymbol{V}_2$ are orthogonal, so are $\boldsymbol{V}_1'$ and $\boldsymbol{V}_2'$. ($\boldsymbol{A}$ and $\boldsymbol{B}$ are orthogonal if $\boldsymbol{A} \cdot \boldsymbol{B}^{*} = 0$.)

4.11 Show that the (Jones) matrix (in the rectangular eigenwave representation) of a birefringent plate with a retardation Γ that is rotated by an angle ψ from the x axis is

$$W(\Gamma,\psi) = \begin{vmatrix} \cos^2\psi\exp(-i\Gamma/2) + \sin^2\psi\exp(+i\Gamma/2) & -i\sin2\psi\sin\Gamma/2 - \\ i\sin2\psi\sin\Gamma/2 & \sin^2\psi\exp(-i\Gamma/2) + \cos^2\psi\exp(+i\Gamma/2) \end{vmatrix}$$

Derive an expression for the intensity transmission through a system consistent of a polarizer $\parallel$ to $\hat{x}$, a Faraday rotator θ, wave plate with retardation Γ rotated an angle ψ from the x axis, and a crossed output polarizer ($\parallel$ to $\hat{y}$).

4.12

a. Show that $(AB)^+ = B^+ A^+$.

b. Show that if an optical element is represented by a unitary matrix, the intensity of an incident wave of arbitrary polarization is preserved in passage through the element.

c. SHow that the matrix representing a train of arbitrary retardation plates is unitary.

4.13

a. Show that in an isotropic medium we can take the general solution of the wave equation of a monochromatic field

$$\nabla^2 \boldsymbol{E} + k^2 \boldsymbol{E} = 0$$

as

$$\boldsymbol{E}(\boldsymbol{r}) = \int_{-\infty}^{\infty}\int_{\infty}^{\infty} \mathrm{d}k_x \mathrm{d}k_y \boldsymbol{A}(\boldsymbol{k}) \mathrm{e}^{-\mathrm{i}(k_x x + k_y y + \sqrt{k^2 - k_x^2 - k_y^2} z)}$$

where $\boldsymbol{A}(\boldsymbol{k})$ is an arbitary vector lying on a plane normal to $\boldsymbol{k}$ and $k^2 = \omega^2 \mu\varepsilon$.

b. Show that if $\boldsymbol{E}(\boldsymbol{r})$ is specified at some plane S, say the plane $z = 0$, as $\boldsymbol{E}(x, y, 0)$, then

$$\boldsymbol{A}(\boldsymbol{k}) = \left(\frac{1}{2\pi}\right)^2 \int\int_S \mathrm{d}x \mathrm{d}y \boldsymbol{E}(x, y, 0) \mathrm{e}^{\mathrm{i}(k_x x + k_y y)}$$

where $\boldsymbol{k} = \hat{\boldsymbol{x}} k_x + \hat{\boldsymbol{y}} k_y + \hat{\boldsymbol{z}} \sqrt{k^2 - k_x^2 - k_y^2}$. [*Hint*: Compare Equations (1) and (2) with $z = 0$ to the integral Fourier transform relationships.]

c. Assume that at $z = 0$ the field is given by

$$E(x,y,0) = E_0 \begin{cases} -a/2 \leqslant x \leqslant a/2 \\ -a/2 \leqslant y \leqslant a/2 \end{cases}$$

and is zero everywhere else. Find the spreading angle of the beam far away to the right of the aperture. [*Hint*: Each $\boldsymbol{k}$ signifies a direction of propagation so that $|A(\boldsymbol{k})|^2$ can be viewed as the distribution function of directions $\boldsymbol{k}$ of the beam to the right of the aperture.]

4.14 Consider light propagating through a sequence of $\lambda/2$ retardation plates ($\Gamma = \pi$) as shown:

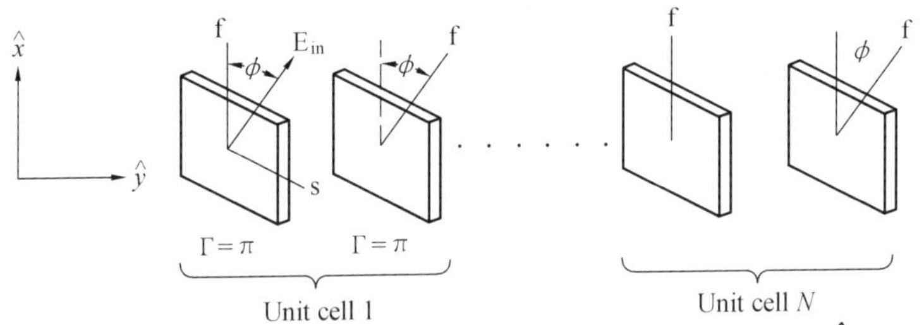

Each unit cell consists of two plates whose surfaces are normal to $\hat{y}$— one with its f (fast) axis ‖ to $\hat{z}$ and one rotated by ϕ about $\hat{x}$. Find the effect of propagation through N cells on a beam initially polarized as shown. Solve the problem first by simple considerations, if possible, then formally.

4.15 Show that if we define

$$\boldsymbol{v}_g \equiv \nabla k\omega\boldsymbol{k}$$

$$\boldsymbol{v}_e \equiv \frac{\boldsymbol{E} \times \boldsymbol{H}}{\frac{1}{2}[\boldsymbol{E} \cdot \varepsilon \boldsymbol{E} + \boldsymbol{H} \cdot \mu \boldsymbol{H}]}$$

in a crystal, then

$$\boldsymbol{v}_g = \boldsymbol{v}_e$$

a. Recall that ε is a tensor.

b. After giving the problem a real try, you may consult *Optical Waves in Crystals*, A. Yariv and P. Yeh, New York: Wiley, P. 79, 1983.

4.16 Derive the transfer matrix of a polarizer whose transmission direction is rotated by α from the laboratory x axis.

4.17 Prove relation (4.6.35).

4.18 The electric field at some point in a medium and the position x of an electron are given by

$$e_x(\boldsymbol{r}, t) = R_e[E_x e^{i(\omega t + \phi_E)}]$$

$$x(\boldsymbol{r}, t) = R_e[X e^{i(\omega t + \phi_e)}]$$

Plot the instantaneous power flow $e e_x v_x$ during one complete oscillation cycle for the case $\phi E - \phi e = 0, -\pi/2, +\pi/2$. What is the (cycle) average power flow for each case ($v_x = \mathrm{d}x/\mathrm{d}t$)?

References

1. Ramo, S., J. R. Whinnery, and T. Van Duze, *Fields and Waves in Communication Electronics*. New York: Wiley, 1965.

2. Born, M., and E. Wolf, *Principles of Optics*. New York: Macmillan, 1964.

3. Yariv, A., *Quantum Electronics*, 2d ed. New York: Wiley, 1975.

4. Yariv, A., and P. Yeh, *Optical Waves in Crystals*. New York: Wiley, 1983.

5. Jones, R. C., "New calculus for the treatment of optical systems," *J. Opt. Soc. Am.* 31:488,1941.

6. Yariv, Ammon, "Operator algebra for propagation problems involving phase conjugation and nonreciprocal elements," *Appl. Opt.* 26:4538, 1987.

7. Lohman, A. W., and D. Medlovic, "Temporal filtering with time lenses," *Appl. Opt.* 31:6212, 1992.

8. Papoulis, A. "Pulse compression, fiber communication, and diffraction: a unified approach," *J. Opt. Soc. Am.* 11:3, 1994.

9. Goodman, J. W., *Introduction to Fourier Optics*, 2d ed. San

Francisco: McGraw Hill, 1995, Ch. 3.

10. See, for example, M. Born and E. Wolf, *Principles of Optics*, 6th ed. Oxford: New York: Pergamon, 1986.

New Words and Expressions

propagation	*n*. 传播
plane	*adj*. 平面的
homogeneous	*adj*. 均一的,均匀的
isotropic	*adj*. 等方性的,各向同性的
dissipation	*n*. 消散,消耗
birefringence	*n*. 双折射
polarization	*n*. 偏振
index ellipsoid	*n*. 折射率椭球
coherent optics	*n*. 相干光学
real part	*n*. 实部
derivative	*n*. 微商,导数
displacement vector	*n*. 置换向量
dipole	*n*. 偶极
permeability of vaccum	*n*. 真空中的磁导率
susceptibility	*n*. (物)磁化系数
dipole moment	*n*. 偶极数
approximation	*n*. (数)近似值
uniform plane wave	*n*. 均匀平面波
longitudinal	*adj*. 纵向的
self-consistent	*adj*. 自相容的
harmonic	*adj*. 谐振
arbitrary constant	*n*. 任意,常数
index of refraction	*n*. 折射率
revolution	*n*. 旋转
circular symmetry	*adj*. 圆对称
hormal	*n*. (数)法线

uniaxial crystal	*n*. 单轴晶体
origin	*n*. (数)原点
unidirectional	*adj*. 单向的
superposition	*n*. 重叠
orthogonalize	*vt*. 使正交
eigen wave	*n*. 特征波
biaxial	*n*. 双轴
calcite	*n*. 方解石
emerging beam	*n*. 新兴光束
phase retardation	*n*. 相位延迟
azimuth	*n*. 方位角
unitary	*adj*. 单一的,一元的
conjugate	*v*. 变化,变换
polarizer	*n*. 偏光器,起偏镜
magnetic moment	*n*. 磁矩
precess	*vi*. 产生进动
resonance	*n*. 共振,谐振
nonreciprocal	*adj*. 单向的
back-reflected radiation	*n*. 背反射,辐射
term	*vt*. 把…称为
convolution integral	*n*. 卷积积分
algorithm	*n*. 运算法则
monochromatic	*adj*. 单色,单频的

NOTES

1. This power goes into an increase in the potential energy stored by the dipoles as well as into supplying the dissipation that may accompany the change in p.

该能量一方面利用偶极子使势能存储增加,另一方面补充了随 P 变化而造成的能量消耗。

2. One of the most important consequences of the dielectric

anisotropy of crystals is the phenomenon of birefringence in which the phase velocity of an optical beam propagating in the crystal depends on the direction of polarization of its e vector.

晶体介电各向异性的最重要结果之一就是双折射现象,在这一现象中,晶体中传播的光束的相速度依赖于向量 **E** 的偏振方向。

3. To determine how the light propagates in the retardation plate, we need to resolve it into a linear combination of the fast and slow eigenwaves of the crystal.

为确定光线在延迟镜中是如何传播的,我们需要将其分解为晶体中快慢特征波的线性组合。

4. The Jones matrix representations of the polarizers oriented so as to transmit light with electric field vectors parallelive to the x and y laboratory axes, respectively, are given by (4.6.14).

为了使光在传播时其电场矢量分别平行于 x、y 实验轴,已定向的偏光器的 JONES 的矩阵描述由(4.6.14)式给出。

5. This difference is due to the fact that the individual atomic magnetic moments precess in a unique sense about the z axis and thus interact differently (have slightly displaced resonances) with the two cp waves.

这种差别是由于单个原子的磁力矩的进动只与 z 轴有关,进而相互作用与两个 cp 波是不同的(有微弱的置换谐振)。

6. It states that given an "input" field with an arbitrary complex amplitude $u_1(x, y)$ at some plane, which without loss of generality we take as $z=0$, we can write down the field amplitude at any other plane z by first deriving the fourier integral transform $u_1(k_x, k_y)$ of $u_1(x, y)$ and then using it in the integral of Equation (4.7.12).

这表明,对于给定的在某些面上具有任意复振幅 $u_1(x, y)$的输入光场。(不失一般性,我们取 $z=0$)那么就能够得出在其他任意 z 值平面上的场振幅。其过程如下:首先,推导出 $u_1(x, y)$的傅里叶积分变换 $u_1(k_x, k_y)$,而后将其带入方程(4.7.12)的积分式中。

PART TWO
SEMICONDUCTOR LASERS

5

Semiconductor Lasers—Theory and Applications

5.1 Introduction

The semiconductor laser invented in 1961 [1 ~ 3] is the first laser to make the transition from a research topic and specialized applications to the mass consumer market. It is by economic standards and the degree of its applications, the most important of all lasers.

The main features that distinguish the semiconductor laser are

1. Small physical size (300 μm × 10 μm × 50 μm) that enables it to be incorporated easily into ether instruments.

2. Its direct pumping by low-power electric current (15 mA at 2 volts is typical), which makes it possible to drive it with conventional transistor circuitry.

3. Its efficiency in converting electric power to light. Actutal operating efficiencies exceed 50 percent.

4. The ability to modulate its output by direct modulation of the pumping current at rates exceeding 20 GHz. This is of major importance in high-data-rate optical communication systems.

5. The possibility of integrating it *monolithically* with electronic field

effect transistors, microwave oscillators, bipolar transistors, and optical components in Ⅲ Ⅴ semiconductors to form integrated optoelectronic circuits.

6. The semiconductor-based manufacturing technology, which lends itself to mass production.

7. The compatibility of its output beam dimensions with those of typical silicabased optical fibers and the possibility of tailoring its output wavelength to the low-loss, low-dispersion region of such fibers.

From the pedagogic point of view, understanding how a modern semiconductor laser works requires, in addition to the basic theory of the interaction of radiation with electrons that was developed in Chapter 5, an understanding of dielectric wave guiding [4, 5] (Section 13. 1) and elements of solid--state theory of semiconductors [6,7]. The latter theory will be taken up in the next few sections.

5.2 Some Semiconductor Physics Background

In this section we will briefly develop some of the basic background material needed to understand semiconductor lasers. The student is urged to study the subject in more detail, using any of the numerous texts dealing with the wave mechanics of solids (Reference [6], for example).

The main difference between electrons in semiconductors and electrons in other laser media is that in semiconductors all the electrons occupy, thus share, the whole crystal volume, while in a conventional laser medium, ruby, for example, the Cr^{3+} electrons are localized to within 1 or 2 Å of their parent Cr^{3+} ion and electrons on a given ion, for the typical Cr doping levels used, do not communicate with those on other ions.

In a semiconductor, on the other hand, because of the spatial overlap of their wavefunctions, no two electrons in a crystal can be placed in the same quantum state, i.e., possess the same eigenfunction. This is the so-called *Pauli exclusion principle*, which is one of the more important axiomatic foundations of quantum mechanics. Each electron thus must possess a unique spatial wavefunction and an associated eigenenergy (the total energy associated with the state). If we plot a horizontal line, as in Figure 5.1, for each allowed electron energy (eigenenergy), we will discover that the energy levels cluster within bands that are separated by "energy gaps" ("forbidden" gaps). A schematic description of the energy level spectrum of electrons in a crystals is shown in Figure 5.1.

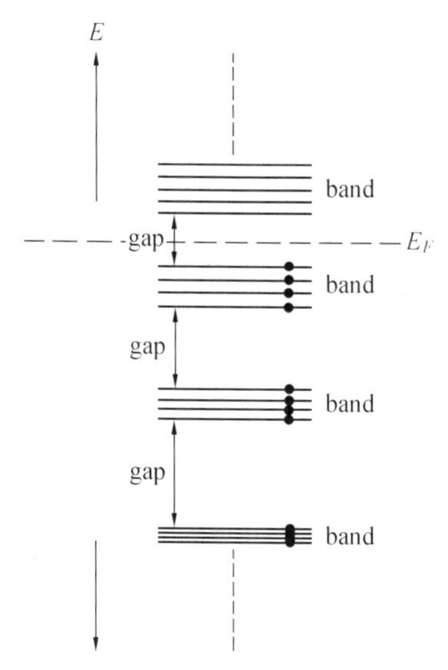

Figure 5.1 The energy levels of electrons in a crystal. In a given material these levels are usually occupied, in the ground state, up to some uppermost level. The energy E_F that marks in the limit of $T \to 0$, the transition from fully occupied electron states ($E < E_F$) to empty states ($E < E_F$), is called the *Fermi energy*. It does not, except accidentally, correspond to an eigenenergy of an electron in the crystal.

The manner in which the available energy states are occupied determines the conduction properties of the crystal. In an insular the uppermost occupied band is filled up with electrons while the next highest band is completely empty. The gap between them is large enough, say, ~3 eV, so that thermal excitation across the gap is negligible. If we aplly an electric field to such an idealized crystal, no current will flow, since the electronic motion in a filled band is completely balanced and for each electron moving with a velocity $\boldsymbol{v}$ there exists another one with $-\boldsymbol{v}$.

If the gap between the uppermost filled band—the valence band—and the next highest—the conduction band—is small, say, <2 eV, then thermal excitation causes partial transfer of electrons from the valence band to the conduction hand and the crystal can conduct electricity. Such crystals are called *semiconductors*. Their degree of conductivity can be controlled not only by the temperature but also by "doping" them with impurity atoms.

The wavefunction of an electron in a given band, say, the valence band, is characterized by a vector $\boldsymbol{k}$ and a corresponding (Bloch) wavefunction

$$\psi_v(\boldsymbol{r}) = uv\boldsymbol{k}(\boldsymbol{r})e^{i\boldsymbol{k}\cdot\boldsymbol{r}} \tag{5.2.1}$$

The function u_{v_k} possesses the same periodicity as the lattice. The factor $\exp(i\boldsymbol{k}\cdot\boldsymbol{r})$ is responsible for the wave nature of the electronic motion and is related to the de Broglie wavelength λ_e of the electron by①

$$\lambda_e = \frac{2\pi}{k} \tag{5.2.2}$$

the vector $\boldsymbol{k}$ can only possess a prescribed set of values (i.e., it is quantized), which is obtained by requiring that the total phase shift $\boldsymbol{k}\cdot\boldsymbol{r}$ across a crystal with dimensions L_x, L_y, L_z be some multiple integer of 2π.

$$k_i = \frac{2\pi}{L_i}s \quad s = 1,2,3,\ldots \tag{5.2.3}$$

① This is true if the value of $\boldsymbol{k}$ is taken in the extended (i.e. not reduced) $\boldsymbol{k}$ space.

where $i = x, y, z$. We con thus divide the total volume in $\boldsymbol{k}$ space into cells each with a volume

$$\Delta V_k \equiv \Delta k_x \Delta k_y \Delta k_z = \frac{(2\pi)^3}{L_x L_y L_z} = \frac{(2\pi)^3}{V} \tag{5.2.4}$$

and associate with each such differential volume a quantum state (two states when we allow for the two intrinsic spin states of each electron). The number of such states within a spherical shell (in $\boldsymbol{k}$ space) of radial thickness $\mathrm{d}k$ and radius k is then given by the volume of the shell divided by the volume (5.2.4) ΔV_k per state

$$\rho(k)\mathrm{d}k = \frac{k^2 V}{\pi^2}\mathrm{d}k \tag{5.2.5}$$

so that $\rho(k)$ is the number of states per unit volume of $\boldsymbol{k}$ space. (A factor of 2 for spin was included to account for the fact that an electron in a given (spatial) state is also in a spin "up" or "down" state.

The energy, measured from the bottom of the band, of an electron $\boldsymbol{k}$ in, say, the conduction band (indicated henceforth by a subscript c) is

$$E_c(\boldsymbol{k}) = \frac{\hbar^2 k^2}{2m_c} \tag{5.2.6}$$

where m_c is the effective mass of an electron in the conduction band. In the simplest and idealized case, which is the one we are considering here, the energy depends only on the magnitude k of the electron propagation vector and not its direction.

We often need to perform electron counting, not in $\boldsymbol{k}$ space but as a function of the energy. The density of states function $\rho(E)$ (the number of electronic states per unit energy interval per unit crystal volume) is determined from the conservation of states relation

$$\rho(E)\mathrm{d}E = \frac{1}{V}\rho(k)\mathrm{d}k$$

which with the use of (5.2.5) and (5.2.6) leads to

$$\rho_c(E) = \frac{1}{2\pi^2}\left(\frac{2m_c}{\hbar^2}\right)^{3/2} E^{1/2}$$

or

$$\rho_c(\omega) = \hbar\,\rho_c(E) = \frac{1}{2\pi^2}\left(\frac{2m_c}{\hbar}\right)^{3/2}\omega^{1/2} \qquad (5.2.7)$$

where $\hbar\,\omega = E$. A similar expression but with m_c replaced by m_v, the effective mass in the valence band, applies to the valence band.

Figure 5.2 depicts the energy – k relationship of a direct gap semiconductor, i.e., one where the conduction hand minimum and the valence band maximum occur at the same value of $\boldsymbol{k}$. The dots represent allowed (not necessarily occupied) electron energies. Note that, following (5.2.3), these states are spaced uniformly along the k axis.

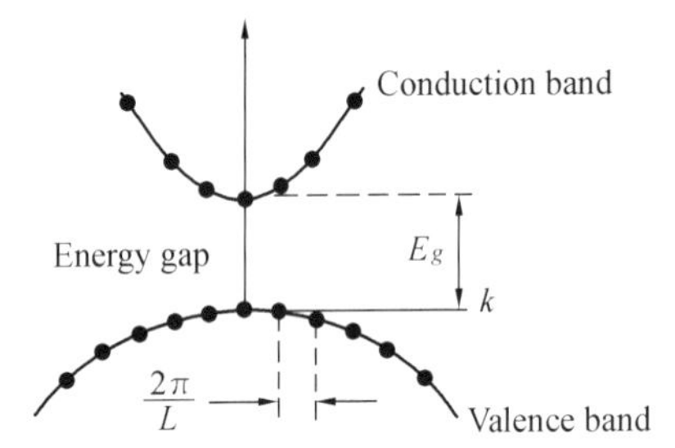

Figure 5.2 A typical energy band structure for a direct gap semiconductor with $m_c < m_v$. The uniformly spaced dots correspond to electron states.

The Fermi-Dirac Distribution Law

The probability that an electron state at energy E is occupied by an electron is given by the Fermi-Dirac law [6,7]

$$f(E) = \frac{1}{e^{(E-E_F)/kT} + 1} \qquad (5.2.8)$$

where E_F is the Fermi energy and T is the temperature. For electron energies well below the Fermi level such that $E_F - E \gg kT, f(E) \to 1$ and the electronic states are fully occupied, while well above the Fermi level $E - E_F \gg kT$, $f(E) \propto \exp(-E/kT)$ and approaches the Boltzmann distribution. At $T = 0 f(E) = 1$, for $E < E_F$, and $f(E) = 0$, for $E > E_F$ so that all levels below the Fermi level are occupied while those above it are empty. In thermal equilibrium a single Fermi energy applies to both the valence and conduction bands. Under conditions in which the thermal equilibrium is disturbed, such as in a *p-n* junction with a current flow or a bulk semiconductor in which a large population of conduction electrons and holes is created by photoexcitation. separate Fermi levels called *quasi-Fermi levels* are used for each of the bands. The concept of quasi-Fermi levels in excited systems is valid whenever the carrier scattering time within a band is much shorter than the equilibration time between bands. This is usually true at the large carrier densities used in *p-n* junction lasers.

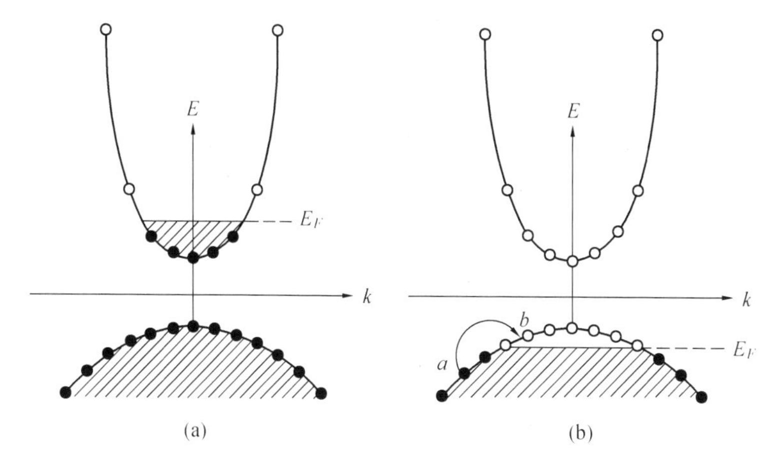

Figure 5.3 (a) Engrgy band of a degenerate n-type semiconductor at 0 K. (b) A degenerate p-type semiconductor at 0 K. The cross-hatching represents regions in which all the electron states are filled. Empty circles indicate unoccupied states (holes).

In very highly doped semiconductors, the Fermi level is forced into either (1) the conduction hand for donor impurity doping or (2) into the valence band for acceptor impurity doping. This situation is demonstrated by Figure 5.3. According to (5.2.8) at 0 K, all the states below E_F are filled while those above it are unoccupied as shown in the figure. In this respect the degenerate semiconductor behaves like a metal in which case the conductivity does not disappear at very low temperatures. The unoccupied states in the valence band [unshaded area in Figure 5.3(b)] are called *holes*, and they are treated exactly like electrons except that their charge, corresponding to an electron deficiency, is positive and their energy increases downward in the diagram. The number of holes in the semiconductor depicted by Figure 5.3 (b) is the number of electron states falling within the unshaded area at the top of the valence band. The process of exciting an electron from state a to state b [Figure 5.3(b)]in the valence band can also be viewed as one whereby a hole is excited from b to a. The advantage of this point of view is the symmetry in the language and mathematical description that it brings to the discussions of current flow the to electrons in the conduction hand and those in the valence hand.

To better appreciate the role of the quasi-Fermi level, consider a nonthermal equilibrium situation in which electrons are excited into the conduction hand of a degenerate p-type semiconductor at a very high rate. This can be done by injecting electrons into the p region across a p-n junction or by subjecting the semiconductor to an intense light beam with $h\nu > E_g + E_{Fc} + E_{Fv}$, so that for each absorbed photon an electron is excited into the conduction band from the valence band. This situation is depicted in Figure 5.4. Following this excitation, electrons relax, by emitting optical and acoustic phonons, to the bottom of he conduction band in times of $\sim 10^{-12}$s while their relaxation across the ga back to the valence band—a process referred to as electron-hole recombination—is characterized by a time constant of

$$\tau \sim 3 - 4 \times 10^{-9}\ \mathrm{s}$$

It is important in analyzing the process of light amplification in semiconductors to determine the quasi-Fermi level E_{Fc} for a given rate of excitation. Assuming that the relaxation to the bottom of the band into

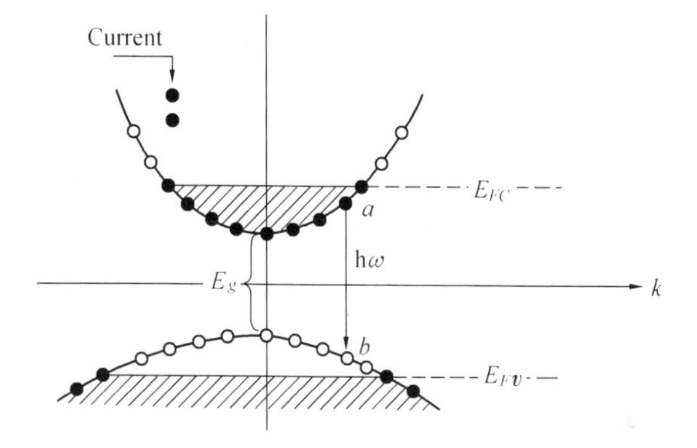

Figure 5.4 Electrons are injected at a rate of I/eV per unit volume (I = total current) into the conduction band of a semiconductor.

which the carriers are excited is instantaneous, we have

$$\frac{Nc}{\tau}=\frac{I}{eV} \tag{5.2.9}$$

where N_c is the density (m^{-3}) of electrons in the conduction band, I the injection current (in amperes), τ is the electron relaxation time back to the valence band (electron-hole recombination time), and V is the volume into which the electrons are confined following injection. The density of electrons with energies between E and $E+dE$ is the product of $\rho_c(E)$—the density of allowed electron states—and the occupation probability $f_c(E)$ of these states.

Using (5.2.7) and (5.2.8)

$$N_c=\frac{I\tau}{eV}=\int_0^\infty \rho_c(E)f_c(E)\mathrm{d}E=$$

$$\frac{1}{2\pi^2}\left(\frac{2m_c}{\hbar^2}\right)^{3/2}\int_0^\infty \frac{E^{1/2}}{\mathrm{e}^{(E-E_{F_c})/kT}+1}\mathrm{d}E \tag{5.2.10}$$

For a given injection current I the only unknown quantity in (5.2.10) is the conduction quasi-Fermi level E_{F_c}. We can thus invert, in practice by numerical methods, (5.2.10) and solve it for $E_{F_c}(\mathrm{T})$ as a function of I, or equivalently of N_c. We shall make use, later, of this fact. At $T=$

0 the integral is replaced by

$$\int_0^{E_{Fc}} E^{1/2} \mathrm{d}E = \frac{2}{3} E_{Fc}^{3/2}$$

yielding

$$E_{Fc}(T = 0) = (3\pi^2)^{2/3} \frac{\hbar^2}{2m_c} N_c^{2/3} \tag{5.2.11}$$

Another fact that we need before proceeding to the subject of optical gain in semiconductors is that when an electron makes a transition (induced or spontaneous) between a conduction band state and one in the valence band, the two states involved must have the same $\boldsymbol{k}$ vector. This is due to the fact that according to quantum mechanics the rate of such a transition is always proportional to an integral over the crystal volume that involves the product of the initial state wavefunction and the complex conjugate of that of the final stat. Such an integral would, according to (5.2.1), be vanishingly small except when the condition

$$\boldsymbol{k}_f = \boldsymbol{k}_i \tag{5.2.12}$$

is satisfied. In band diagrams such as that of Figure 5.4, the transitions are consequently described by vertical arrows.

5.3 Gain and Absorption in Semiconductor (laser) Media

Consider the semiconductor material depicted in Figure 5.5 in which by virtue of electron pumping a nonthermal equilibrium steady state is obtained in which *simultaneously* large densities of electrons and holes coexist in the *same* space. These are characterized by quasi-Fermi levels E_{F_c} and E_{F_v}, respectively, as shown.

Let an optical beam at a (radian) frequency ω_0 travel through the crystal. This beam will induce downward $a \to b$ transitions that lead to amplification as well as $b \to a$ absorbing transitions. Net amplification of the beam results if the rate of $a \to b$ transitions exceeds that of $b \to a$.

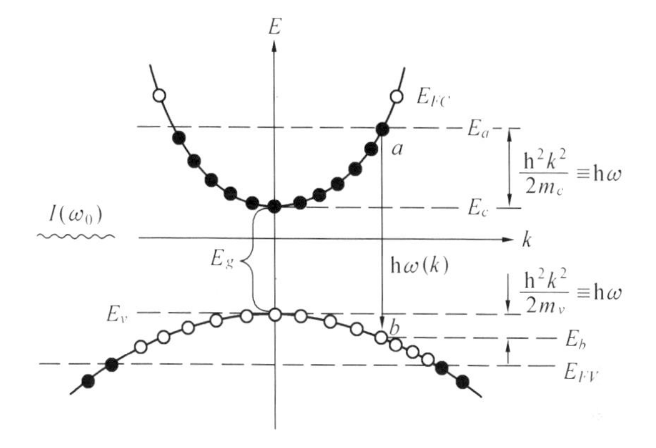

Figure 5.5 An optical beam at ω_0 with intensity $I(\omega_0)$ is incident on a pumped semiconductor medium characterized by quasi-Fermi levels E_{F_c} and E_{F_v}. A single level pair *a*-*b* with the same ***k*** value is shown. The induced transition $a \to b$ contributes one photon to the beam.

As discussed in the previous section, only transitions in which the upper and lower electron states have the same ***k*** vector are allowed. The pair of levels *a* and *b* in Figure 5.5 are thus characterized by some ***k*** value. Let us consider a group of such levels with nearly the same ***k*** value and hence with nearly the same transition energy

$$\hbar\,\omega(\boldsymbol{k}) = E_g + \frac{\hbar^2 k^2}{2m_c} + \frac{\hbar^2 k^2}{2m_v} \tag{5.3.1}$$

(In the following the ***k*** dependence of ω will be omitted but understood.) The density of such level pairs whose ***k*** values fall within a spherical shell of thickness dk is, according to (5.3.5), $\rho(k)\mathrm{d}k/V$.

Before proceeding let us remind ourselves of some results developed in connection with conventional laser media. The gain constant $\gamma(\omega_0)$ is given by (5.5.7) as

$$\gamma(\omega_0) = -\frac{k}{n^2}\chi''(\omega_0) \qquad k = \frac{2\pi n}{\lambda} \tag{5.3.2}$$

where $\chi''(\omega_0)$, the imaginary part of the electric susceptibility, is

$$\chi''(\omega_0) = \frac{(N_1 - N_2)\lambda_0^3}{8\pi^3 t_{\text{spont}} \Delta v n} \frac{1}{1 + 4(v - v_0)^2/(\Delta v)^2} \quad (5.3.3)$$

Combining the last two equations and defining the "relaxation time" T_2 by $T_2 = (\pi \Delta v)^{-1}$ leads to

$$\gamma(\omega_0) = \frac{(N_1 - N_2)\lambda_0^2}{4n^2 t_{\text{spont}} \Delta v n} \frac{T_2}{\pi[1 + (\omega - \omega_0)^2 T_2^2]} \quad (5.3.4)$$

In semiconductors T_2 is the mean lifetime for coherent interaction of k electrons with a monochromatic field and is of the order of the phonon-electron collision time. Numerically $T_2 \sim 10^{-12}$ s. Given an electron in an upper state "a." the lower state "b" with the same $\boldsymbol{k}$ value may be occupied by another electron. The downward rate of transitions is thus proportional to

$$R_{a \to b} \propto F_c(E_a)[1 - f_v(E_b)]$$

i.e., to the product of the probabilities $f_c(E_a)$ that the upper (conduction) state is occupied and the probability $(1 - f_v)$ that the lower (valence) state is empty. The functions $f_{v,c}(E)$ are given, according to (5.3.8), by

$$f_c(E) = \frac{1}{e^{(E - E_{F_c})/kT} + 1} \quad (5.3.5)$$

$$f_v(E) = \frac{1}{e^{(E - E_{F_v})/kT} + 1} \quad (5.3.6)$$

allowing for the fact that under pumping conditions $E_{F_c} \neq E_{F_v}$.

In translating to the case of semiconductors the results that were developed for conventional lasers, the population inversion density $(N_2 - N_1)$ is thus replaced by the effective inversion due to electrons and holes within $\mathrm{d}k$.

$$N_2 - N_1 \to \frac{\rho(k)\mathrm{d}k}{V} \{f_c(E_a)[1 - f_v(E_b)] - f_v(E_b)[1 - f_c(E_a)]\} =$$

$$\frac{\rho(k)\mathrm{d}k}{V}[f_c(E_a) - f_v(E_b)] \quad (5.3.7)$$

$$E_a - E_b \equiv \hbar\,\omega = E_g + \frac{\hbar^2 k^2}{2m_c} + \frac{\hbar^2 k^2}{2m_v} \qquad (5.3.8)$$

Equation (5.3.7) is of central importance and is a capsule statement of the different between the population inversion in a conventional laser medium where the occupation probability obeys Boltzmann statistics and that of a semiconductor medium governed by Fermi-Dirac statistics.

Returning to the gain expression (5.3.4), we use (5.3.7) to rewrite it as

$$\mathrm{d}\gamma(\omega_0) = \frac{\rho(k)\mathrm{d}k}{V}(f_c - f_v)\frac{\lambda_0^2}{4n^2\tau}\left(\frac{T_2}{\pi[1 + (\omega - \omega_0)^2 T_2^2]}\right)$$

where $\omega \equiv \omega(k)$ is the transition frequency at k as in (5.3.1). The different designation $\mathrm{d}\gamma(\omega_0)$ is to remind us that only electrons with $\boldsymbol{k}$ vectors within $\mathrm{d}k$ included here. We have also replaced, to agree with popular usage, the term sontaneous lifetime (t_{spont}) by the recombination lifetime τ for an electron in the conduction band with a hole in the valence band. To obtain the gain constant, we must add up the contributions from all the electrons

$$\gamma(\omega_0) = \int_0^\infty \frac{\mathrm{d}k\rho(k)}{V}[f_c(\omega) - f_v(\omega)]\frac{\lambda_0^2}{4n^2\tau}\left(\frac{T_2}{\pi[1 + (\omega - \omega_0)^2 T_2^2]}\right) \qquad (5.3.9)$$

We will find it easier to carry out the indicated integration in (5.3.9) in the ω domain [$\hbar\,\omega$ being the separation $E_a(\boldsymbol{k}) - E_b(\boldsymbol{k})$]. From (5.3.1)

$$\hbar\,\omega = E_g + \frac{\hbar^2}{2m_r}k^2 \qquad (5.3.10)$$

$$\frac{1}{m_r} = \frac{1}{m_v} = \frac{1}{m_c} \quad (m_r \equiv \text{reduced effective mass}) \qquad (5.3.11)$$

Using the relations

$$\mathrm{d}\omega = \frac{\hbar^2}{2m_r}k\mathrm{d}k$$

$$k = (\hbar\omega - E_g)^{1/2}\left(\frac{2m_r}{\hbar^2}\right)^{1/2}$$

the expression (5.3.9) for $\gamma(\omega_0)$ be comes

$$\gamma(\omega_0) = \int_0^\infty (\hbar\omega - E_g)^{1/2}\left(\frac{2m_r}{\hbar^2}\right)^{1/2} \frac{m_r \lambda_0^2 T_2 [f_c(\omega) - f_v(\omega)]}{\pi^2 \hbar 4n^2 \tau\pi[1 + (\omega - \omega_0)^2 T_2^2]} d\omega \tag{5.3.12}$$

In most situations we can replace the normalized function

$$\frac{T_2}{\pi[1 + (\omega - \omega_0)^2 T_2^2]} \rightarrow \delta(\omega - \omega_0)$$

which is merely a statement of the fact that its width $\Delta\omega \sim T_2^{-1}$ is narrower another spectral features of interest. In this case the integration (5.3.12) leads to

$$\gamma(\omega_0) = \frac{\lambda_0^2}{8\pi^2 n^2 \tau}\left(\frac{2m_c m_v}{\hbar(m_v + m_c)}\right)^{3/2}\left(\omega_0 - \frac{E_g}{\hbar}\right)^{1/2}[f_c(\omega_0) - f_v(\omega_0)] \tag{5.3.13}$$

The condition for net gain $\gamma(\omega_0) > 0$ is thus

$$f_c(\omega_0) > f_v(\omega_0) \tag{5.3.14}$$

which is the equivalent, in a semiconductor, of the conventional inversion condition $N_2 > N_1$. Using (5.3.5) and (5.3.6), the gain condition (5.3.14) becomes

$$\frac{1}{e^{(E_a - E_{F_c})/kT} + 1} > \frac{1}{e^{(E_b - E_{F_c})/kT} + 1} \tag{5.3.15}$$

Recalling that $E_a - E_b = \hbar\omega_0$, (5.3.15) is satisfied provided

$$\hbar\omega_0 < E_{F_c} - E_{F_v} \tag{5.3.16}$$

so that only frequencies whose photon energies $\hbar\omega_0$ are smaller than the quasi-Fermi levels separation are amplified. Condition (5.3.16) was first derived by Basov, etal. [1], Bernard and Duraffourg [8]. The general features of the gain dependence $\gamma(\omega_0)$ on the frequency ω_0 are illustrated by Figure 5.6. The gain is zero at $\hbar\omega < E_g$, since no electronic transitions exist at these energies. The gain becomes zero again at the frequency where $\hbar\ \omega_0 = E_{F_c} - E_{F_v}$. At higher frequencies the semiconductor absorbs.

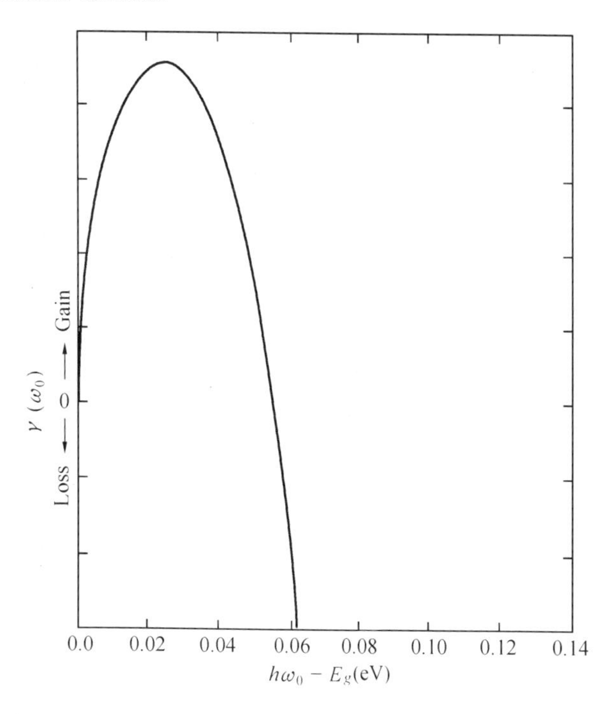

Figure 5.6 A typical polt of gain $\gamma(\omega_0)$ as a function of frequency for a fixed pumping level N. (After Reference [9].)

Figure 5.7 shows calculated plots based on (5.3.12) with the density of the (injected) electrons as a parameter. The curves are based on the following physical constants of GaAs: $mc = 0.067 m_e$, $m_v = 0.48 m_e$, $T_2 \sim 0.5$ ps, $\tau \simeq 3 \times 10^{-9}$ s, $E_g = 1.43$ eV. We note that the minimum density to achieve transparency ($\gamma = 0$) is $N_{tr} \sim 1.55 \times 10^{18}$ cm^{-3}. The peak gain corresponding to a given inversion density N_c is plotted in Figure 5.8.

If follows from Figure 5.8 that semiconductor media are capable of

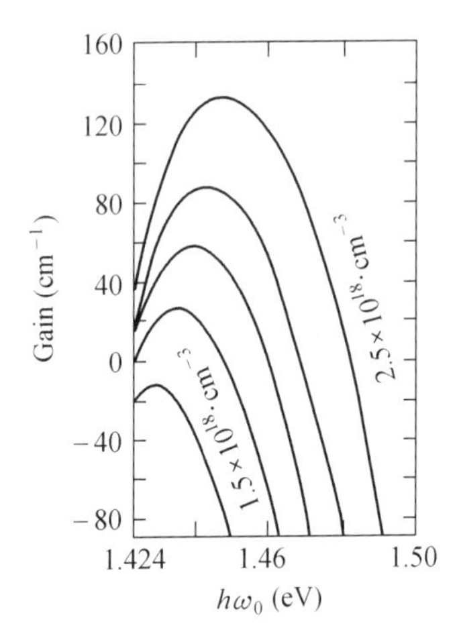

Figure 5.7 A plot based on (5.3.12) of the photon energy dependence of the optical gain (or loss = negative gain) of GaAs with the injected carrier density as a parameter. (After Reference [9].)

achieving very large gain ranging up to a few hundred cm^{-1}. In a laser the amount of gain that actually prevails is clamped by the phenomenon of saturation (see Section 5.6) to a value equal to the loss. In a typical semiconductor laser this works out to $20\ cm^{-1} < \gamma < 80\ cm^{-1}$. In this region we can approximate the polt of Figure 5.8 by a linear relationship

$$\gamma_{max} = B(N - N_{tr}) \tag{5.3.17}$$

The constant B fitting the data of Figure 5.8 is $B \sim 1.5 \times 10^{-16}\ cm^2$ and is typical of GaAs/GaAlAs lasers at 300 K. The gain constant B increases with the decrease of the temperature T. This is due to the narrowing of the transition regions of the Fermi functions $f_c(\omega)$ and $f_v(\omega)$ in (5.3.12). At 77 K, $B \sim 5 \times 10^{-16}\ cm^2$. Figure 5.8 shows that the semiconductor diode is capable of producing extremely large incremental gains, with only moderate increases of the inversion density, hence the current, above the transparency value ($N_{tr} \sim 1.55 \times 10^{18}\ cm^{-3}$ in the figure). It is thus possible to obtain oscillation in a semiconductor laser with active regions that are only a few tens of microns long.

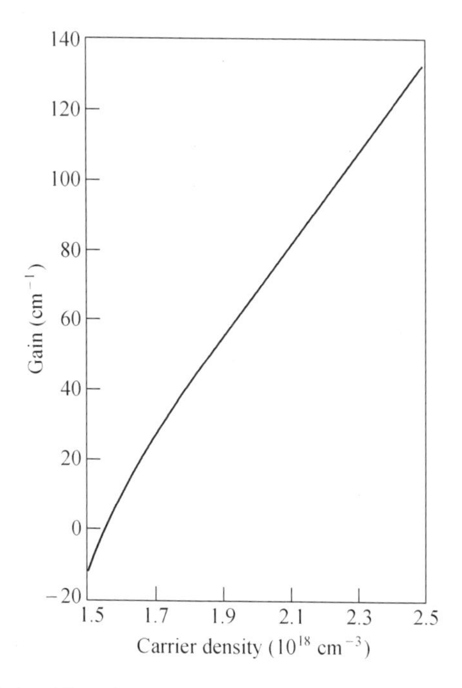

Figure 5.8 A plot of the peak gain γ_{max} of Figure 5.7 as a function of the inversion density at $T = 300$ K.

Commercial diode lasers have typical lengths of ~250 μm.

For additional background material on semiconductor lasers, the student is advised to consult References [10 ~ 12].

5.4 GaAs/$Ga_{1-x}Al_xAs$ Lasers

The two most important classes of semiconductor lasers are those that are based on Ⅲ-Ⅴ semiconductors. The first system is based on GaAs and $Ga_{1-x}Al_xAs$. The active region in this case is GaAs or $Ga_{1-x}Al_xAs$. The subscript x indicates the fraction of the Ga atoms in GaAs that are replaced by Al. The resulting lasers emit (depending on the active region molar fraction x and its doping) at $0.75\ \mu m < \lambda < 0.88\ \mu m$. This spectral region is convenient for the short-haul (< 2 km) optical communication in silica fibers.

The second system has $Ga_{1-x}In_xAs_{1-y}P_y$ as its active region. The lasers emit in the $1.1\ \mu m < \lambda < 1.6\ \mu m$ depending on x and y. The region near 1.55 μm is especially favorable, since, as shown in Figure 3.19, optical fibers are available with losses as small as 0.15 dB/km at this wavelength, making it extremely desirable for long-distance optical communication.

In this section we will consider $GaAs/Ga_{1-x}Al_xAs$ lasers. A generic laser of this type, depicted in Figure 5.9, has a thin (0.1 ~ 0.2 μm) region of GaAs sandwiched between two regions of GaAlAs. It is consequently called a *double heterostructure* laser. The basic layered structure is grown epitaxially on a crystalline GaAs substrate so that it is uninterrupted crystalographically.

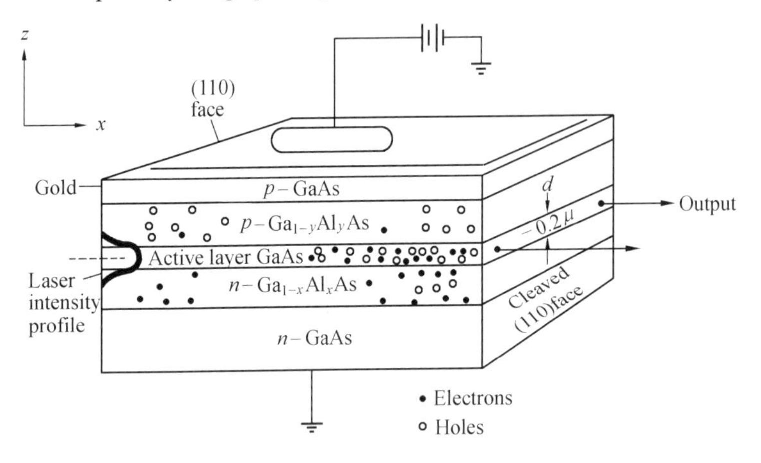

Figure 5.9 A typical double heterostructure GaAs-GaAlAs laser. Electrons and holes are injected into the active GaAs layer from the n and p GaAlAs. Frequencies near $v = E_g/h$ are amplified by stimulating electron-hole recombination.

The favored crystal growth techniques are liquid-phase epitaxy and chemical vapor deposition using metallo-organic reagents (MOCVD) [11, 13, 14]. Another important technique—molecular beam epitaxy [11,

13, 15, 16]—uses atomic beams of the crystal constituents in ultra-high vacuum to achieve extremely fine thickness and doping control.

The thin active region is usually undoped while one of the bounding $Ga_{1-x}Al_xAs$ layers is doped heavily n-type and the other p-type. The difference

$$n_{GaAs} - n_{Ga_{1-x}Al_xAs} \simeq 0.62x$$

between the indices of refraction of GaAs and the ternary crystal with a molar fraction x gives rise to three-layered dielectric waveguide of the type illustrated in Figure 13-1. At this point the student should review the basic modal concepts discussed in Chapter 13. The lowest-order (fundamental) mode has its energy concentrated mostly in the GaAs (high index) layer. The index distribution and a typical modal intensity plot for the lowest-order mode are shown in Figure 5.10. When a positive bias is applied to the device, electrons are injected from the n-type $Ga_{1-x}Al_xAs$ into the active GaAs region while a density of holes equal to that of the electrons in the active region is caused by injection from the p side.

The electrons that are injected into the active region are prevented from diffusing out into the p region by means of the potential barrier due to the difference ΔE_g between the energy gaps of GaAs and $Ga_{1-x}Al_xAs$. The x-dependence of the energy gap of $Ga_{1-x}Al_xAs$ is approximated by [13]

$$Eg(x < 0.37) = (1.424 + 1.247x)\text{eV}$$

and is plotted in Figure 5.11.

The total discontinuity ΔE_g of the energy gap at a GaAs/GaAlAs interface is taken up mostly (60 percent) by the conduction band edge. i.e., $\Delta E_c = 0.6\Delta E_g$, while 40 percent is left to the valence band, $\Delta E_v = 0.4\Delta E_g$, so that both holes and electrons are effectively confined to the active region. This double confinement of injected carriers as well as of the optical mode energy to the same region is probably the single most

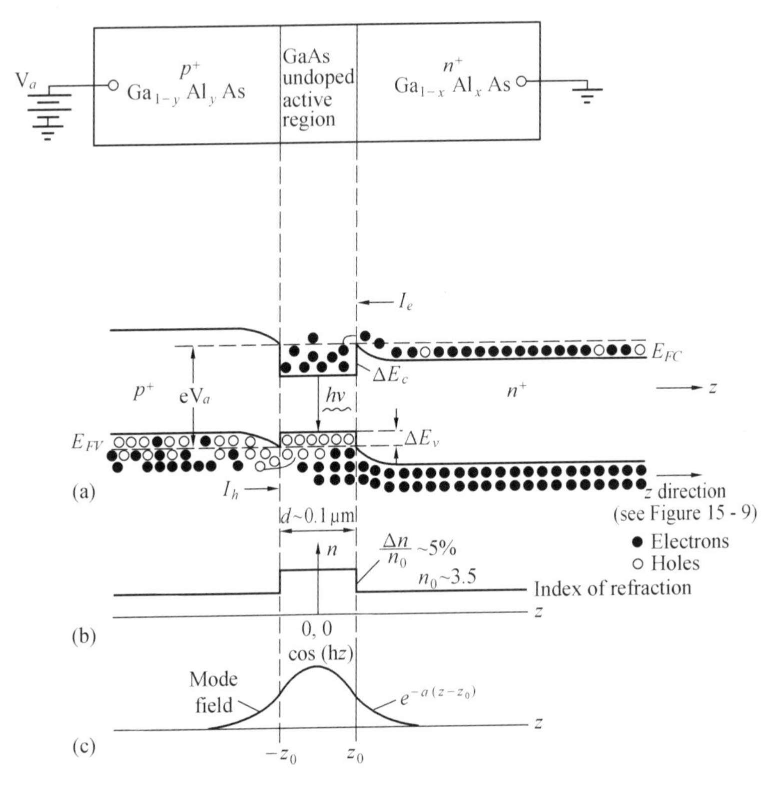

Figure 5.10 (a) The energy band edges of a strongly forward-biased (near-flattened) double heterrostructure GaAs/GaAlAs laser diode. Note trapping of electrons (holes) in the potential well formed by the conduction (valence) band edge energy discontinuity $\Delta E_c(\Delta E_v)$. (b) The spatial (z) profile of the index of refraction which is responsible for dielectric waveguiding in the high index (GaAs) layer. (c) The intensity profile of the fundamental optical mode in a slab waveguide.

important factor responsible for the successful realization of low-thresh old continuous semiconductor lasers [17 ~ 19]. Under these conditions we expect the gain experienced by the mode to vary as d^{-1}, where d is the thickness of the active (GaAs) layer, since at a given total current, the carrier density, hence the gain, will be proportional to d^{-1}. To quantify the last statement, we start with the basic definition of the modal gain

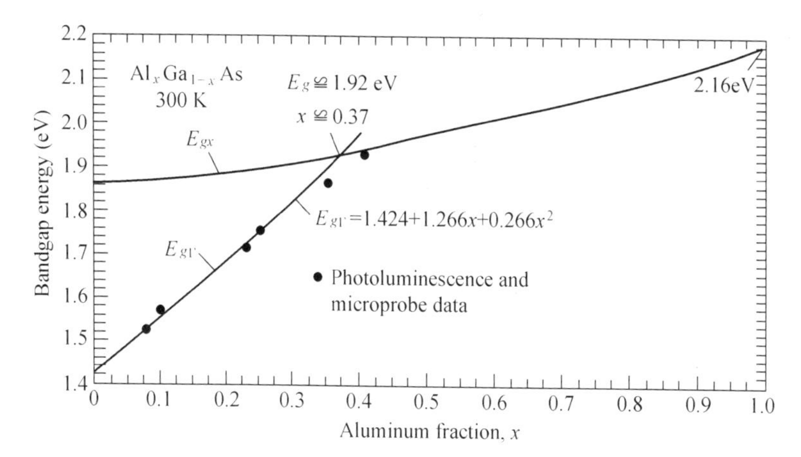

Figure 5.11 The magnitude of the energy gap in $Ga_{1-x}Al_xAs$ as a function of the molar fraction x. For $x > 0.37$ the gap is indirect. (After Reference [11].)

$$g = \frac{\text{power generated per unit length (in } x)}{\text{power carried by beam}} = \frac{-\int_{-\infty}^{-d/2} \alpha_n |E|^2 dz + \int_{-d/2}^{d/2} \gamma |E|^2 dz - \int_{d/2}^{\infty} \alpha_p |E|^2 dz}{\int_{-\infty}^{\infty} |E|^2 dz} \tag{5.4.1}$$

where γ is the gain constant experienced by a plane wave in a medium whose inversion density is equal to that of the active medium. γ is given by (5.4.12) and (5.4.17). α_n is the loss constant of the unpumped n-$Ga_{1-x}Al_xAs$ and is due mostly to free electron absorption. α_p is the loss (by free holes) in the bounding p-$Ga_{1-y}Al_yAs$ region. We note that as $d \to \infty$, $g \to \gamma$.

It is convenient to rewrite (5.4.1) as

$$g = \gamma \Gamma_a - \alpha_n \Gamma_n - \alpha_p \Gamma_p \tag{5.4.2}$$

$$\Gamma_a = \frac{\int_{-d/2}^{d/2} |E|^2 dz}{\int_{-\infty}^{\infty} |E|^2 dz} \tag{5.4.3a}$$

$$\Gamma_n = \frac{\int_{-\infty}^{-d/2} |E|^2 \mathrm{d}z}{\int_{-\infty}^{\infty} |E|^2 \mathrm{d}z} \tag{5.4.3b}$$

$$\Gamma_p = \frac{\int_{d/2}^{\infty} |E|^2 \mathrm{d}z}{\int_{-\infty}^{\infty} |E|^2 \mathrm{d}z} \tag{5.4.3c}$$

$$\Gamma_a + \Gamma_a + \Gamma_p = 1$$

Γ_a is very nearly the fraction of the mode power carried within the active GaAs layer, while Γ_n and Γ_p are, respectively, the fraction of the power in the n and p regions. As long as $\Gamma_a \sim 1$, i.e., most of the mode energy is in the active region, the gain g is inversely proportional to the active region thickness d since decreasing d, for example, increases the optical intensity for a given total beam power and, consequently, the rate of stimulated transitions. As d decreases, an increasing fraction of the mode intensity is carried outside the active region as can be seen from the modal waveguide solution plotted in Figure 5.12[11]. The resulting decreiae of the confinement factor Γ_a eventually dominates over the d^{-1}-dependence and the gain begins to decrease with further decrease of d [22]. A plot of the threshold current dependence on d is depicted in Figure 5.13. The bottoming out and eventual increase of J_{th} for $d \tilde{<} 0.1$ μm is due to the decrease of the confinement factor Γ_v and the increase of the relative role of the losses in the p and n GaAlAs bounding layer as Γ_n and Γ_p increase, i.e., as an increasing fraction of the mode intensity is carried within these lossy unpumped regions as shown in Figure 5.12.

Numerical Example: Threshold Current Density in Double Heterostructure Lasers

Consider the case of a GaAs/GaAlAs laser of the type illustrated in Figure 5.10. We will use the following parameters: $\tau \sim 4 \times 10^{-9}$ s, $L =$

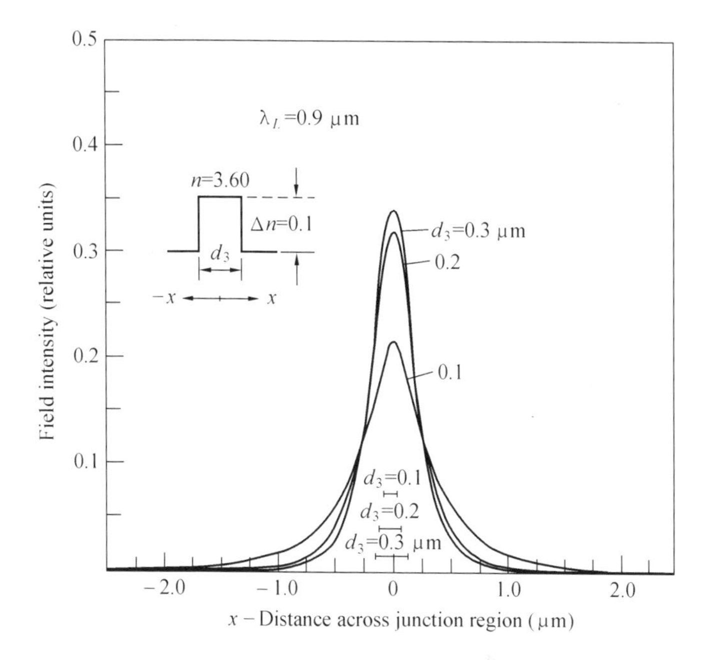

Figure 5.12 Calculated near field intensity distribution of the step discontinuity waveguide for various values of the guiding layer thickness. (After Reference [11].)

500 μm. The threshold gain condition is (5.4.2)

$$\gamma \Gamma_a = \alpha_n \Gamma_n + \alpha_p \Gamma_p - \frac{1}{L}\ln R + \alpha_s \qquad (5.4.4)$$

where the term α_s accounts for scattering losses (mostly at heterojunction interfacial imperfections). The largest loss term in lasers with uncoated faces is usually $L \ln R$. In our case, taking $R = 0.31$ as due to Fresnel reflectivity at a GaAs ($n = 3.5$) air interface, we obtain

$$-\frac{1}{L}\ln R = 23.4 \text{ cm}^{-1}$$

The rest of the loss terms are assumed to add up to ~ 10 cm^{-1} so that taking. $\Gamma_s \sim 1$ the total gain needed is 33.4 cm^{-1}. This requires, according to Figure 5.4, an injected carrier density of $N \sim 1.7 \times 10^{18}$ cm^{-3}. Under steady-state conditions the rate at which carriers are

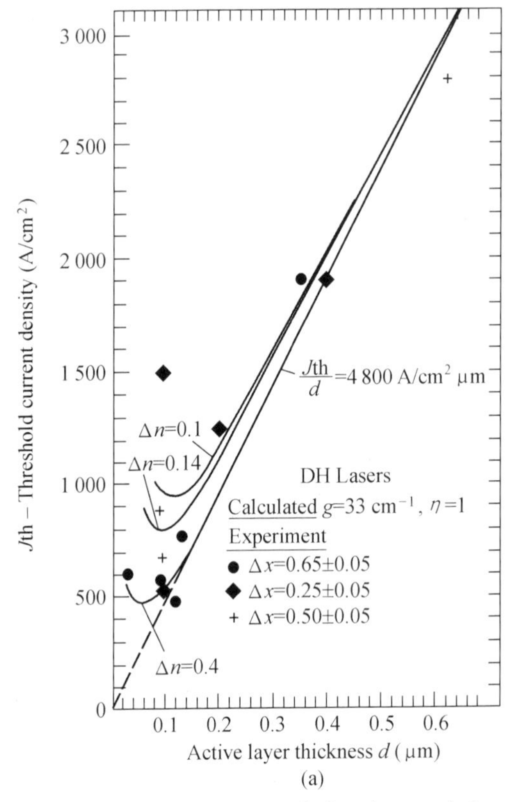

(a)

injected into the active region must equal the electron-hole recombination rate

$$\frac{J}{e} = \frac{Nd}{\tau}$$

Using the above data we obtain

$$\frac{J}{d} = \frac{eN}{\tau} \sim 6.8 \times 10^3 \text{ A/(cm}^2 \cdot \mu\text{m)}$$

This value of J/d is in reasonable agreement with the measured value of $\sim 5 \times 10^3$ in Figure 5.13. If we use this value to estimate the lowest threshold current density which from Figure 5.13 occurs when $d \sim$

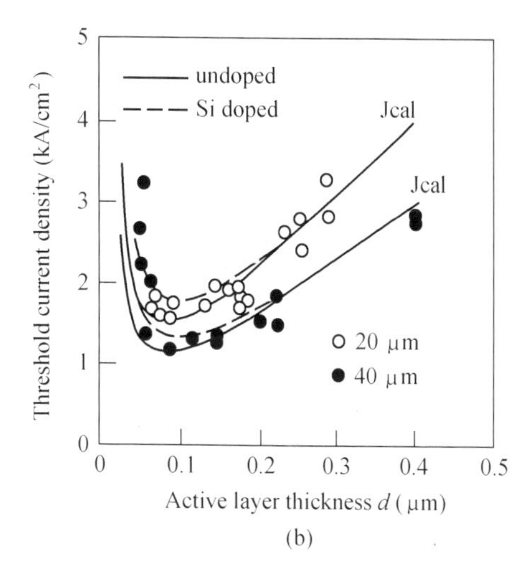

Figure 5.13 (a) Calculated and experimental values of the threshold current density as a function of the active layer thickness d for broad-area 500-μm-long AlGaAs DH diode lasers of "undoped" active layers. Notable exceptions are the experimental data for $\Delta x \simeq 0.25$, which were obtained from diodes with heavy-Ge-doped active layers. (After Reference [11].) (b) Calculated and experimental values of the threshold current density as a function of active layer thickness d for stripe-geometry (20-and 40-μm-wide stripe contacts) 300-μm-long AlGaAs DH diode lasers ($\Delta x = 0.25$) of "undoped" and low-Si-doped active layers ($x = 0.05$). (After Reference [21].)

0.08 μm, we obtain

$$J_{\min} = 0.68 \times 10^4 \times 0.08 = 544 \text{ A/cm}^2$$

again, close to the range of observed values.

The successful epitaxial growth of $Ga_{1-x}Al_xAs$ on top of GaAs (and vice versa), which is the main reason for the success of double heterostructure lasers, is due to the fact that their lattice constants are the same, to within a fraction of a percent, over the range $0 \leqslant x \leqslant 1$. This can be seen from the plot of Figure 5.14, which shows the lattice constant corresponding to various compositions of Ⅲ-Ⅴ semiconductors as a function of the band gap energy. We note that the line connecting the

AlAs ($x = 1$) and the GaAs ($x = 0$) is nearly horizontal, which corresponds to a (very nearly) constant lattice constant over this compositional range.

5.5 Some Real Laser Structures

The double heterostructure lasers discussed in Section 5.5 lack the means for confining the current and the radiation in the lateral (y) direction. The outcome is that typical broad area lasers can support more than one transverse (y) mode, resulting in unacceptable mode hopping as well as spatial and temporal instabilities. To overcome these problems, modern semiconductor lasers employ some form of transverse optical and carrier confinement. A typical and successful example of this approach is th buried heterostructure laser [20] shown in Figure 5.15. To fabricate these lasers, the first three layers: n-$Ga_{1-x}Al_xAs$, and p-$Ga_{1-y}Al_yAs$ are grown on an n-GaAs crystalline substrate by one of the epitaxial

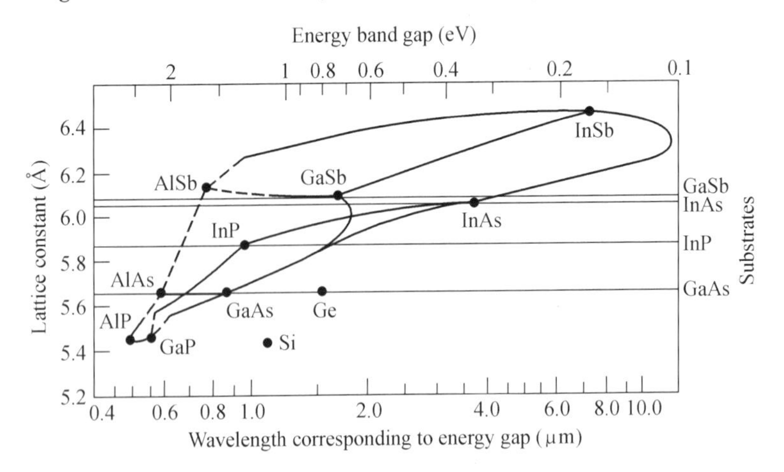

Figure 5.14 Ⅲ-Ⅴ compounds: Lattice constants versus energy band gaps and corresponding wavelengths. The solid lines correspond to direct-gap materials and the dashed lines to indirect-gap materials. The binary-compound substrates that can be used for lattice-matched growth are indicated on the right. [After Reference 11].

techniques described above. The structure is then etched through a mask down to the substrate level, leaving stand a thin (~ 3 μm) rectangular mesa composed of the original layers. A "burying" $Ga_{1-z}Al_zAs$ layer is then regrow on both sides of the mesa, resulting in the structure shown in Figure 5.15.

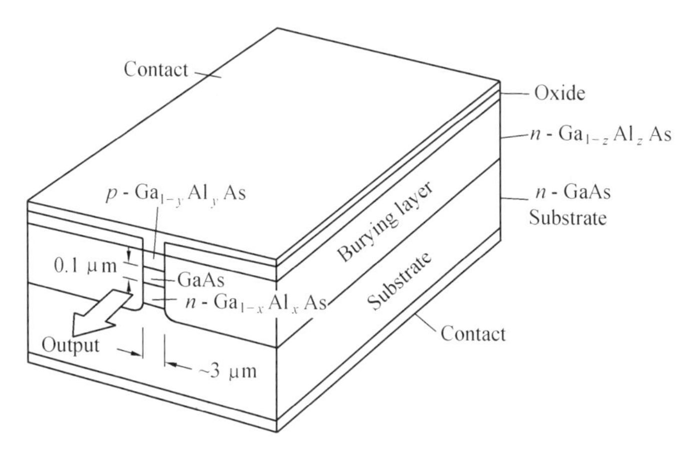

Figure 5.15 A buried heterostructure laser [20].

The most important feature of the buried heterostructure laser is that the active GaAs region is surrounded on *all* sides by the lower index GaAlAs, so that electro-magnetically the structure is that of a rectangular dielectric waveguide. The transverse dimensions of the active region and the index discontinuties (i.e., the molar fractions x, y, and z) are so chosen that only the lowest-order transverse mode can propagate in the laser waveguide. Another important feature of this laser is the confinement of the injected carriers at the boundaries of the active region due to the energyband discontinuity at a GaAs/GaAlAs interface as discussed in the last section. These act as potential barriers inhibiting carrier escape out of the active region. GaAs semiconductor lasers utilizing this structure have been fabricated, see chapter 16, with threshold currents of less than 1 milliampere [38]; more typical lasers have thresholds of

~ 20 milliamperes. Typical power vs. current polt of a commercial laser is shown in Figure 5.16, while the far field angular intensity distribution is shown in Figure 5.17.

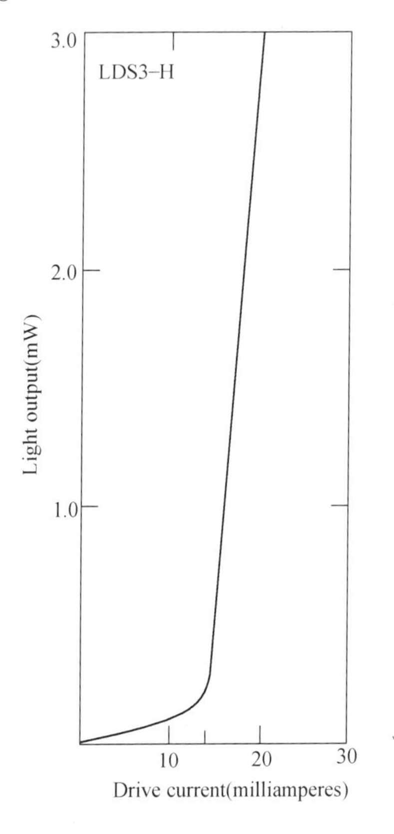

Figure 5.16 Power versus current plot of a low-threshold (~ 14 milliamperes) commercial DH GaAs/GaAlAs laser. (After Reference [23].)

Quaternary GalnAsP Semiconductor Lasers

Optical fiber communication over long distances (say > 10 km) uses, almost exclusively, lasers emitting in spectral regions near 1.3 μm and 1.55 μm. The 1.3 μm lasers are important because the group

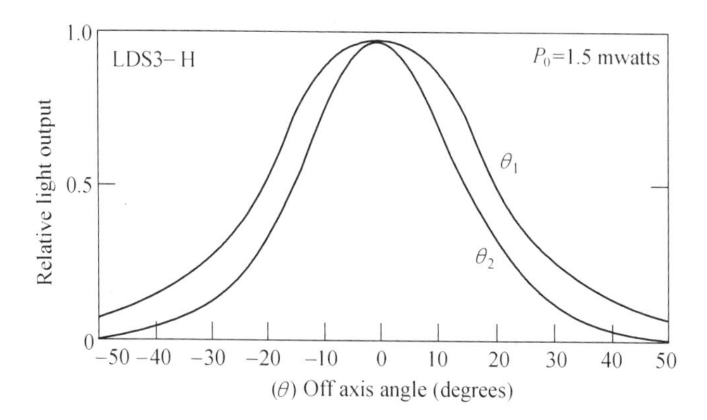

Figure 5.17 Far-field angular intensity distribution of a low-threshold commercial DH GaAs/GaAlAs laser. (After Reference [23.])

velocity dispersion of silica-based fibers at this wavelength is very small. The first-order group velocity dispersion parameter D is plotted in Figure 3-10 and is zero at $\lambda \approx 1.3\ \mu m$, so that optical pulses at this wavelength region around $1.55\ \mu m$ is where the optical absorption coefficient of silica fibers reaches a minimum, which makes it a favorite for long-haul links. Lasers in these wavelength regions [24] are fabricated using active layers of $GA_{1-x}In_xAs_{1-y}P_y$. From Figure 5.14, we find that such lasers spanning the $0.9\ \mu m < \lambda < 1.7\ \mu m$ region can be lattice-matched to InP, which possesses a lower index of refraction to produce dielectric waveguides in which the InP epitaxial layers act as cladding layers to the quaternary $Ga_{1-x}In_xAs_{1-y}P_y$ active layer. The quaternary layer plays in this system the role played by GaAs in the GaAs/GaAlAs laser depicted in Figure 5.9. A typical quaternary laser systems employ active regions with thicknesses in the 50 Å → 100 Å range. These are the so-called *quantum well lasers*. These lasers possess lower threshold currents and have a larger modulated bandwidth compared to earlier generations employing "thick" (~ 1 000 Å) active regions. They are discussed in detail in Chapter 16. Recent experiments [26, 39, 41, 42] have demonstrated propagation without repeaters at distances of ~ 150 km in optical fibers at 1.55 μm.

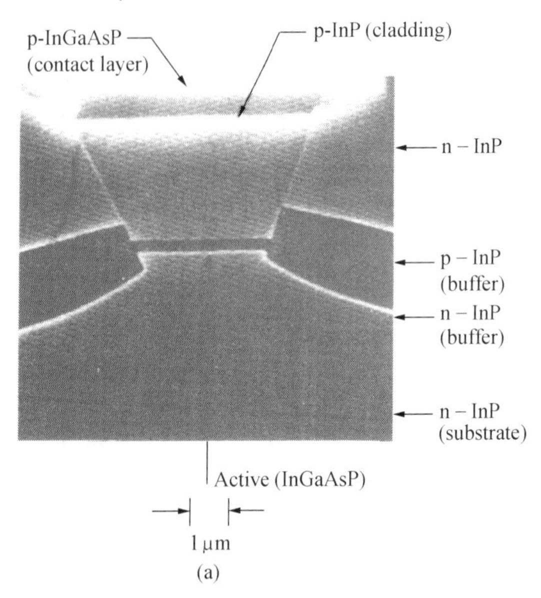

(a)

Power Output of Injection Lasers

The considerations of saturation and power output in an injection laser are basically the same as that of conventional lasers, which were described in Sections 5.6 and 6.4. As the injection current is increased above the threshold value, the laser oscillation intenssity builds up. The resulting stimulated emission shortens the lifetime of the inverted carriers to the point where the magnitude of the inversion is clamped at its threshold value. Taking the probability that an injected carrier recombine radiatively within the active region as η_i,[①] we can write thefollowing expression for the power emitted by stimulated emission:

$$P_e = \frac{(I - I_t)\eta_i}{e}hv \tag{5.5.1}$$

① The reason for a quantum efficiency η_i that is less than unity is, mostly, the existence of a leakage current component that bypasses the active *p-n* junction region.

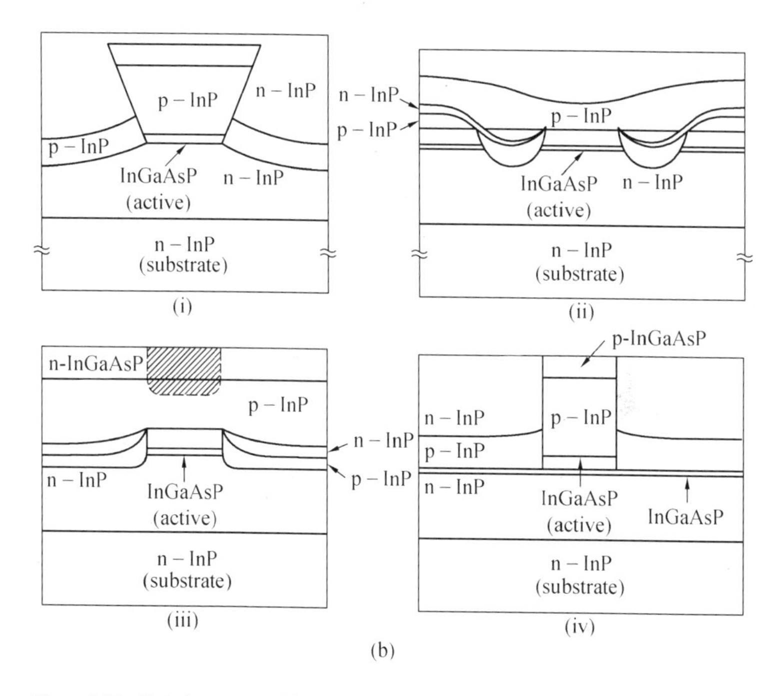

Figure 5.18 Typical structures of buried active region InP/GaInAsP diode lasers. (a) SEM [40, 41]. (b) Drawing of different structures [39].

Part of this power is dissipated inside the laser resonator, and the rest is coupled out through the end reflectors. These two powers are, according to (5.5.4), proportional to the effective internal loss $\alpha \equiv \alpha_n \Gamma_n + \alpha_p \Gamma_p + \alpha_x$ and to $-L^{-1} \ln R$, respectively. We can thus write the output power as

$$P_o = \frac{(I - I_t)\eta_i h\nu}{e} \frac{(1/L)\ln(1/R)}{\alpha + (1/L)\ln(1/R)} \tag{5.5.2}$$

The external differential quantum efficiency η_{ex} is defined as the ratio of the photon output rate that results from an increase in the injection rate (carriers per second) to the increase in the injection rate

$$\eta_{ex} = \frac{d(P_o/h\nu)}{d[(I - I_t)/e]} \tag{5.5.3}$$

Using (5.5.2) we obtain

$$\eta_{ex}^{-1} = \eta_i^{-1} = \left(\frac{\alpha L}{\ln(1/R)} + 1\right) \qquad (5.5.4)$$

By plotting the dependence of η_{ex} on L we can determine η_i, which in GaAs is around 0.9 ~ 1.0.

Since the incremental efficiency of converting electrons into useful output photons is η_{ex}, the main remaining loss mechanisms degrading the conversion of electrical to optical power is the small discrepancy between the energy eV_{appl} supplied to each injected carrier and the photon energy $h\upsilon$. This discrepancy is due mostly to the series resistance of the laser diode. The efficiency of the laser in converting electrical power input to optical power is thus

$$\eta = \frac{P_o}{VI} = \eta_i \frac{I - I_l}{I} \frac{h\upsilon}{eV_{appl}} = \frac{\ln(1/R)}{\alpha L + \ln(1/R)} \qquad (5.5.5)$$

In practice $eV_{appl} \sim 1.4\ E_g$ and $h\upsilon \simeq E_g$, Values of $\eta \sim 30$ percent at 300 K have been achieved.

We conclude this section by showing in Figures 5 ~ 16 and 5 ~ 17 typical plots of the power output versus current and the far field of commercial low-threshold GaAs semiconductor lasers.

5.6 Direct-Current Modulation of Semiconductor Lasers

Since the main application of semiconductor lasers is as sources for optical communication systems, the problem of high-speed modulation of their output by the high-data-rate information is one of great technological importance.

A unique feature of semiconductor lasers is that, unlike other lasers that are modulated externally (see Chapter 9), the semiconductor laser can be modulated directly by modulating the excitation current. This is especially important in view of the possibility of monolithic integration of the laser and the modulation electronic circuit, as will be discussed in Section 5.7. The following treatment follows closely that of Reference [27].

If we denote the photon density inside the active region of a semiconductor laser by P and the injected electron (and hole) density by N, then we can write

$$\frac{dN}{dt} = \frac{I}{eV} - \frac{N}{\tau} - A(N - N_{tr})P$$

$$\frac{dP}{dt} = A(N - N_{tr})P\Gamma_a - \frac{P}{\tau_p} \qquad (5.6.1)$$

where I is the total current, V the volume of the active region, τ the spontaneous recombination lifetime, τ_p the photon lifetime as limited by absorption in the bounding media, scattering and coupling through the output mirrors.

The term $A(N - N_{tr})P$ is the net rate per unit volume of induced transitions. N_{tr} is the inversion density needed to achieve transparency as defined by (5.6.17), and A is a temporal growth constant that by definition is related to the constant B defined by (5.6.17) by the relation $A = Bc/n$. Γ_a is the filling factor defined by (5.6.3), and its presence here is to ensure that the total number, rather than the density variables used in (5.6.1), of electrons undergoing stimulated transitions is equal to the number of photons emitted. The contributing of spontaneous emission to the photon density is neglected since only a very small fraction ($\sim 10^{-4}$) of the spontaneously emitted power enters the lasing mode.

By setting the left side of (5.6.1) equal to zero, we obtain the steady-state solutions N_o and P_o

$$0 = \frac{I_0}{eV} - \frac{N_0}{\tau} - A(N_0 - N_{tr})P_0$$

$$0 = A(N_0 - N_{tr})P_0\Gamma_a - \frac{P_0}{\tau_p} \qquad (5.6.3)$$

We consider the case where the current is made up of dc and ac components

$$I = I_0 + i_1 e^{i\omega_m t} \tag{5.6.3}$$

and detine the samll-signal modulation response n_1 and p_1 by

$$N = N_0 + n_1 e^{i\omega_m t} \quad P = P_0 + p_1 e^{i\omega_m t} \tag{5.6.5}$$

where N_0 and P_0 are the dc solutions of (5.6.2).
Using (5.6.3), (5.6.5), and the result $A(N0 - N_{tr}) = (\tau_p \Gamma_a)^{-1}$ from (5.6.2)
in (5.6.1) leads to the small-signal algebraic equations

$$- i\omega_m n_1 = -\frac{i_1}{eV} + \left(\frac{1}{\tau} + AP_0\right) n_1 + \frac{1}{\tau_p \Gamma_a} p_1$$

$$i\omega_m p_1 = AP_0 \Gamma_a n_1 \tag{5.6.5}$$

Our main interest is in the modulation response $p_1(\omega_m)/i_1(\omega_m)$ so that from (5.6.5) we obtain

$$p_1(\omega_m) = \frac{-(i_1/eV)AP_0\Gamma_a}{\omega_m^2 - i\omega_m/\tau - i\omega_m AP_0 - AP_0/\tau_p} \tag{5.6.6}$$

A typical measurement of $p_1(\omega_m)$ is shown in Figure 5.19(b). The response curve is flat at small frequencies, peaks at the "relaxation resonance frequency" ω_g, and then drops steeply. The expression for the peak frequency is obtained by minimizing the magnitude of the denominator of (5.6.6)

$$\omega_g = \sqrt{\frac{AP_0}{\tau_p} - \frac{1}{2}\left(\frac{1}{\tau} + AP_0\right)^2} \tag{5.6.7}$$

In a typical semiconductor laser with $L = 300\ \mu m$, we have from (4.7.3) $\tau_p \simeq (n/c)(\alpha - (1/L))\ln R - 1 \sim 10^{12}$ s, $\tau \sim 4 \times 10^{-9}$ s, and $AP_0 \sim 10^9\ s^{-1}$ so that to a very good accuracy

$$\omega_R = \sqrt{\frac{AP_0}{\tau_p}} \tag{5.6.8}$$

The last result is extremely useful, since it suggests that to increase ω_R

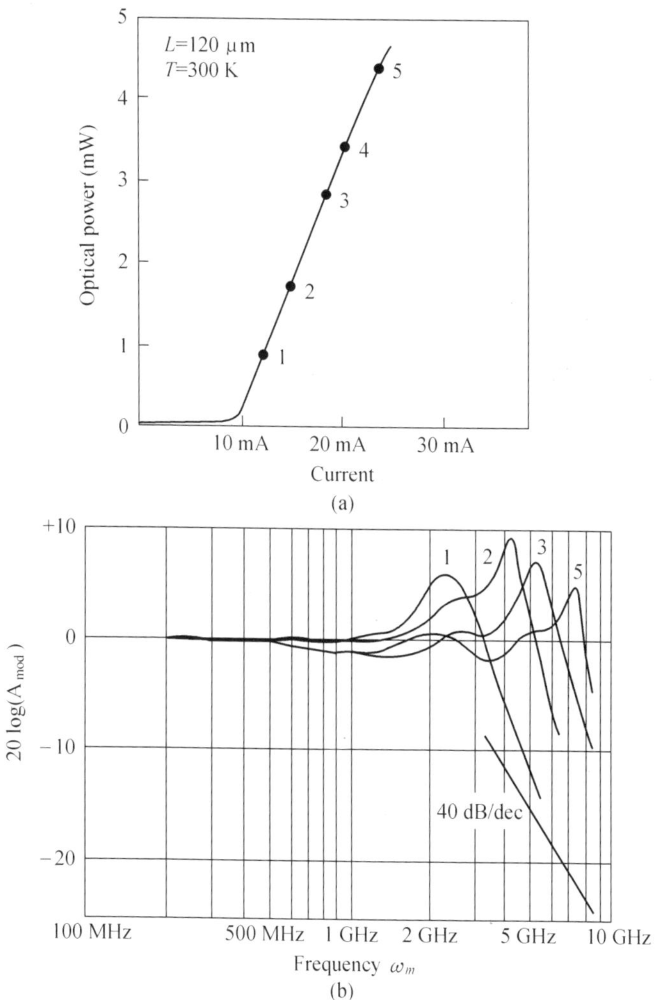

(a)

(b)

and thus increase the useful linear region of the modulation response $p_1(\omega_m)/i_1(\omega_m)$, we need to increase the optical gain coefficient A, decrease the photon lifetime τ_p, and operate the laser at as high internal photon density P_0 as possible. The observed linear dependence of the modulation resonance frequency ω_R on the square root of the power output $\sqrt{P}$ is demonstrated in Figure 5.19(c) for laser of varying lengths. A

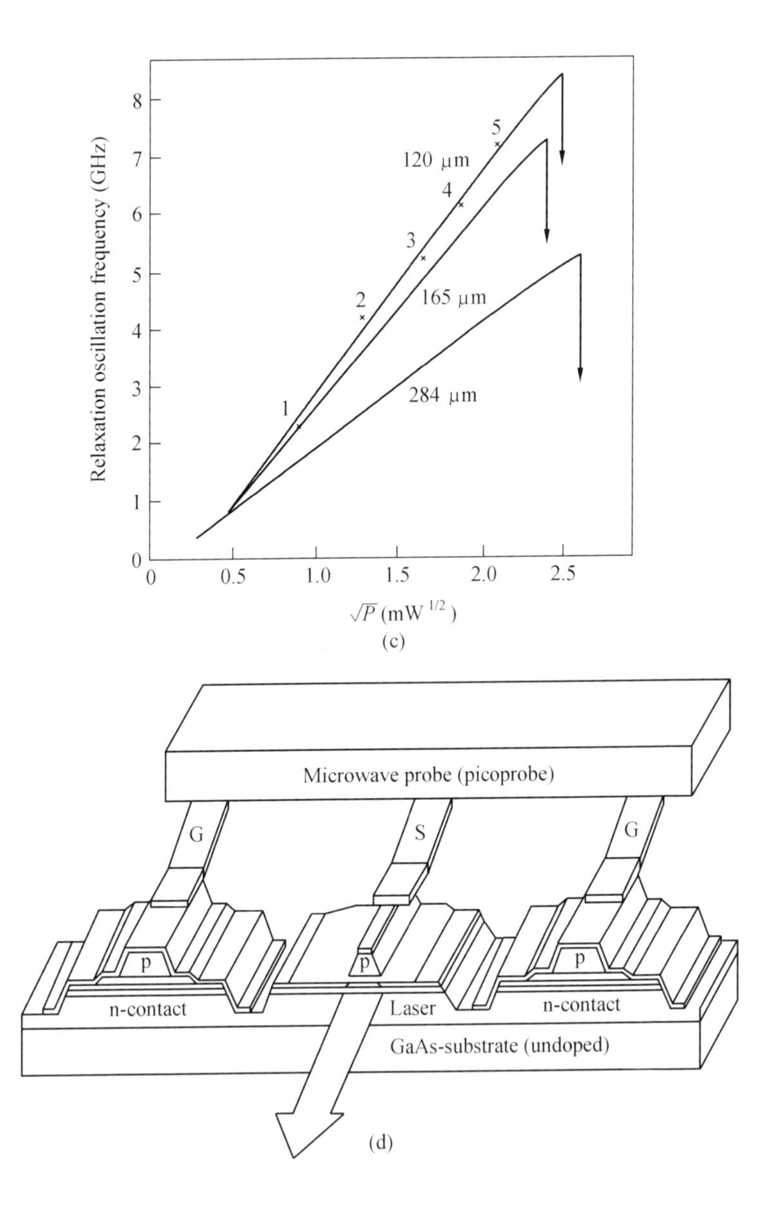

Relaxation oscillation frequency (GHz)
8
7
6
5
4
3
2
1
0
0
0.5
1.0
1.5
2.0
2.5
√P (mW 1/2)
120 μm
165 μm
284 μm
1
2
3
4
5
(c)
Microwave probe (picoprobe)
G
S
G
p
p
p
n-contact
Laser
n-contact
GaAs-substrate (undoped)
(d)

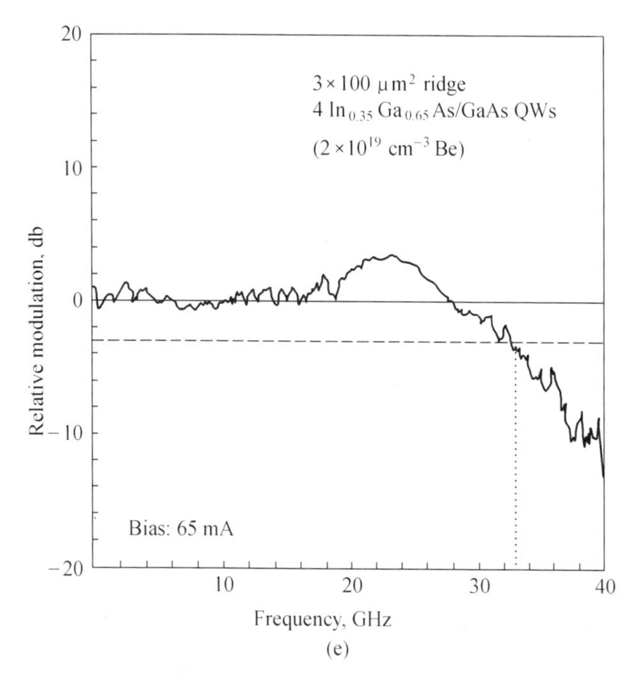

Figure 5.19 (a) *CW* light output power versus current characteristic of a laser of length = 120μm. (b) Modulation characteristics of this laser at various bias points indicated in the plot. (c) Measured relaxation oscillation resonance frequency of lasers of various cavity lengths as a function of $\sqrt{P}$, where P is the cw output optical power. The points of catastrophic damage are indicated by downward pointing arrows. (After Reference [27].) (d) Current feed network for microwave modulation of high-speed lasers. (e) The corresponding frequency response (after Reference [41].)

detailed discussion of the optimum strategy for maximizing ω_R is given in Reference [27]. Figure 5.19(d) shows the microwave current feeding electrodes for high-frequency modulation, and 5.19(e) the corresponding frequency response.

It is somewhat tedious but straightforward to show that (5.6.8) can also be written as

$$\omega_R = \sqrt{\frac{1 + A\tau_p \Gamma_a N_{tr}}{\tau\tau_p}\left(\frac{I_0}{I_{th}} - 1\right)} \tag{5.6.9}$$

Numerical Example: Modulation Bandwidth in GaAs/AgalAs Lasers

Here, using (5.6.8), we will estimate the uppermost useful modulation frequency ω_R of a typical GaAs/GaAlAs laser. We shall assume a typical laser emitting 5×10^{-3} watt from a single face with an active area cross section of $3\mu m \times 0.1\mu m$, a facet reflectivity of $R = 0.31$ and an index of refraction $n_0 = 3.2$. Solving for P_0 from the relationship

$$\frac{(1 - R)P_0 ch\nu}{n_0} = \frac{\text{power}}{\text{area}}$$

we obtain $P_0 = 1.21 \times 10^{15}$ photons/cm^3 for the photon density in the laser cavity. The constant A has a typical value of 2×10^{-6} cm^3/s. [This can be checked against the relationship $A = Bc/n_0$, where B is the spatial gain parameter of (5.6.17).] The photon lifetime τ_p is obtained from (4.7.3)

$$\tau_p = \frac{n_0}{c}\left(\alpha_{ab} - \frac{1}{L}\ln R\right)^{-1}$$

which for $L = 120$ μm, $\alpha_{ab} = 10$ cm^{-1}, and $R = 0.31$ yields $\tau_p \sim 1.08 \times 10^{-12}$ s. Combining these results gives

$$\nu_R \equiv \frac{\omega_R}{2\pi} = \frac{1}{2\pi}\sqrt{\frac{AP_0}{\tau_p}} = \frac{1}{2\pi}\sqrt{\frac{2 \times 10^{-6} \times 1.2 \times 10^{15}}{1.08 \times 10^{-12}}} = 7.53 \times 10^9 \text{ Hz}$$

This value is in the range of the experimental data shown in Figure 5.19, which was obtained on a laser with characteristics similar to that used in our example. The square root law dependence of ω_R on the photon density (or power output) predicted by (5.6.8) is verified by the data of Figure 5.19(c).

5.7 Gain Suppression and Frequency Chirp in Current-Modulated Semiconductor Lasers

In Section 5.7, we solved for the modulation of the power output (or, equivalently, the photon density, inside the laser resonator), which is due to a modulation of the current flowing through the laser. The current is taken as

$$I(t) = I_0 + i_1(\omega_m) exp(i\omega_m t) \tag{5.7.1}$$

while the photon density inside the laser, which is proportional to the power output, is

$$P(t) = P_0 + p_1(\omega_m)\exp(i\omega_m t) \tag{5.7.2}$$

We also take the carrier density in the active regions (the inverted population) as

$$N(t) = N_0 + n_1(\omega_m)\exp(i\omega_m t) \tag{5.7.3}$$

Ideally we would like $p_1(\omega_m)/i_1(\omega_m)$, the frequency modulation response, to be a constant independent of ω_m and, above threshold, we will expect that $n_1(\omega_m) = 0$, indicating perfect gain clamping. As we shall find out. neither expectation is realized fully. As a matter of fact. if we solve Equation (5.7.5) for $n_1(\omega_m)$, the result is

$$n_1(\omega_m) = - i\left(\frac{i_1}{\mathrm{eV}}\right)\frac{\omega_m}{\omega_m^2 - \dfrac{AP_0}{\tau_p} - i\omega_m\left(\dfrac{1}{\tau} + AP_0\right)} \tag{5.7.4}$$

We thus find that, under dynamic conditions, the carrier density, hence the gain, is not clamped at the threshold value N_0 but has an oscillating component whose amplitude $n_1(\omega_m)$ is given by (5.7.4). Its general feature are depicted in Figure 5.20. We note a peak at ω_R, which is also the resonance frequency for the amplitude modulation response $p_1(\omega_m)/i_1(\omega_m)$, as given by (5.7.7). Since the index of refraction of a semiconductor medium depends on the carrier density, the modulation of the latter is accompained by a modulation of the index of refraction, leading to a frequency modulation (FM) of the output optical field. This parasitic FM modulation, most of the time undesired, has a number of important consequences. The most important of these is the resulting spectral broadening of the laser field that, in dispersive, $D \neq 0$, fibers, leads to an increase in the spreading of optical pulses with distance.

Before embarking on the analysis of the parasitic frequency modulation, we need to introduce two new physical concepts: (1) the gain suppression effect and (2) the amplitude-phase coupling effect.

Gain suppression. The gain experienced in an inverted semiconductor laser medium by an optical wave is invariably lower the higher the optical intensity. This is due partly to optical gain saturation and partly to gain suppression. The first effect reflects the drop of the total electron population density—N in (5.7.1)—with the increase in P. This effect is accounted for properly by the coupled rate equations (5.7.1). The second mechanism reducing the gain takes place even when the total density N is constant and reflects the reduction of the density of *resonant carriers* (electrons and holes) in the *immediate vicinity* of points a and b in Figure 5.5 which contribute to the gain. This is due both to spectral "hole burning," as discussed in Section 5.7 and illustrated in

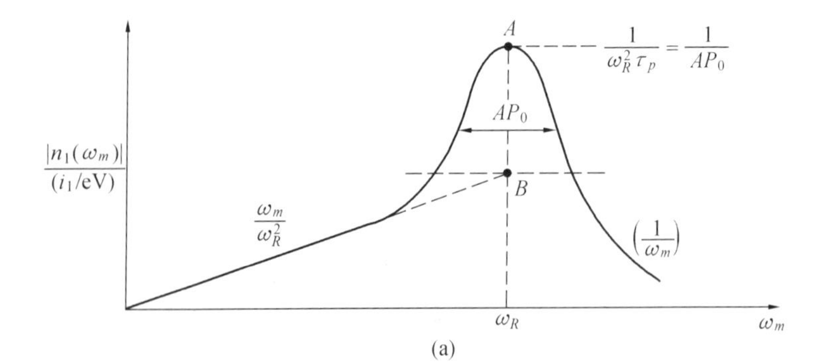

(a)

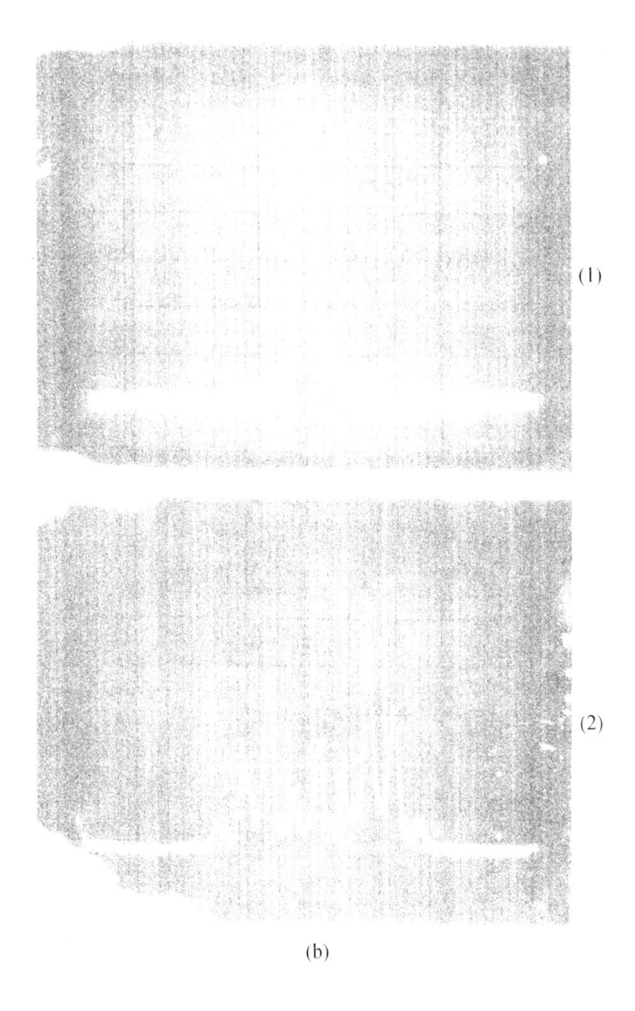

(b)

Figure 5.20 (a) A theoretical plot of the carrier density modulation n_1 as a function of the current modulation frequency ω_m. (b)(1) A scanning Fabry-Perot spectrum of a GaInAsP ($\lambda = 1.31\ \mu m$) DFB laser with no current modulation. (2) The spectrum of the same laser when the current is modulated at f_m = 550 MHz, horiz, scale = 1 GHz/div. (Courtesy of H. Blauvelt. P. C. Chen, and N. Kwong of ORTEL Corporation, Alhambra, California)

Figure 6.9(d), and to an increase in the effective electron temperature① by the optical field. Under dynamic nonthermal equilibrium conditions, this temperature may differ from the lattice temperature. Such an increase causes, according to the discussion in Section 5.1 and 5.2, the electrons (and holes) to spread to higher energies, reducing in the process the density of resonant carriers that contribute to the gain. This effect asserts itself with a time constant of $< 10^{-12}$s, characteristic of electron-electron and electron-phonon collisions, and for system applications involving modulation at $\omega_m/2\pi < 3 \times 10^{10}$ Hz, it can be considered as responding instantaneously to the optical field.

Our main departure from the analyses of Section 5.5 is to take the optical gain constant as

$$G(N,P) = G(N)(1 - \in P) \approx G(N_{th}) + A(N - N_{th}) - \in G(N_{th})P \tag{5.7.5}$$

where gain suppression is represented by the factor $(1 - \in P)$. The constant $\in$ is called the *gain suppression factor*. We can view (5.7.5) as a Taylor expansion about the threshold point $N = N_{th}$, $P = 0$. The numerical value of $\in$ can be estimated theoretically in a given system but more often is evaluated experimentally [42].

We now rewrite the rate equations (5.7.1) as

$$\begin{aligned} \frac{\mathrm{d}N}{\mathrm{d}t} &= \frac{1}{eV} - \frac{N}{\tau} - G(N,P)P \\ \frac{\mathrm{d}P}{\mathrm{d}t} &= \Gamma_a G(N,P)P - \frac{P}{\tau_p} \end{aligned} \tag{5.7.6}$$

at steady state $\mathrm{d}/\mathrm{d}t = 0$

$$0 = \frac{I_0}{eV} - \frac{N_0}{\tau} - [G(N_{th}) + A(N_0 - N_{th}) - \in G(N_{th})P_0]P_0$$

$$0 = \Gamma_a[G(N_{th}) + A(N_0 - N_{th}) - \in G(N_{th})P_0]P_0 - \frac{P}{\tau_p}$$

① By electron "temperature" we mean the temperature used in the Fermi function (5.1.8).

The value of $G(N_{th})$ is obtained from the second equation evaluated at threshold ($N_0 = N_{th}, P_0 = 0$)

$$G(N_{th}) = \frac{1}{\Gamma_a \tau_p}$$

We use the last result to simplify the last two equations

$$0 = \frac{I_0}{eV} - \frac{N_0}{\tau} - \frac{P_0}{\Gamma_a \tau_p}$$

$$0 = A(N_0 - N_{th}) - \frac{\in P_0}{\Gamma_a \tau_p} \tag{5.7.7}$$

Performing a "small signal" expansion as in Equations (5.7.3) and (5.7.4) leads to

$$i\omega_m n_1 = \frac{i_1}{eV} - \left(\frac{1}{\tau} + AP_0\right) n_1 - \frac{(1 - \in P_0)}{\Gamma_a \tau_p} p_1$$

$$i\omega_m p_1 = \Gamma_a A P_0 n_1 - \frac{\in P_0}{\tau_p} p_1 \tag{5.7.8}$$

We note that for the case $\in = 0$, these equations reduce to Equations (5.7.5). Solving (5.7.8) for $p1(\omega_m)$ yields

$$p_1(\omega_m) = \frac{-\Gamma_a A P_0 \left(\frac{i_1}{eV}\right)}{\omega_m^2 - i\omega_m \left(\frac{1}{\tau} + AP_0 + \frac{\in P_0}{\tau_p}\right) - \left(\frac{AP_0}{\tau_p} + \frac{\in P_0}{\tau \tau_p}\right)} \tag{5.7.9}$$

Under typical conditions such as those of the numerical example above, we have

$$P_0 = 1.2 \times 10^{21} \text{ photons/m}^3$$

$$A = 2 \times 10^{-12} \text{ m}^3/\text{s} \quad \tau = 4 \times 10^{-9} \text{ s}$$

$$\tau_p = 10^{-12} \text{ s} \quad \in = 10^{-23} \text{ m}^3$$

so that $\in P_0/\tau_p = 1.2 \times 10^{10} \text{ s}^{-1} \gg AP_0$.

Using the inequalities $\in P_0/\tau_p \gg AP_0$ and $\in/\tau p \gg A$, we find that the peak modulation response occurs at

$$\omega_R \approx \sqrt{\frac{AP_0}{\tau_p} - \frac{\in^2 P_0^2}{2\tau_p^2}}$$

A comparison of this last result to (5.7.8) shows that when $\in \neq 0$, i.e., gain suppression is considered, the modulation resonant frequency, ω_R, does not increase indefinitely with P_0 but reaches a maximum value at a photon density of

$$(P_0)_{\max} = \frac{A\tau_p}{\in^2} \tag{5.7.10}$$

The maximum value of ω_R, which obtains when $P_0 = (P_0)_{\max}$, is

$$(\omega_R)_{\max} = \frac{A}{\sqrt{2}\in} \tag{5.7.11}$$

Using the typical numerical constants given following (5.7.9) and in the numerical example of Section 5.7, we estimate

$$(P_0)_{\max} = \frac{A\tau_p}{\in^2} = \frac{2 \times 10^{-12} \times 10^{-12}}{10^{-46}} = 1 \times 10^{22} \text{ photons/m}^3$$

which, using the expressions given in the example of Section 5.7, corresponds to a power output of ~ 80 mwper facet. The corresponding maximal resonant modulation frequency is

$$\left(\frac{\omega_R}{2\pi}\right)_{\max} \approx \left(\frac{1}{2\pi}\right)\frac{A}{\sqrt{2}\in} = \frac{2 \times 10^{-12}}{2\pi\sqrt{2} \times 10^{-23}} = 2.24 \times 10^{10} \text{ Hz}$$

The last result points out the role of the gain suppression in placing a practical upper limit on the modulation bandwidth that can be achieved by current modulation of semiconductor lasers. To modulate at significantly higher frequencies, $\omega_m > \omega_R$, one needs to employ external modulators.

Amplitude-phase coupling

Amplitude-phase coupling which causes a second effect that is attendant upon the carrier density modulation ($n_1(\omega_m) \neq 0$) is the frequency modulation ("chirp") of the laser output field. The amplitude $n_1(\omega_m)$ of the carrier density fluctuation is obtained by solving (5.7.8)

$$n_1(\omega_m) = \frac{-\left(i\omega_m + \frac{\in P_0}{\tau_p}\right)\left(\frac{i_1}{eV}\right)}{\omega_m^2 - \left(\frac{AP_0}{\tau_p} + \frac{\in P_0}{\tau\tau_p}\right) - i\omega_m\left(\frac{1}{\tau} + AP_0 + \frac{\in P_0}{\tau_p}\right)} \tag{5.7.12}$$

A comparison of (5.7.12) to (5.7.9) shows that

$$n_1(\omega_m) = \frac{\left(i\omega_m + \frac{\in P_0}{\tau_p}\right)}{\Gamma_a AP_0} p_1(\omega_m) \tag{5.7.13}$$

Since for our assumed exp ($i\omega_m t$) time dependence, $i\omega_m = d/dt$, we use (5.7.13) to write

$$\Delta N(t) = \frac{1}{\Gamma_a A}\left[\frac{1}{P_0}\frac{\mathrm{d}P}{\mathrm{d}t} + \frac{\in}{\tau_p}\Delta P(t)\right] \tag{5.7.14}$$

where $N(t) = N_0 + \Delta N(t)$, $P(t) = P_0 + \Delta P(t)$. Equation 5.7.14 applies to any arbitrary photon density modulation, not necessarily harmonic.

Our next task is to find the effect of the carrier density modulation $\Delta N(t)$ on the laser frequency. The index of retraction of the gain medium is complex

$$n_0(t) = n_0'(t) + in_0''(t) \tag{5.7.15}$$

where the time dependence reflects the dependence of the real index n_0' and the gain (or loss), which involves n_0'', on the time-modulated carrier density $\Delta N(t)$.

The imaginary part of the index n_0 is related to the spatial exponential gain constant of the laser medium since the spatial dependence of the field is

$$\mathscr{E}(z) \propto \boldsymbol{E}_0\exp\left[-\frac{i\omega}{c}(n_0' + n_0'')z\right] = \boldsymbol{E}_0\exp\left(-\frac{i\omega n_0'}{c}z\right)\exp\left(\frac{\omega n_0''}{c}z\right)$$

Since our rate equations (5.7.6) and (5.7.8) are in the time domain, we need to convert the spatial growth parameter $\omega n_0''/c$ to a temporal one. We use

$$\frac{\mathrm{d}|\mathscr{E}|}{\mathrm{d}t}=\frac{\partial|\mathscr{E}|}{\mathrm{d}z}\frac{\mathrm{d}z}{\mathrm{d}t}\cong\left(\frac{\omega n_0''}{c}z\right)\frac{c}{n_0'}|\mathscr{E}|=\frac{\omega n_0''}{n_0'}|\mathscr{E}|$$

It thus follows that the exponential gain constant $G(N,P)$ of (5.7.6) is related to n''_0 by

$$G=\frac{2\omega n_0''}{n_0'}=\frac{4\pi v n_0''}{n_0'} \tag{5.7.16}$$

where the factor of 2 accounts for the fact that G is the temporal growth constant of the photon density ($\propto$ optical intensity) i.e., of $|\mathscr{E}|^2$.

$$\frac{\partial n_0''}{\partial N}=\frac{n_0'}{4\pi v}\frac{\partial G}{\partial N}=\frac{n_0'}{4\pi v}A \tag{5.7.17}$$

where we used (5.7.5) to substitute $\partial G/\partial N=A$, A modulation $\Delta N(t)$ of the carrier density causes, according to (5.7.17), a corresponding perturbation

$$\Delta n_0''=\frac{n_0'}{4\pi v}A\Delta N(t) \tag{5.7.18}$$

Our next task is to obtain the dependence of the real index perturbation $\Delta n_0'$ on$\Delta N(t)$. Now $\Delta n_0'$ and $\Delta n_0''$ are related through Kramers-Kroning relations, which were discussed in Section 5.4. It has, however, proved useful to relate $\Delta n_0'$ and $\Delta n_0''$ by means of a parameter, the so-called phase-amplitude coupling constant, the α parameter [36,43].

$$\alpha=\frac{\Delta n_0'}{\Delta n_0''} \tag{5.7.19}$$

The α parameter is a function of the carrier density N_0 of the semiconductor laser medium, and typical values for it range between 3 and 5 [36,44]. Combining (5.7.19) and (5.7.18), results in

$$\Delta n_0''(t)=\frac{\alpha n_0' A}{4\pi v}\Delta N(t) \tag{5.7.20}$$

A pertubation $\Delta n_0'$ due to carrier density modulation causes a perturbation Δv of the laser frequency

$$\frac{\Delta v}{v} = -\frac{\Delta n_0'}{n_0'}\Gamma_a = \frac{\alpha F_a A}{4\pi v}\Delta N(t) \qquad (5.7.21)$$

The first equality of (5.7.21) follows directly from the basic Fabry-Perot resonance frequency relation. The factor $\Gamma_a \approx V_{active}/V_{mode}$ accounts for (possibly) partial filling of the resonator by the active medium. Substitution in the last equation for $\Delta N(t)$ from (5.7.14) [36,37] gives

$$\Delta v(t) = -\frac{\alpha}{4\pi}\left[\frac{1}{P_0}\frac{dP}{dt} + \frac{\in}{\tau_p}\Delta P(t)\right] \qquad (5.7.22)$$

for the laser frequency chirp. The contribution involving dP/dt is called the *transient chirp*, while that which is proportional to $\Delta P(t)$ is termed the *adiabatic chirp*. The latter involves the gain suppression parameter $\in$ and is usually dominant at low ($\leqslant 10^8$ Hz) frequencies, while the transient term dominates at typical microwave ($> 10^9$ Hz) frequencies.

The Field Spectrum of A Chirping Laser (36,37)

In the development leading to (5.7.22), we showed that any modulation of the power of a semiconductor laser by means of current modulation causes a frequency chirp. In the following, we will derive the spectrum of the output optical field of a laser with sinusoidal power modulation in order to obtain an appreciation of the expected order of magnitude of the effect, especially the amount of spectral spread. We take the optical field of the laser as

$$\mathscr{E}(t) = \left[E_0 + \frac{s}{2}E_0\sin\omega_m t\right]\exp\{i[2\pi v_0 t + \phi(t)]\} \qquad (5.7.23)$$

where ω_m is the modulation frequency, and v_0 is the average optical frequency.

The phase $\phi(t)$ in (5.7.23) is related to the chirp $\Delta v(t)$ of (5.

7.22) by the general result[①]

$$\phi(t) = 2\pi\int_0^t \Delta v(t')\mathrm{d}t' = -\frac{\alpha}{2}\left[\frac{1}{P_0}P(t) + \frac{\in}{\tau_p}\int_0^t \Delta P(t')\mathrm{d}t'\right] \tag{5.7.24}$$

If $\phi(t)$ were a constant, which would be the case when $\alpha = 0$, the spectrum of the topical field $\mathscr{E}(t)$, would consists of a carrier of amplitude E_0 at v_0 and two sidebands with amplitudes $s/4$. When $\phi(t)$ is not constant, additional sidebands appear and the optical spectrum broadens. This chirped-induced broadening is of special concern in applications that involve high data rate communication in fibers, since, as shown in Section 2.9, the temporal broadening of pulses in dispersive fibers is directly poportional to the product of the spectral width of the light and the propagation length. Any increase in the laser's spectral width would thus limit the rate at which the data can be transmitted in such a fiber. In the case considered here, we can write the photon density $P(t)$ at

$$P(t) \propto \langle |\mathscr{E}(t)|^2\rangle = E_0^2\left(1 + \frac{s^2}{4}\right) + sE_0^2\sin\omega_m t = P_0 + P_1\sin\omega_m t \tag{5.7.25}$$

where $\langle\ \rangle$ indicates averaging over a few optical periods. Substitution of the last expression (5.7.24) leads to

$$\phi(t) = -\frac{\alpha}{2}\left[\begin{matrix}\text{transient} & \text{abiabatic}\\ \dfrac{P_1}{P_0}\sin\omega_m t & -\dfrac{\in P_1}{\omega_m\tau_p}\cos\omega_m t\end{matrix}\right]$$

where we left out time-independent terms that correspond to unimportant fixed phase shifts. At high modulating frequencies such that $\omega_m \gg \in P_0/\tau_p$ (this corresponds, using the numerical data used earlier in this section, to $\omega_m/2\pi \gg 2\times10^9$ Hz), the first term ("transient") in the

① The basic definition of the frequency $v(t)$ of a field $A\exp[i\phi(t)]$ is $v(t) = (1/2\pi)(\mathrm{d}\phi/\mathrm{d}t)$ and corresponds to the number of optical cycles per second at time t.

square brackets dominates, so that the optical phase can be taken as

$$\phi(t) = -\frac{m\alpha}{2}\sin\omega_m t$$

$$m \equiv \frac{P_1}{P_0} = \frac{s}{1 + s^2/4} \approx s \qquad (5.7.26)$$

when s (and m) $\ll 1$. The total optical field, Equation (5.7.23), assumes the form

$$\mathscr{E}(t) = \left[E_0 + \frac{m}{2}E_0\sin(\omega_m t)\right]\exp\left\{i\left[\omega_0 t - \frac{m\alpha}{2}\sin(\omega_m t)\right]\right\} \qquad (5.7.27)$$

where we used $s \approx m (s \ll 1)$.

We can use the Bessel function identity

$$e^{i\delta\sin x} = \sum_{n=-\infty}^{\infty} J_n(\delta)e^{inx} \qquad (5.7.28)$$

to rewrite (5.7.27)

$$\frac{\mathscr{E}(t)}{E_0} = \sum_{n=-\infty}^{\infty} J_n(\delta)\exp[i(\omega_0 + n\omega_m)t] - i\frac{m}{4}\sum_{n=-\infty}^{\infty} J_n(\delta) \times \exp\{i[\omega_0 + (n+1)\omega_m]t\} + i\frac{m}{4}\sum_{n=-\infty}^{\infty} J_n(\delta)\exp\{i[\omega_0 + (n-1)\omega_m]t\}$$

Some of the sidebands are

at ω_0, $\quad E_0\left[J_0(\delta) + i\frac{m}{2}J_1(\delta)\right]\exp(i\omega_0 t)$

at $\omega_0 + \omega_m$, $E_0\left[J_1(\delta) - i\frac{m}{4}J_0(\delta) + i\frac{m}{4}J_2(\delta)\right]\exp[i(\omega_0 + \omega_m)t]$

at $\omega_0 - \omega_m$, $E_0\left[-J_1(\delta) + i\frac{m}{4}J_0(\delta) + i\frac{m}{4}J_2(\delta)\right]\exp[i(\omega_0 - \omega_m)t]$

(5.7.29)

at $\omega_0 + 2\omega_m$, $E_0\left[J_2(\delta) - i\frac{m}{4}J_1(\delta) + i\frac{m}{4}J_3(\delta)\right]\exp[i(\omega_0 + 2\omega_m)t]$

at $\omega_0 - 2\omega_m$, $E_0\left[J_2(\delta) - i\frac{m}{4}J_1(\delta) + i\frac{m}{4}J_3(\delta)\right]\exp[i(\omega_0 + 2\omega_m)t]$

where $\delta = -m\alpha/2$ is the phase modulation index of the optical field.

The amplitudes of the sidebands at $\omega_0 \pm n\omega_m$ in this case have the same magnitude. This is a consequence of the form of (5.7.27) but is not generally true, so that the optical sidebands, in general, for m and δ not zero, are not symmetric about ω_0. This is considered in Problem 5.11. An experimental graph showing the spectrum of the output field of a laser whose current is modulated is given in Figure 5.20(b). The spectrum can be fit well with an adiabatic chirp spectrum (i.e., phase modulation 90° out of phase with current), corresponding to a field

$$\mathscr{E}(t) = \left[E_0 + \frac{m}{2} E_0 \sin\omega_m t \right] \exp[\mathrm{i}2\pi v_0 t + \delta \cos\omega_m t] \tag{5.7.30}$$

with $m = 0.2$ and $\delta = 3.3$.

In the transient case, we found [see (5.7.27)] that the phase modulation index δ, the amplitude of the phase excursion, is equal to $m\alpha/2$ and that it can be determined from a fit to the experimental sideband distribution. Since the intensity modulation index m can be determined straightforwardly from a spectral analysis of the laser intensity, the combination of the field spectrum, obtainable with a scanning Fabry-Perot etalon, and the intensity spectrum, obtained from a spectral analysis of the detected photocurrent, can be used to determine the amplitude-phase coupling constant α[36].

5.8 Integrated Optoelectronics

In one of its rare moments of cooperative spirit, nature has endowed the Ⅲ-Ⅴ semiconductors based on GaAs/GaAlAs and InP/GaInAsP with a double gift. These are, as discussed above, the materials of choice for semiconductor lasers, but in addition it is possible to use them, especially In GaAs/GaAs and GaAs/GaAlAs, as base materials for electronic circuits in a manner similar to that in silicon.①

Iw was pointed out in 1971 [30] that it should be possible to bring together monolithically in a Ⅲ-Ⅴ semiconductor the two principal actors of the modern communication era—the transistor and the laser—in new integrated optoelectronic circuits. This new technology is now taking its first tentative steps from the laboratory to applications.

The basic philosophy, as well as an example of an integrated optoelectronic device, is shown in Figure 5.21, which shows a buried heterostructure GaAs/GaAlAs laser, sim lar to that illustrated in Figure 5.15, fabricated monolithically on the same crystal as a field-effect transistor (FET). The output current of the FET (see arrows) supplies the electron injection to the active region of the laser. This current and thus the laser power output can be controlled by a bias voltage applied to the gate electrode.

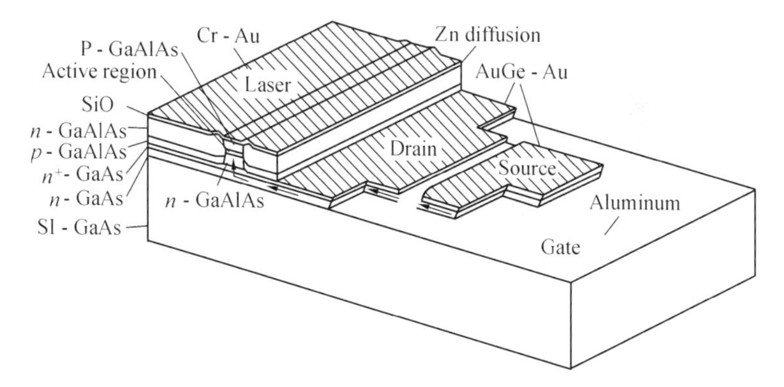

Figure 5.21 A GaAs n-channel field-effect transistor integrated monolithically with a buried heterostructure GaAs/GaAlAs laser. The application of gate voltage is used to control the bias current of the laser. This voltage can oscillate and modulate the light at frequencies > 10 GHz. (After Reference [31].)

① A completely new electronic technology based on GaAs/GaAlAs is now emerging [29]. It takes advantage of the large mobility of electrons in GaAs for very high switching speeds.

An example of a feasibility model of an integrated optoelectronic optical repeater, which incorporates a detector, a FET current preamplifier, a FET laser driver, and a laser, is shown in Figure 5.22. The main reason for the accelerating drive toward an integrated optoelectronic circuit technology [33] derives from the reduction of parasitic reactance that are always associated with conventional wire interconnections, plus the compatibility with the integrated electronic circuits technology that makes it possible to apply the advanced techniques of the latter to this new class of devices. More recent examples of optoelectronic integrated circuits are demonstrated in Figures 5.23 and 5.24.

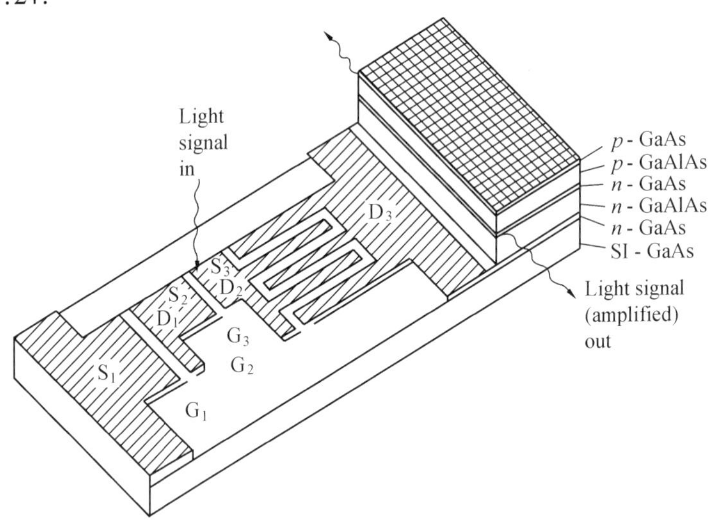

Figure 5.22 A monolithically integrated optoelectronic repeater containing a detector, transistor current source, a FET amplifier, and a laser on a single crystal GaAs substrate. (After Reference [32].)

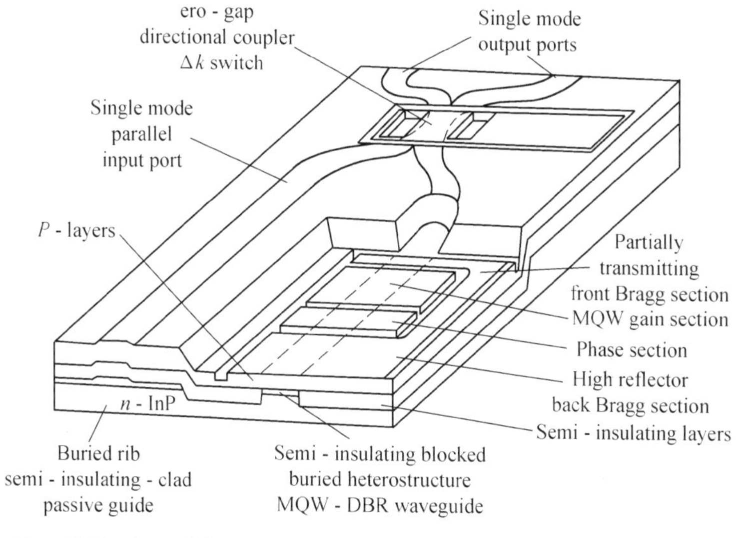

Figure 5.23 A monolithic circuit containing a tunable multisection In GaAsP/InP 1.55 μm laser employing multiquantum well gain section, a passive waveguide for an external input optical wave, and a directional coupler switch for combining the laser output field and that of the external input at the output ports. (After Reference [34].)

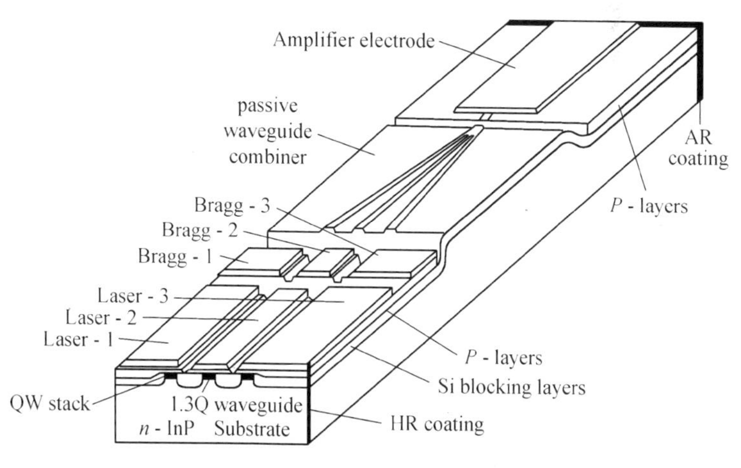

Figure 5.24 An optoelectronic integrated circuit composed of three ~ 1.5 μm InGaAs/InP distributed feedback lasers each tuned to a slightly different wavelength. The three wavelengths are fed into a single waveguide and amplified in a single amplifying section. (After Reference [35].)

Problems

5.1 Derive Equations (5.8.12).

5.2 Derive Equations 5.8.7 and 5.8.8.

5.3 Assume a fiber with $L = 10$ km and a group velocity dispersion parameter of 10 psec/nm-km (see Section 3.4). Calculate the maximum data rate through the fiber in bits/s if we use a semiconductor laser with characteristics similar to those used in the example of Section 5.8. For the purpose of this calculation, assume that a data rate of N bits/s is equivalent to a current modulation frequency of $\omega_m/2\pi = N$.

5.4 Find the frequency ω_R that maximizes $p_i(\omega_m)$ as given by equation (5.8.9) and, using the approximations given, derive (5.8.11).

5.5 Evaluate and plot:

(a) The gain $\gamma(\omega)$ of an inverted GaAs crystal under the following conditions:

$$N_{\text{elec}} = N_{\text{hole}} = 3 \times 10^{18}\ \text{cm}^{-3}$$

$$m_c = 0.07\ m_{\text{electron}}$$

$$m_h = 0.4\ m_{\text{electron}}$$

$$T = 0\ \text{K}$$

$$E_R = 1.45\ \text{eV}$$

$$T_2 = \infty$$

(b) Comment qualitatively on the changes in $\gamma(\omega)$ as the temperature is raised.

(c) What is the effect of a finite T_2 on $\gamma(\omega)$?

5.6 Consider the effect on the modulation response $p_1(\omega_m)/i_1(\omega_m)$ of the inclusion of a nonlinear gain term bP in the rate equations (5.8.1)

$$\frac{dN}{dt} = \frac{I}{eV} - \frac{N}{\tau} - A(1 - bP)(N - N_{tr})P$$

$$\frac{dP}{dt} = A(1 - bP)(N - N_{tr})P\Gamma_a - \frac{P}{\tau_p}$$

where $bP \ll 1$. Show that the main effect is a damping of the resonance peak at ω_R.

5.7 Solve for the carrier density modulation $N = N_0 + N_1 e^{i\omega_m t}$ in semiconductor lase whose current is modulated at

$$I = I_0 + I_1 e^{i\omega_m t} \tag{1}$$

$$\omega_m = \text{modulation frequency} \ll \omega_{opt} \tag{2}$$

(See Section 5.8.)

5.8 Assume $\epsilon = \epsilon_0 - aN$, a is a constant and that the instantaneous frequency of the semiconductor laser obeys

$$\frac{\Delta v}{v} = \frac{\Delta \epsilon}{\epsilon} \tag{3}$$

find the form of the laser optical field due to the current modulation. What is the (phase) modulating index of the field?

5.9 Using the data of Figure 5.7, what is the total current needed to render the active medium of a semiconductor laser transparent? Assume an active volume of $300 \times 2 \times 0.2$ (μm^3) and a recombination lifetime of $\tau = 3 \times 10^{-9}$ seconds.

5.10 If the thickness of the active region in Problem 5.9 were reduced to 100 Å, can we obtain enough gain from a semiconductor laser to overcome a distributed loss constant of $\alpha = 20\ \text{cm}^{-1}$ and $R = 0.9$? What will be the transparency current? What will be the threshold current? Assume a mode height normal to the interfaces of $t = 4000$ A and

$$\Gamma_a \sim \frac{\text{d(active region)}}{\text{t(mode height)}}$$

5.11 Plot the optical spectrum of a wave with simultaneous AM and FM modulation

$$\mathscr{E}(t) = E_0\left(1 + \frac{m}{2}\sin\omega_m t\right)\exp\{i[\omega_0 t + \delta\sin(\omega_m t + \alpha)]\}$$

for $\alpha = 0, \pi/6, \pi/4, \pi/2$.

References

1. Basov, N. G., O. N. Krokhin, and Y. M. Popov, "Production of negative temperature states in *p-n* junctions of degenerate semiconductors," *J. E. T. P.* 40:1320, 1961.

2. Hall, R. N., G. E. Fenner, J. D. Kingsley, T. J. Soltys, and R. O. Carlson. "Coherent light emission from GaAs junctions," *Phys. Rey. Lett.* 9:366, 1962

3. Nathan, M. I., W. P. Dumke, G. Burns, F. H. Dills, and g. Lasher, "Stimulated emission of radiation from GaAs *p-n* junctions" *Appl. Phys. Lett.* 1:62, 1962.

4. Yariv, A., and R. C. C. Leite, "Dielectric waveguide mode of light propagation in *p-n* junctions," *Appl. Phys. Lett.* 2:55, 1963.

5. Anderson, W. W., "Mode confinement in junction lasers," *IEEE J. Quant. Elec.* QE-1:228, 1965.

6. Kittel, C., *Introduction to Solid State Physics*, 5th ed. New York: Wiley, 1982.

7. Yariv, A., *Introduction to the Theory and Applications of Quantum Mechanics*. New York: Wiley, 1982.

8. Bernard, M. G., and G. Duraffourg, "Laser conditions in semiconductors." *Phys. Status Solidi* 1:699, 1961.

9. Vahala, K., L. C. Chiu, S. Margalit, and A. Yariv, "On the linewidth enhancement factor α in semiconductor injection lasers," *Appl, Phys. Lett.* 42:631, 1983.

10. Stern, F., "Semiconductor laser: Theory," *Laser Handbook*, F. T. Arecchi and E. O. Schultz Du Bois, eds. Amsterdam: North Holland, 1972.

11. Kressel, H., and J. K. Butler, *Semiconductor Lasers and Herterojunction LEDS*. New York: Academic Press, 1977.

12. Yariv, A., *Quantum Electronics*, 2d ed. New York: Wiley, 1975, p. 219.

13. Casey, H.C., and M. B. Panish, *Heterostructure Lasers*. New York: Academic Press, 1978.

14. Dupuis, R. D., and P. D. Dapkus, *Appl. Phys. Lett*. 31: 466, 1977.

15. Cho, A. Y., and J. R. Arthur, in *Progress in Solid State Physics*, J. O. Mc Caldin and G. Somoraj, eds., vol. 10. Elmsford, N.Y.: Pergamon Press, p. 157.

16. Tsang, W. T. and A. Y. Cho, "Growth of GaAs/GaAlAs by molecular beam epitoxy," *Appl. Phys. Lett*. 30:293, 1977.

17. Hayashi, J., M. B. Panish, and P. W. Foy, "A low-threshold room-temperature injection laser," *IEEE J. Quant. Elec*. 5: 211, 1969.

18. Kressel, H., and H. Nelson, "Close confinement gallium arsenide *p-n* junction laser with reduced optical loss at room temperature," *RCA Rev*. 30: 106, 1969.

19. Alferov, Zh. I., et al., *Sov. Phys.—Semicond*., 4: 1573, 1971.

20. Tsukada, T., "GaAs-$Gal_{1-x}Al_xAs$ buried heterostructure injection lasers," *J. Appl. Phys*. 45:4899, 1974.

21. Chinone, N., H. Nakashima, I. Ikushima, and R. Ito, "Semiconductor lasers with a thin active layer ($<0.1\mu m$) for optical communication," *Appl. Opt*. 17:311, 1978.

22. Botez, D., and G. J. Herskowitz, "Components for optical communication systems: A review," *Proc. IEEE* 68: 1980.

23. Ortel Corp., Alhambra, Calif. Product Data Sheets.

24. Hsieh, J. J., J. A. Rossi, and J. P. Donnelly, "Room

temperature CW operation of GaIn AsP/InP double heterostructure diode lasers emitting at 1.1 μm," *Appl. Phys. Lett*. 28:709,1976.

25. Yu, K. L., Koren, U., T. R. Chen, and A. Yariv, "A Groove GaIn AsP Laser on Semi-Insulating InP," *IEEE J. Quant. Elec*. QE-8:817, 1982.

26. Suematsu, Y., "Long wavelength optical fiber communication," *Proc. IEEE*. 71:692, 1983.

27. Lau, K. T., N. Bar-Chaim, I. Ury, and A. Yariv, "Direct amplitude modulation of semiconductor GaAs lasers up to X-band frequencies," *Appl. Phys. Lett*. 43:11,1983.

28. Lau, K. Y., Ch. Harder, and A. Yariv, "Direct modulation of semiconductor lasers at $f > 10$ GHz," *Appl. Phys. Lett*. 44:273-275, 1984.

29. See, for example, Bailbe, J. P., A. Marty, P. H. Hiep, and G. E. Rey, "Design and fabrication of high speed GaAlAs/GaAs heterojunction transistors," *IEEE Trans. Elect. Dev*. ED-27: 1160, 1980.

30. Yariv, A., "Active integrated optics," in *Fundamental and Applied Laser Physics*, Proc. ESFAHAN Symposium, Aug. 29, 1971, M. S. Feld, A. Javan, N. Kurnit, eds. New York: Wiley, 1972.

31. Ury, I., K. Y. Lau, N. Bar-Chaim, and A. Yariv, "Very high frequency GaAlAs laser-field effect transistor monolithic integrated circuit," *Appl. Phys. Lett*. 41:126,1982.

32. Yust, M., et al., "A monolithically integrated optical repeater," *Appl. Phys. Lett*. 10:795,1979.

33. Yust, M., et al., "A monolithically integrated optical repeater," *Appl. Phys. Lett*. 10:795, 1979.

34. Hernandez-Gil, F., T. L. Koch, U. Koren, R. P. Gnall, C. A. Burrus, "Tunable MQW-DBR laser with a monolithically integrated InGaAs/InP Directional coupler switch," paper PD 17, Conference on

Laser Engineeing and Optics (CLEO) 1989.

35. Koren, U., et al., "Wavelength division multiplexing light source with integrated quantum well tunable lasers and optical amplifiers," *Appl. Phys. Lett.* 54:21 (May 1989).

36. Harder, C., K. Vahala, and A. Yariv, "Measurement of the linewidth enhncement factor α of semiconductor lasers," *Appl, Phys. Lett.* 42:428(1983).

37. Koch, T., and J. Bowers, "Nature of wavelength chirping in directly modulated semiconductor lasers," *Electr. Lett.* 20:1038-1040, 1984.

38. Derry, P., et al., "Ultra low threshold graded-index separate confinement single quantum well buried heterostructure (Al, Ga) as lasers with high reflectivity coatings," *Appl. Phys. Lett.* 50: 1773, 1987.

39. A comprehensive book dealing with long wavelength lasers is G. P. A grawal and N. K. Dutta, *Long Wavelength Semiconductor Lasers*, New York: Von Nostrand, 1986.

40. Hirao, M., S. Tsuji, K. Mizushi, A. Doi, and M. Nakamura, *J. Opt. Comm.* 1:10,1980.

41. Ralston, J. D., S. Weisser, K. Eisele, R. E. Sah, E. C. Larkins, J. Rosenzweig, J. Fleissner, and K. Bender, "Low-bias-current direct modulation up to 33 Ghz in In GaAs/AlGaAs pseudomorphic MQW Ridge-waveguide lasers," *IEEE Phot. Tech.* 6: 1076,1994.

42. Coldren, L. A., and S. W. Corzine, *Diode Lasers and Photonic Integrated Circuits*, New York: Wiley, 1995, p. 195.

43. Henry C. H., *Line Broadening of Semicoductor Lasers in Coherence Amplification and Quautum Effects*, in *Semiconductor Lasers* ed. Y. Yamamoto New York: Wiley-Interscience, 1991, Chap.2.

44. Vahala, K., L. C. Chiu, S. Margalit, and A. Yariv, "On the

Linewidth Factor α in Semiconductor Injection Lasers," *Appl. Phys. Lett.* 42:631-633(April 1983).

New Words and Expressions

conventional transistor circuitry	*n.* 常规晶体管电路
oscillator	*n.* 振荡器
bipolar transistors	*n.* 双极晶体管
circuits	*n.* 电路
pedagogic	*adj.* 教育学的
dielectric	*n.* 电介质,绝缘体 *adj.* 非传导性的
schematic	*adj.* 示意性的
bloch	布洛赫,美国物理学家
eigen energy	*n.* 本征能量,特征能量
lattice	*n.* 格子
quantum	*n.* 量,额,[物]量子,量子论
propagation	*n.* (声波,电磁辐射等)传播
dope	*n.* 粘稠物,浓液,涂料 *vt.* 上涂料,预测出 *vi.* 服麻醉品
quasi-Fermi	*adj.* 类似的,准的
equilibrium	*n.* 平衡,均衡,保持平衡的能力
photon	*n.* [物]光子
conjugate	*v.* 使变化,变化
susceptibility	*n.* 易感性,感受性,感性,[物]磁化系数
diode	*n.* 二极管
incremental	*adj.* 增加的
heterostructure	*n.* [电子]异质结构
ternary	*adj.* 三重的
molar	*adj.* 质量的,[化][物]摩尔的

epitaxial	*adj*. [晶](晶体)取向附生的,外延的
vice versa	*adv*. 反之亦然
propagate	*v*. 繁殖,传播,宣传
quaternary	*n*. 四,四个一组 *adj*. 四进制的
cladding	*n*. 覆层
saturation	*n*. 饱和(状态),浸润,浸透,饱和度
algebraic	*adj*. 代数的,关于代数学的
tedious	*adj*. 单调乏味的,沉闷的,冗长乏味的
resonance	*n*. 共鸣,谐振,共振,极短命的不稳定基本粒子
adiabatic	*adj*. [物]绝热的,隔热的

NOTES

1. The compatibility of its output beam dimensions with those of typical silica-based optical fibers and the possibility of tailoring its output wavelength to the low-loss, low-dispersion region of such fibers.

输出光柱的尺寸对典型的基于硅的光纤的兼容性和对输出波长到光纤中低消耗、低发散位置的可能性。

2. The manner in which the available energy states are occupied determines the conduction properties of the crystal.

可利用能级被占用的方式决定着晶体的传输特性。

3. In thermal equilibrium a single Fermi energy applies to both the valence and conduction bands.

在热平衡状态下,单费尔米能量应用到价电层和传导层。

4. Another fact that we need before proceeding to the subject of optical gain in semiconductors is that when an electron makes a transition (induced or spontaneous) between a conduction band state and one in the valence band, the two states involved must have the same k vector.

在半导体中我们进行这个光学获得主题之前,我们需要的另一因素是当电子在传导层和价电层状态之间传输时,这两层必须有相同的 k 矢量。

5. Figure 5.8 shows that the semiconductor diode is capable of producing extremely large incremental gains, with only moderate increases of the inversion density, hence the current, above the transparency value.

图 5.8 显示了半导体二极管有足够的能力产生极大的增量,随着翻转密度的稳定增加,电流高于透过值。

6. The double confinement of injected carriers as well as of the optical mode energy to the same region is probably the single most important factor responsible for the successful realization of low-threshold continuous semiconductor lasers.

对相同的位置输入载体和光学模式能量的两种限制,可能是对低阈值连续性半导体激光器成功认识的惟一的重要因素。

7. The most important feature of the buried heterostructure laser is that the active GaAs region is surrounded on all sides by the lower index GaAlAs, so that electromagnetically the structure is that of a rectangular dielectric waveguide.

埋沟异质结构激光器最重要的特点是 GaAs 激活区域被低折射率 GaAlAs 在各个方向所包围,以至于其从电磁特性来看这种结构相当于矩形介质波导。

6

Advanced Semiconductor Lasers: Quantum Well Lasers, Distributed Feedback Lasers, Vertical Cavity Surface Emitting Lasers

6.1 Introduction

During the last few years a new type of a semiconductor laser. the quantum well (QW) laser, has come to the fore①. It is similar in most respects to the conventional double heterostructure laser of the type shown in Figure 15-10 except for the thickness of the active layer. In the quantum well it is ~ 50 – 100 Å, while in conventional lasers it is ~ 1 000 Å. This feature leads to profound differences in performance. The main advantage to derive from the thinning of the active region is almost too obvious to state – a decrease in the threshold current that is nearly proportional to the thinning. This reduction can be appreciated directly from Figure 15-7. The carrier density in the active region needed to render the active region transparent is ~ 10^{18} cm^{-3}. It follows that just to reach transparency we must maintain a total population of $N_{\text{transp}} \sim V_a \times$

① A basic, first-year knowledge of quantum mechanics is assumed in this chapter.

10^{18} electrons(holes) in the conduction(valence) band of the active region where V_a(cm^3) is the volume of the active region. The injection current to sustain this population is approximately

$$I_{\text{transp}} \sim \frac{eV_a \times 10^{18}}{\tau} \tag{6.1.1}$$

and is proportional to the volume of the active region. A thinning of the active region thus reduces V_a and I_{transp} proportionately. In a properly designed laser, the sum of the free carrier, scattering, and mirror(output) coupling can be made small enough so that the increment of current, above the transparency value, needed to reach threshold is small in comparison to I_{transp}. The reduction of the transparency current that results from a small V_a thus leads to a small threshold current[4,5,6,7,8].

Quantum well active regions are also used as the amplifying medium in distributed feedback lasers and in vertical cavity surface emitting lasers. Both of these classes of important semiconductor lasers are discussed in separate sections of this chapter.

6.2 Carriers in Quantum Wells (Advanced Topic)

The essential difference between the gain of a pumped quantum well semiconductor medium and that of a bulk semiconductor laser has to do with the densities of states in both of these media. The density of states of a bulk semiconductor was derived in Section 5.1 and is given by Equation(5.1.7). It was used in (5.2.9) to derive an expression for the gain $\gamma(\omega_0)$ of that medium. In this section we will repeat this procedure using the density of states function of a quantum well. This is one of the few places in this book where we need to turn to quantum mechanics. The student without a quantum mechanical background but with a good electromagnetic preparation can simply think of the electron using the de Broglie picture as a wave obeying, not Maxwell's equations but the Schrödinger equation. The running wave solutions of this equation are of the form of modes $\psi_i(x, z) = u_i(\boldsymbol{r})\exp(-iE_i t/\hbar)$ where $\hbar$ is Planck's constant divided by 2π, E_i is the quantized energy of an electron in the state i, while $u_i(\boldsymbol{r})$ is the eigen function.

We consider the electron in the conduction band of a QW to be free (with an effective mass m_c) to move in the x and y directions, but to be confined in the z (normal to the junction planes) as shown in Figure 6.1. The potential barrier ΔE_c was given in Section 5.3 for the GaAs/$Ga_{1-x}Al_xAs$ system as $\Delta E_c \sim 0.75\times$ (eV).

For the sake of simplification, we shall take ΔE_c as infinite. (This is a close approximation for barriers $\gtrsim 100$ Å and $x \gtrsim 0.3$.) The wavefunction $u(\boldsymbol{r})$ of the electrons in this well obeys the time independent Schrödinger equation[2].

$$V(z)u(\boldsymbol{r}) - \frac{\hbar^2}{2m_c}\left(\frac{\partial^2}{\partial z^2} + \frac{\partial^2}{\partial x^2} + \frac{\partial^2}{\partial y^2}\right)u(\boldsymbol{r}) = Eu(\boldsymbol{r}) \qquad (6.2.1)$$

E is the energy of the electron while $V(z) = E_c(z)$ is the potential energy function confining the electrons in the z direction. We will measure the energy relative to that of an electron at the bottom of the conduction band in the GaAs active region as shown in Figure 6.1. The eigenfunction $u(\boldsymbol{r})$ can be separated into a product.

$$u(\boldsymbol{r}) = \psi_k(\boldsymbol{r}_\perp)u(z) \qquad (6.2.2)$$

which. when substituted in (6.1.1), leads to

$$\left[V(z) - \frac{\hbar^2}{2m_c}\frac{\partial^2}{\partial z^2}\right]u(z) = E_z u(z) \qquad (6.2.3)$$

where E_z is a separation constant to be determined. Since we agreed to take the height of $V(z)$ the well region as infinite, $u(z)$ must vanish at $z = \pm L_z/2$.

$$u_l(z) = \begin{cases} \cos l\dfrac{\pi}{L_z}z & l = 1,3,5\cdots\cdots \\ \sin l\dfrac{\pi}{L_z}z & l = 2,4,6\cdots\cdots \end{cases} \qquad (6.2.4)$$

$$E_z = l^2 \frac{\hbar^2 \pi^2}{2m_c L_z^2} = l^2 E_{1c} \equiv E_{lc} \qquad l = 1,2,3,\cdots \tag{6.2.5}$$

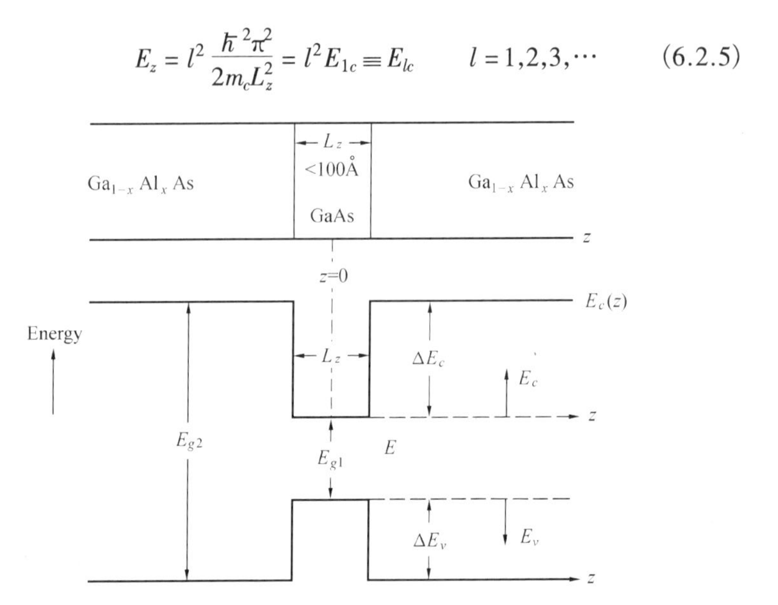

Figure 6.1 The layered structure and the band edges of a GaAlAs/GaAs/GaAlAs quantum well

Using(6.1.3,6.1.4,and 6.1.5)in (6.1.1)leads to

$$\Psi(\boldsymbol{r}_\perp)\cdot H(\boldsymbol{r}_\perp) = (E - E_z)\Psi(\boldsymbol{r}_\perp) \tag{6.2.6}$$

We can take $\psi(\boldsymbol{r}_\perp)$as a two-dimensional Bloch wavefunction(see 5.1.1).

$$\psi(\boldsymbol{r}_\perp) = \mathrm{u}_{\boldsymbol{k}_\perp}(\boldsymbol{r}_\perp)\mathrm{e}^{(\mathrm{i}\boldsymbol{k}_\perp \boldsymbol{r}_\perp)} \tag{6.2.7}$$

where $\mathrm{u}_{\boldsymbol{k}_r}(\boldsymbol{r}_\perp)$ possesses the crystal periodicity. The wavefunction $\Psi(\boldsymbol{r}_\perp)$obeys the Schrödinger equation

$$H(\boldsymbol{r}_\perp) = \frac{\hbar}{2m_c}\boldsymbol{\Psi}(\boldsymbol{r}_\perp) \tag{6.2.8}$$

and from Equations(6.2.6,7,8)

$$E_c(\boldsymbol{k}_\perp, l) = \frac{\hbar^2 k_\perp^2}{2m_c} + l^2 \frac{\hbar^2 \pi^2}{2m_c L_z^2} = \frac{\hbar^2 k_\perp^2}{2m_c} + E_{lc} \qquad l = 1,2,3,\ldots \tag{6.2.9}$$

where the zero energy is taken as the bottom of the conduction band.

$U_{k_\perp}(\boldsymbol{r}_\perp)$ possesses the lattice (two-dimensional) periodicity. Similar results with $m_c \to m_v$ apply to the holes in the valence band. We recall that the hole energy E_v is measured downward in our electronic energy diagrams so that

$$E_v(\boldsymbol{k}_\perp, l) = \frac{\hbar^2 k_\perp^2}{2m_v} + l^2 \frac{\hbar^2 \pi^2}{2m_v L_z^2} = \frac{\hbar^2 k_\perp^2}{2m_v} + E_{lv} \quad l = 1,2,3,\ldots \tag{6.2.10}$$

measured (downward) from the top of the valence band. The complete wavefunctions are then

$$\psi_c(\boldsymbol{r}) = \sqrt{\frac{2}{L_z}} \Psi \boldsymbol{k}_{\perp c}(\boldsymbol{r}_\perp) CS\left(l \frac{\pi}{L_z} z\right) \tag{6.2.11}$$

for electrons and

$$\psi_v(\boldsymbol{r}) = \sqrt{\frac{2}{L_z}} \Psi \boldsymbol{k}_{\perp v}(\boldsymbol{r}_\perp) CS\left(l \frac{\pi}{L_z} z\right)$$

for holes. We defined $CS(x) \equiv \cos(x)$ or $\sin(x)$ in accordance with (6.2.4).

The lowest-lying electron and hole wavefunctions are

$$\psi_c(\boldsymbol{r})_{\text{ground state}} = \sqrt{\frac{2}{L_z}} \psi_{\boldsymbol{k}_{\perp c}}(\boldsymbol{r}_\perp) \cos\left(\frac{\pi}{L_z} z\right)$$

$$\psi_v(\boldsymbol{r})_{\text{ground state}} = \sqrt{\frac{2}{L_z}} \psi_{\boldsymbol{k}_{\perp v}}(\boldsymbol{r}_\perp) \cos\left(\frac{\pi}{L_z} z\right) \tag{6.2.12}$$

and are shown along with the next higher level in Figure 6.2. In a real semiconducting quantum well, the height ΔE_c of the confining well (see Figure 5.10) is finite, which causes the number of confined states, i.e., states with exponential decay in the z direction outside the well to be finite. The mathematical procedure for solving (6.2.3) is similar to that used in Section 3.1 to obtain the TE modes of a dielectric waveguide that obeys a Schrödinger-like equation (3.3.1). As a matter of fact, to determine the number of confined eigenmodes as well as their eigenvalues we use a procedure identical to that of Figure 3.4.

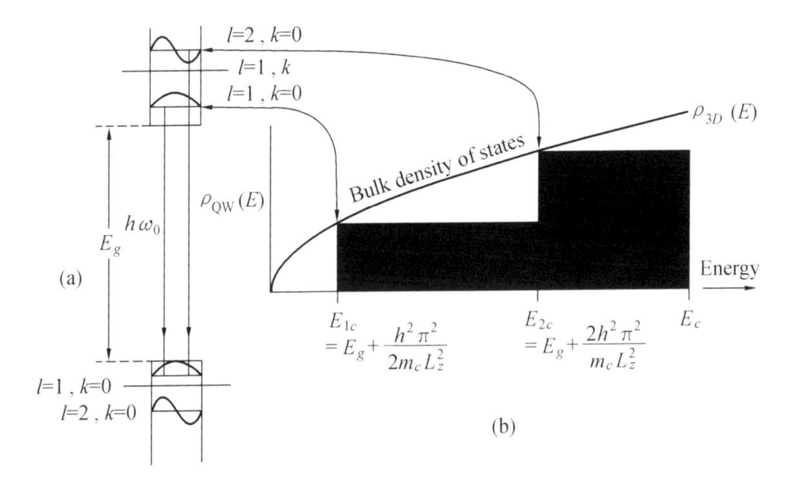

Figure 6.2 (a) The first two $l = 1$, $l = 2$ quantized electron and hole states and their eigenfunctions in an infinite potential well. (b) A plot of the volumetric density of states $(1/AL_z)/[\mathrm{d}N(E)/\mathrm{d}E]$ (i.e., the number of states per unit area (A) per unit energy divided by the thickness z of the active region) of electrons in a quantum well and of a bulk semiconductor. (Courtesy of M. Mittelstein, The California Institute of Technology, Pasadena, California)

Waveguide that obeys a Schrödinger-like equation (3.3.1). As a matter of fact, to determine the number of confined eigenmodes as well as their eigenvalues we use a procedure identical to that of Figure 3.4.

The Density of States

The considerations applying here are similar to those of Section 5.1. Since the electron is "free" in the x and y directions, we apply two-dimensional quantization by assuming the electrons are confined to a rectangle L_xL_y. This leads, as in Equation (5.1.3), to a quantization of the components of the $\boldsymbol{k}$ vectors.

$$k_x = n\frac{\pi}{L_x} \qquad n = 1,2,\ldots, \qquad k_y = m\frac{\pi}{L_y} \qquad m = 1,2,\ldots$$

The area in $\boldsymbol{k}_\perp$ space per one eigenstate is thus $A_k = \pi/L_xL_y \equiv \pi/A_\perp$. We will drop the subnotation from now on so that $k \equiv k_\perp$. The number of states with transverse values of k' less than some given k is obtained by dividing the area $\pi k^2/4$ by A_k (the factor 1/4 is due to the fact that $\boldsymbol{k}_\perp$ and $-\boldsymbol{k}_\perp$ describe the same state). The result is

$$N(k)=\frac{k^2 A_\perp}{2\pi}$$

where a factor of two for the two spin orientations of each electron was included.

The number of states between k and $k+\mathrm{d}k$ is

$$\rho(k)\mathrm{d}k=\frac{\mathrm{d}N(k)}{\mathrm{d}k}\mathrm{d}k=A_\perp\frac{k}{\pi}\mathrm{d}k \qquad (6.2.13)$$

and is the same for the conduction or valence band. The total number of states with total energies between $E+\mathrm{d}E$ is

$$\frac{\mathrm{d}N(E)}{\mathrm{d}E}\mathrm{d}E=\frac{\mathrm{d}N(k)}{\mathrm{d}k}\frac{\mathrm{d}k}{\mathrm{d}E}\mathrm{d}E \qquad (6.2.14)$$

The number of states per unit energy per unit area is thus

$$\frac{1}{A_\perp}\frac{\mathrm{d}N(E)}{\mathrm{d}E}=\frac{k}{\pi}\frac{\mathrm{d}k}{\mathrm{d}E}$$

from (6.2.9) with $\boldsymbol{k}_\perp\rightarrow\boldsymbol{k}$, $l=1$, and limiting the discussion to the conduction band, the relation between the electron energy at the lowest state $l=1$ and its $\boldsymbol{k}$ value is

$$k=\sqrt{\frac{2m_c}{\hbar^2}}(E-E_{1c})^{1/2}$$

so that the two-dimensional density of states (per unit energy and unit area) is

$$\rho_{\mathrm{QW}}(E)=\frac{1}{A_\perp}\frac{\mathrm{d}N(E)}{\mathrm{d}E}=\frac{m_c}{\pi\hbar^2} \qquad (6.2.15)$$

Recall that this is the density of electron states. The actual density of electrons depends on the details of occupancy of these states as is discussed in the next section. An expression similar to (6.2.15) but with $m_c\rightarrow m_v$ applies to the valence band. In the reasoning leading to (6.2.15), we considered only one transverse $u(z)$ quantum state with a fixed l

quantum number (see Equation 6.2.15). But once $E > E_{2c}$, as an example, an electron of a given total energy E can be found in either $l = 1$ or $l = 2$ state so that the density of states at $E = E_{2c}$ doubles. At $E = E_{3c}$ it triples. and so on. This leads to a staircase density of states function. The total density of states thus increases by $m_c/\pi\hbar^2$ at each of the energies E_{lc} of (6.2.15), which is expressed mathematically as

$$\rho_{QW}(E) = \sum_{n=1}^{\text{all states}} \frac{m_c}{\pi\hbar^2} H(E - E_{nc}) \qquad (6.2.16)$$

where $H(x)$ is the Heaviside function that is equal to unity when $x > 0$ and is zero when $x < 0$.

The first two steps of the staircase density of states are shown in Figure 6.2. In the figure we plotted the volumetric density of states of the quantum well medium ρ_{QW}/L_z so that we can compare it to the bulk density of states in a conventional semiconductor medium. It is a straightforward exercise to show that in this case the QW volumetric density of states equals the bulk value $\rho_{3D}(E)$ at each of the steps, as shown in the figure.

The selection rules. Consider an amplifying electron transition from an occupied state in the conduction band to an unoccupied state in the valence band. The states $l = 1$ in the conduction band have the highest electron population. (The Fermi law Equation (15.1.8) shows how the electron occupation drops with energy.) The same argument shows that the highest population of holes is to be found in the $l = 1$ valence band state. It follows that, as far as populations are concerned, the highest optical gain will result from an $l = 1$ to $l = 1$ transition. The gain constant is also proportional to the (square of) the integral involving the initial and final states and the polarization direction x, y, or z of the optical field, Since the lowest lying electron and hole wavefunctions have, according to Equation (6.2.12), a z dependence that is proportional to $\cos(\pi z/L_z)$ and since

$$\int_{-L_z/2}^{L_z/2} z\cos^2 \frac{\pi z}{L_z} \mathrm{d}z = 0$$

it follows that the optical field must be x or y polarized. The optical E vector thus must lie in the plane of the quantum well. A field polarized along the z direction does not stimulate any transitions between the two lowest lying levels and thus does not exercise gain(or loss).

6.3 Gain in Quantum Well Lasers (3)

To obtain an expression for the gain of an optical wave confined (completely) within a quantum well medium, we follow a procedure identical to that employed in the case of a bulk semiconducting medium that was developed in Section 5.2. An amplifying transition at some frequency $\hbar\omega_0$ is shown in Figure 6.2. The upper electron state and the lower hole state(the unoccupied electron state in the valence band) have the same l and $\boldsymbol{k}$ values(see discussion of selection rules in Section 6.1) so that the transition energy is

$$\hbar\omega = E_c - E_v = E_g + E_c(\boldsymbol{k}, l) + E_v(\boldsymbol{k}, l) =$$

$$E_g = \left(\frac{1}{m_c} + \frac{1}{m_v}\right)\frac{\hbar^2}{2}\left(k^2 + l^2\frac{\pi^2}{L_z^2}\right) = \tag{6.3.1}$$

$$E_g + \frac{\hbar^2}{2m_r^*}\left(k^2 + l^2\frac{\pi^2}{L_z^2}\right)$$

$$m_r^* = \frac{m_c m_v}{m_c + m_v} \tag{6.3.2}$$

$l = 1, 2, \ldots$ is the quantum number of the z dependent eigenfunction $u_l(z)$ as in Equation(6.2.4). We start again with (5.2.4) but this time in the correspondence of Equation (5.2.7) replace $\rho(k)/V$ by the equivalent quantum well $\rho_{\mathrm{QW}}(\boldsymbol{k})/L_z$ volumetric carrier density

$$N_1(m^{-3}) \longrightarrow \frac{\rho_{\mathrm{QW}}(\boldsymbol{k})}{L_z}\mathrm{d}k f_v(E_v)[1 - f_c(E_c)] = \frac{k}{\pi L_z}(f_v - f_v f_c)\mathrm{d}k$$

$$N_2(m^{-3}) \longrightarrow \frac{\rho_{QW}(\boldsymbol{k})}{L_z} \mathrm{d}k f_c(E_c)[1 - f_c(E_c)] = \frac{k}{\pi L_z}(f_c - f_v f_c)\mathrm{d}k$$

where $\rho_{QW}(\boldsymbol{k})$ is given by (6.2.15) and is independent of $\boldsymbol{k}$. The effective inversion population density due to carriers between k and k + dk is thus

$$N_2 - N_1 \longrightarrow \frac{k\mathrm{d}k}{\pi L_z}[f_c(E_c) - f_v(E_v)] \tag{6.3.3}$$

The division of ρ_{QW} by L_z is due to the need, in deriving the gain constant to use the *volumetric* density of inverted population consistent with the definition of N_1 and N_2 in(5.2.4). E_c and E_v are, respectively, the upper and lower energies of the carriers involved in a transition. We use(5.2.4) and (6.3.3) to write the contribution to the gain due to electrons within dk and in a single say $l = 1$ sub-band as

$$\mathrm{d}(\gamma\omega_0) \frac{k\mathrm{d}k}{\pi L_z}[f_c(E_c) - f_v(E_v)] \frac{\lambda_0^2}{4n^2\tau} \frac{\mathrm{T}_2}{\pi[1 + (\omega - \omega_0)^2 T_2^2]} \tag{6.3.4}$$

where T_2 is the coherence collision time of the electrons and τ is the electron-hole recombination lifetime assumed to be a constant. We find it more convenient to transform from the k variable to the transition frequency ω (see Equation6.3.1). From(6.3.1) it follows that

$$\mathrm{d}k = \frac{\mathrm{m_r^*}}{\hbar k}\mathrm{d}\omega$$

so that (6.3.4) becomes

$$\gamma_{n=1}(\omega_0) = \frac{m_r^* \lambda_0^2}{4\pi \hbar L_z n^2 \tau} \int_0^\infty [f_c(\hbar\omega) - f_v(\hbar\omega)] \frac{T_2 \mathrm{d}\omega}{\pi[1 + (\omega - \omega_0)^2 T_2^2]} \tag{6.3.5}$$

where we used the convention that $f_c(\hbar\omega)$ is the Fermi function at the upper transition (electron) energy E_c, while $f_v(\hbar\omega)$ is the valence band Fermi function at the lower transition energy. To include, as we should, the contributions from all other subbands (l = 2, 3...) we replace, using(6.2.16)

$$\frac{m_r^*}{\pi \hbar^2} \longrightarrow \frac{m_r^*}{\pi \hbar^2} \sum_{l=1}^{\infty} H(\omega - \omega_l) \tag{6.3.6}$$

where $\hbar \omega_l$ is the energy difference between the bottom of the l subband in the conduction band and the l subband in the valence band.

$$\hbar \omega_l = E_g + l^2 \frac{\hbar^2 \pi^2}{2 m_r^* L_z^2} \tag{6.3.7}$$

To get an analytic form for Equation(6.3.5) we will assume that the phase coherence "collision" time T_2 is long enough so that

$$\frac{T_2}{\pi[1 + (\omega - \omega_0)^2 T_2^2]} \longrightarrow \delta(\omega - \omega_0) \tag{6.3.8}$$

which simplifies(6.3.5)to

$$\gamma(\omega_0) = \frac{m_r^* \lambda_0^2}{4\pi \hbar L_z n^2 \tau} [f_c(\hbar \omega_0) - f_v(\hbar \omega_0)] \sum_{l=1}^{\infty} H(\hbar \omega_0 - \hbar \omega_l) \tag{6.3.9}$$

Equations(6.3.5) and (6.3.9) constitute our key result. They contain most of the basic physics of gain in quantum well media. Consider, first, the dependence of the gain on the Fermi functions f_c, f_v, An increase in the pumping current leads to an increase in the density of injected carriers in the active region and with it to an increase in the quasi-Fermi energies E_{Fc} and E_{Fv}, This leads to a larger region of ω_0 where the gain condition(Equation 5.2.14)

$$f_c(\hbar \omega_0) - f_v(\hbar \omega_0) > 0 \tag{6.3.10}$$

is satisfied. This situation is depicted in Figure 6.3. The solid curves (a), (b), and (c) show the modal gain of a typical GaAs quantum well laser at three successively in creasing current densities. The modal gain is equal to the medium gain $\gamma(\omega_0)$ of Eq. (6.3.5) multiplied by the optical confinement factor $\Gamma_a \sim L_z/d$. The dashed curve corresponds to the gain available at infinite current density ($f_v(\hbar \omega_0) = 0, f_c(\hbar \omega_0) = 1$) and thus, the gain in this case according to (6.3.9), is proportional to density-of -states function

$$\rho_{QW}(\hbar\omega_0) = \sum_{l=1}^{\infty} \frac{m_r^*}{\pi\hbar^2} \tag{6.3.11}$$

The first frequency to experience transparency, then gain, as the current is increased, according to the idealized staircase density of states model , is ω_0 where

$$\hbar\omega_0 = E_g + E_{1c} + E_{1v} = E_g + \frac{\hbar^2\pi^2}{2m_r^* L_z^2} \tag{6.3.12}$$

$\hbar\omega_0$ is thus the energy difference between the $l = 1(k = 0)$ conduction band state and the $l = 1(k = 0)$ valence band state. The inversion factor $f_c(\hbar\omega_0) - f_v(\hbar\omega_0)$ is always larger at this frequency than at larger frequencies. As the current is increased, and with it the density of electrons (holes) in the conduction (valence) band, the quasi Fermi levels (E_{F_c}, E_{F_v}) move deeper into their respective bands. There now exists a range of frequencies between the value given by(6.3.12)and $\omega_0 = (E_g + E_{F_c} + E_{F_v})$ where the gain condition(6.3.10)is satisfied. At even higher pumping the contribution from the $l = 2$ sub-band [see Figure 6.3 (b)]adds to that form $l = 1$ and the maximum available gain doubles to $2\gamma(\omega_0)$. Curve (d) in Figure 6.3 shows the gain of a conventional double heterostructure laser. We note that equal increments of current will yield larger increments of gain in the SQW case, that, at low currents, the SQW gain tends to saturate at a constant value γ_0, and that the width of the spectral region that experiences gain is much larger in SQW case compared to *DH* lasers.

The gain expression $\gamma(\omega_0)$ derived as (6.3.9) is that of the quantum well medium. It is the gain experienced by a wave that is completely confined to the quantum well. Since the quantum well thickness if typically 50 Å $< L_z <$ Å. while the mode height is typically $d \sim 1\ 000$ Å, the actual gain experienced by the mode—modal gain—is given as in Equations(5.3.2,3)by

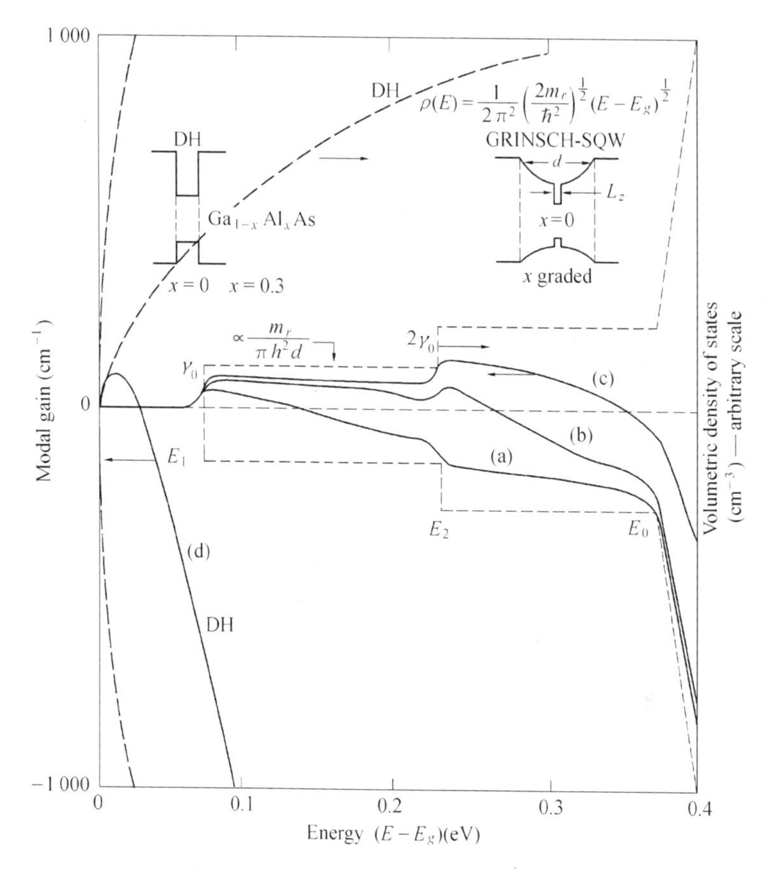

Figure 6.3 Gain (solid curves) and the joint density-of-states function (dashed lines) in a graded index, separate confinement heterostructure single quantum well laser (GRINSGH SQW). and a conventional double heterostructure (DH) laser. The gain curves (a), (b), and (c) are for successively larger injection current densities, and curve (d) applies to the DH with the same current density as the QW laser curve (a). To meaningfully compare the density of states of a quantum well laser and the bulk (DH) laser we divided the former by the width $w = 4L_z$ of the optical confinement distance. This, in addition to rendering the dimensions identical, makes both curves proportional to the maximum (available) modal gain. (Courtesy of D. Mehuys, The California Institute of Technology)

$$g_{\text{modal}} = \gamma \Gamma_a \approx \gamma \frac{L_z}{d} \tag{6.3.13}$$

When we use the expression (6.3.9) for γ, we find that the modal gain at a given areal density of carriers is independent of the thickness L_z of the quantum well and is inversely dependent on the mode height d.

Multiquantum well Laser

The small thickness of the quantum well relative to that of the mode height ($Lz/d \approx 2 \times 10^{-2}$ typically) makes it practical to employ more than one quantum well as the active region. To first approximation. the total electronic inversion is divided equally among the quantum wells, and the total modal gain is the sum of the individual modal gains of each well. The advantage of multiple-quantum well lasers is that, as shown in Figure 6.4, the gain from a single well tends to saturate with carrier density, hence with current, because of the flat-top nature of the density of states. The use of multiple-quantum wells enables each well to operate much within its linear gain-current region, thus extracting the maximum modal gain at a given total injected carrier density. This effect also results in a large differential gain $A \equiv \partial g / \partial N$, which, as shown in Section 5.5, leads to a higher laser modulation band width. This behavior is illustrated by Figure 6.6. We see that the optimal number of wells in a given laser depends on the requisite modal gain which at oscillation, is equal to the laser losses. A laser with an effective loss constant of $d_{\text{eff}} = \alpha - 1/(L \ln R)$ (α = loss constant, R is the mirrors' reflectivity) of, say, 10 cm^{-1}) will have, according to the figure, the lowest threshold current with one ($N = 1$) quantum well.

A theoretical plot of the exponential (modal) gain constant as a function of photon energy, or wavelength, is shown in Figure 6.4. The parameter is the injection current density. The interesting feature is the leveling off of the gain at the lower photon energies with increasing current and the appearance of a second peak at the higher current due to the population of the $l = 2$ well state.

Figure 6.5 shows the layered structure of a single quantum well

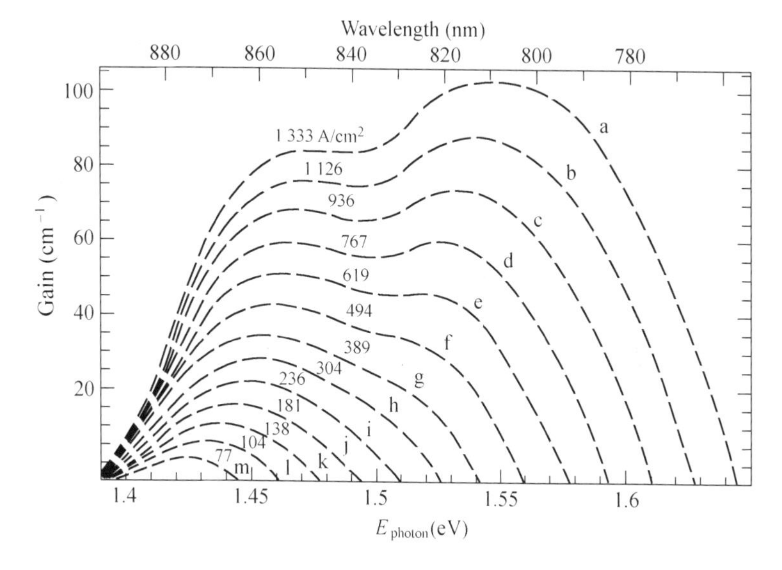

Figure 6.4 A theoretical plot of the exponential (modal) gain constant vs. wavelength of a quantum well laser. (Courtesy of Michael Mittelstein, The California Institute of Technology)

GaAs/GaAlAs laser. The 80 Å wide quantum well is bounded on each side with a graded index region. This graded index (and graded energy gap) region is grown by tapering the Al concentration from 0% to 60% in a gradual fashion as shown. The graded region functions as both a dielectric waveguide and as a funnel for the injected electrons and , not shown, the holes, herding them into the quantum well.

6.4 Distributed Feedback Lasers(9,10,11)

All laser oscillators employ optical feedback. By the word *feedback* we mean a means for ensuring that part of the optical field passing through a given point returns to the point repeatedly. If the delay is equal to an integral number of optical periods, this leads, in the presence of gain, to a sustained self-consistent oscillating mode where the field stimulated by atoms at any moment adds up coherently and *in phase* to those emitted earlier. In the laser resonators studied so far in this book, the feedback

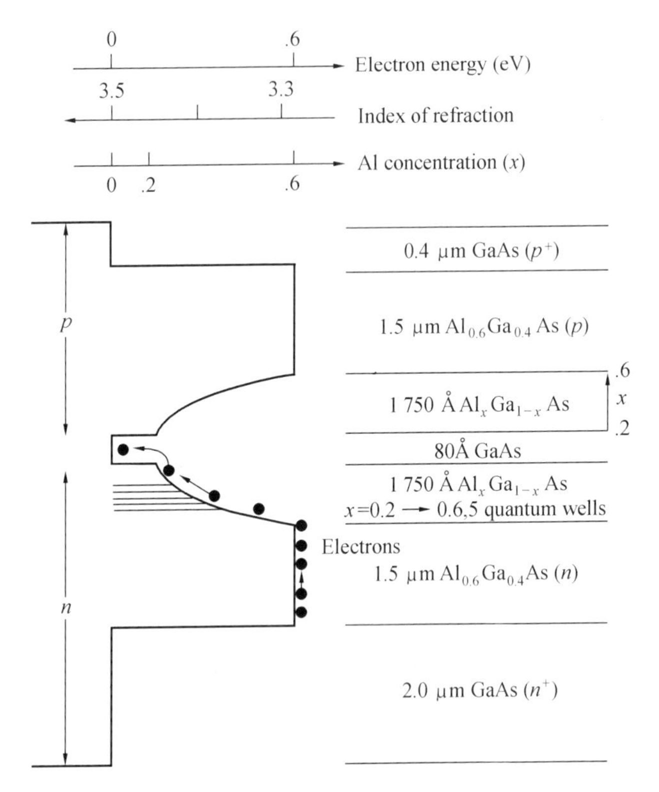

Figure 6.5 Schematic drawing of the conduction band edge and doping profile of a single quantum well, graded index separate confinement heterojunction lasers (GRINSCH). (Courtesy of H. Chen, The California Institute of Technology)

was provided by two oppositely facing reflectors. Feedback can also be achieved in a traveling wave folded-path geometry.

In distributed feedback (DFB) lasers, the reflection feedback of forward into backward waves, and vice versa, takes place not at the end reflectors but continuously throughout the length of the resonator. This coupling is due to a spatially periodic modulation of the index of refraction of the medium or of its optical gain. These lasers enjoy a wavelength stability that is far superior to those of ordinary FabryPerot lasers. This stability is due to the fact that the laser mode prefers to oscillate at a frequenicy such that the spatial period Λ of the index perturbation is equal to some (usually small) integer (l) number of guide half wavelengths:

$$\Lambda = l\frac{\lambda_g}{2}\left(\lambda_g = \frac{2\pi}{\beta}\right) \qquad l = 1,2,3,\ldots$$

where β is the propagation constant of the optical field in the waveguide. This condition. which ensures that reflections from different unit cells of the periodic perturbation add up in phase, is referred to as the *Bragg condition*. This is in analogy with the , formally simsilar, phenomenon of x-ray diffraction from the periodic lattice of crystals. This enables the laser designer, through a choice to "force" the laser to oscillate at any predetermined wavelength, provided that the amplifying medium is capable of providing sufficient gain at that wavelength. This property is especially important in semiconductor lasers used in optical fiber communication. Such lasers are often required to operate within narrow, prescribed wavelength regions to minimize pulse spreading by chromatic (group velocity) dispersion or to avoid crosstalk from other laser beams at different wavelengths sharing the same fiber.

We will start our treatment of the distributed feedback (DFB) laser with a derivation of the relevnt coupled mode equations. The essence of the DFB laser is a spatially periodic waveguide *with gain*. It is thus described by the coupled-mode equations (3.5.1) with the addition of gain terms

$$\frac{dA}{dz} = k_{ab}e^{-2i(\Delta\beta)z}B - \gamma A$$

$$\frac{dB}{dz} = k_{ab}^{*}e^{i2(\Delta\beta)z}A + \gamma B \qquad (6.4.1)$$

$$\Delta\beta = \beta - \beta_0 = \beta - \frac{\pi}{\Lambda} \qquad \text{for } l = 1 \qquad (6.4.2)$$

We shall choose, without loss of generality, a real k so that $k_{ab} \equiv k$. The gain terms, $-\gamma A$ and γB, are chosen such that if. hypothetically. we eliminate the periodic perturbation ($k = 0$), the waves A and B are uncoupled and grow exponentially, each along its direction of propagation as exp ($\gamma \times$ distance). We could, of course, have derived Equations (6.3.) by including *ab initio* gain in the derivation leading to Equations (3.5.1)①. Equations (6.4.2) can be simplified y defining new complex amplitudes $A'(z)$ and $B'(z)$

$$A(z) = A'(z)\exp(-\gamma z)$$
$$B(z) = B'(z)\exp(-\gamma z) \quad (6.4.4)$$

The result is

$$\frac{\mathrm{d}A'}{\mathrm{d}z} = k_{ab}B'e^{-i2(\Delta\beta + i\gamma)z} \quad (6.4.5)$$

$$\frac{\mathrm{d}B'}{\mathrm{d}z} = k_{ab}^{*}A'e^{i2(\Delta\beta + i\gamma)z} \quad (6.4.6)$$

These equations are identical in form to those of (3.5.1), provided we replace in the latter, $\Delta\beta \rightarrow \Delta\beta + \mathrm{i}\gamma$. We can thus use the solution (3.5.2) of Equations (3.5.1) to write down directly the solutions for the complex amplitudes $E_i(z)$ and $\mathrm{E_r(z)}$ of the incident and reflected waves, respectively, *inside* the *amplifying* periodic waveguide. We take the boundary conditions to be those of a *single* right-traveling wave incident from the left with an amplitude $B(\theta)$. The solution of Equations (6.4.5)、(6.4.6) in this case is

$$E_i(z) = B'(z)e[(-i\beta + \gamma)z] =$$
$$B(0)\frac{\mathrm{e}^{-\mathrm{i}\beta_0 z}\{(\gamma - \mathrm{i}\Delta\beta)\sinh[S(L-z)] - S\cosh[S(L-z)]\}}{(\gamma - \mathrm{i}\Delta\beta)\sinh(SL) - S\cosh(SL)} \quad (6.4.7)$$

$$E_r(z) = A'(z)e[(\mathrm{i}\beta - \gamma)z] =$$
$$B(0)\frac{k_{ab}\mathrm{e}^{\mathrm{i}\beta_0 z}\sinh[S(L-z)]}{(\gamma - \mathrm{i}\Delta\beta)\sinh(SL) - S\cosh(SL)} \quad (6.4.8)$$

---

① This can be done formally by replacingthe real dielectric constant $\epsilon(\mathbf{r})$ in (3.3.3) by a complex $\epsilon_c(\mathbf{r})$: $\epsilon_c(\mathbf{r}) = \epsilon_r(\mathbf{r}) + i\epsilon_i(\mathbf{r})$. For the case of a uniform $\epsilon_i(\mathbf{r}) = \epsilon_i$, we obtain $\gamma = \omega\sqrt{\mu/\epsilon_i\epsilon_r}$. Otherwise γ involves a spatial integral (see Problem 6.6)

$$S^2 = |k|^2 + (\gamma - i\Delta\beta)^2 \tag{6.4.9}$$

The fact that S now is complex makes for a major qualitative difference between the behavior of the passive periodic guide(3.5.2)and the periodic guide with gain. To demonstrate this difference, consider the case when the condition

$$(\gamma - i\Delta\beta)\sinh SL = S \cosh SL \tag{6.4.10}$$

is satisfied. It follows from (6.4.7, 8) that both the reflectance, $E_r(0)/E_i(0)$, and the transmittance, $E_i(L)/E_i(0)$, become infinite. *The device acts as an oscillator, since it yields finnite output fields* $E_r(0)$ and $E_i(L)$ with no input $[E_i(0) = 0]$. Condition (6.4.10) is thus the oscillation condition for a distributed feedback laser. For the case of $\gamma = 0$, it follows, from(13.5.2), that, $|E_i(L)/E_i(0)| < 1$, and $|E_r(0)/E_i(0)| < 1$ as appropriate to a passive device with no internal gain.

For Frequencies very near the Bragg frequency $\omega_0(\Delta\beta \cong 0)$ and for sufficiently high-gain constant γ so that(6.4.10) is nearly satisfied, the guide acts as a high-gain amplifier. The amplified output is available either in reflection with a field gain

$$\frac{E_r(0)}{E_i(0)} = \frac{k_{ab}\sinh SL}{(\gamma - i\Delta\beta)\sinh SL - S\cosh SL} \tag{6.4.11}$$

or in transmission with a gain

$$\frac{E_i(L)}{E_i(0)} = \frac{-Se^{i\beta_0 L}}{(\gamma - i\Delta\beta)\sinh SL - S\cosh SL} \tag{6.4.12}$$

The behavior of the incident and reflected field for a high-gain case is sketched in Figure 6.7. Note the qualitative difference between this case and the passive(no gain)one depicted in Figure 3.7.

The reflection power gain, $|E_r(0)/E_i(0)|^2$, and the transmission power gain $|E_i(L)/E_i(0)|^2$, are plotted in Figure 6.8 as a function of $\Delta\beta$ and γ. Each plot contains four infinite gain singularities at which the oscillation condition(6.4.10)is satisfied. These are four longitudinal laser modes. Higher orders exist but are not shown.

Oscillation Condition

The oscillation condition(6.4.10)can be written as

$$\frac{S-(\gamma-\mathrm{i}\Delta\beta)}{S+(\gamma-\mathrm{i}\Delta\beta)}\mathrm{e}^{2SL}=-1 \qquad (6.4.13)$$

In general, one has to resort to a numerical solution to obtain the threshold values of $\Delta\beta$ and γ for oscillation [17]. In some limiting cases, however, we can obtain approximate solutions. In the high-gain $\gamma \gg k$ case, we have from the definition of $S^2 = k^2 + (\gamma - \mathrm{i}\Delta\beta)^2$

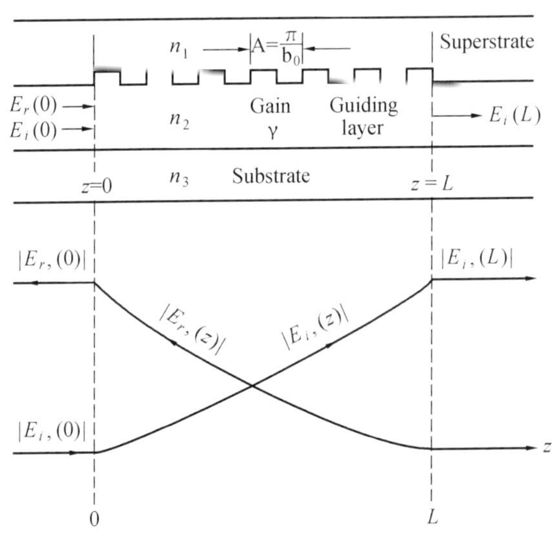

Figure 6.7 Incident and reflected fields inside an amplifying periodic waveguide.

$$S \approx -(\gamma-\mathrm{i}\Delta\beta)\left(1+\frac{k^2}{2(\gamma-\mathrm{i}\Delta\beta)^2}\right) \qquad \gamma \gg k$$

so that

$$S-(\gamma-\mathrm{i}\Delta\beta) \cong -2(\gamma-\mathrm{i}\Delta\beta)$$

$$S+(\gamma-\mathrm{i}\Delta\beta) \cong \frac{-k^2}{2(\gamma-\mathrm{i}\Delta\beta)}$$

and (6.4.13)becomes

$$\frac{4(\gamma-\mathrm{i}\Delta\beta)^2}{k^2}\mathrm{e}^{2SL}=-1 \qquad (6.4.14)$$

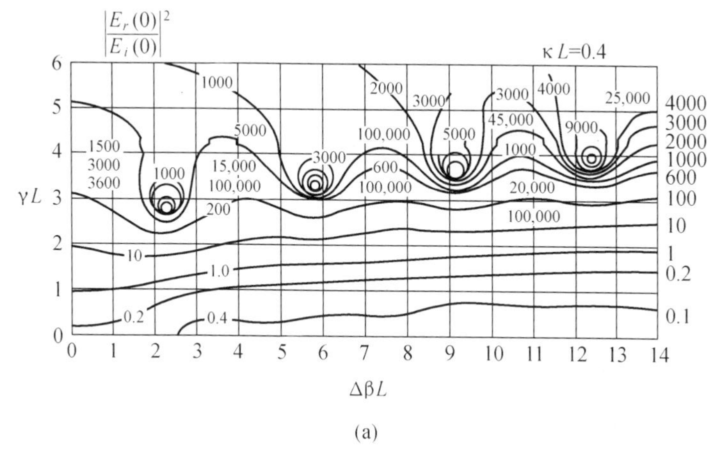

(a)

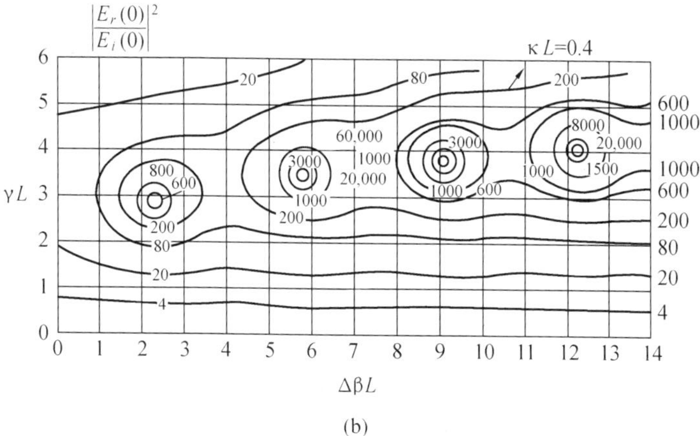

(b)

Figure 6.8 (a) Reflection gain contours in the $\Delta\beta L - \gamma L$ plane. $\Delta\beta$ is defined by (6.4.2) and is proportional to the deviation of the frequency $\omega_0 \equiv \pi c\Lambda n$. The plots are symmetric about $\Delta\beta$ so that only one-half ($\Delta\beta > 0$) of the plots is shown. (b) Transmission gain. [Courtesy of H. W. Yen.]

Equating the phases on both sides of (6.4.14) results in

$$\tan^{-1}\frac{2(\Delta\beta)_m}{\gamma_m} - 2(\Delta\beta)_m L + \frac{(\Delta\beta)_m Lk^2}{\gamma_m^2 + (\Delta\beta)_m^2} = (2m+1)\pi \tag{6.4.15}$$

$$m = 0, \pm 1, \pm 2, \ldots$$

In the limit $\gamma_m \gg (\Delta\beta)_m$, k, the oscillating mode frequencies are given by

$$(\Delta\beta_m) L \cong -\left(m + \frac{1}{2}\right)\pi \tag{6.4.16}$$

and since $\Delta\beta \equiv \beta - \beta_0 \approx (\omega - \omega_0) n_{\text{eff}}/c$

$$\omega_m = \omega_0 - \left(m + \frac{1}{2}\right)\frac{\pi c}{n_{\text{eff}} L} \tag{6.4.17}$$

We note that no oscillation can take place exactly at the Bragg frequency ω_0. The mode frequency spacing is

$$\omega_{m-1} - \omega_m \simeq \frac{\pi c}{n_{\text{eff}} L} \tag{6.4.18}$$

which is approximately the same as in a two-reflector resonator of length L.

The threshold gain value γ_m is obtained from equating the amplitudes in 6.4.14)

$$\frac{e^{2\gamma_m L}}{\gamma_m^2 + (\Delta\beta)_m^2} = \frac{4}{k^2} \tag{6.4.19}$$

indicating an increase in threshold with increasing mode number m. This is also evident from the numerical gain plots (Figures 6.7 and 6.8). An important feature that follows from (6.4.19) is that the threshold gain for modes with the same $|\omega - \omega_0|$, or equivalently the same $|\Delta\beta|$, is the same. Thus two modes will exist with the lowest threshold, one on each side of ω_0. This property of DFB lasers is usually undesirable, and methods for obtaining single-mode operation are discussed in the last part of this section.①. The periodic perturbation in semiconductor DFB lasers is achieved by incorporating a grating, usually in the form of a rippled interface, in the laser structure. This is achieved by interrupting the crystal

---

① High-speed (data rate) optical communication in fibers requires that the optical source put out a single frequency in order to minimize the temporal spread of the optical pulses with distance, which is caused by group velocity dispersion.

growth at the appropriate stage and wet-chemical etching a corrugation into the topmost layer by using an interferometrically produced photoresist mask[10]. Growth of a layer with a different index of refraction, or optical absorption on top of the rippled surface, results in the desired spatial modulation.

A diagram of a distributed feedback laser using a GaAs-GaAlAs structure is shown in Figure 6.9. The waveguiding layer, as well as that providing the gain (active layer), is that of p-GaAs. The feedback is provided by corrugating the interface between the p-$Ga_{93}Al_{07}As$ and p-Ga-Al_3As, where the main index discontinuity responsible for the guiding occurs. Figure 6.12 shows an example of a periodic gain grating. The laser in this example is based on the quaternary $Ga_{1-x}In_xAs_{1-y}P_y$ as the active region and InP as the high-energy gap, low indexcladding layer. The feedback is achieved by growing an extra-absorbing, i.e., low energy gap, layer and then etching through a mask to leave behind a periodic array of absorbing islands.

The increase in threshold gain with the longitudinal mode index m predicted by (3.6.19) and by the plots of Figures 6.7 and 6.8 manifests itself in a high degree of mode discrimination in the distributed feedback laser.

It follows from (6.4.17) and (6.4.19) that the two lowest threshold modes are those with $m = 0$ and $m = -1$ and that they are situated symmetrically on either side of the Bragg frequency ω_0 just outside the bandgap.

To understand why the basic DFB laser of Figure 6.6. in which the index of refraction is spatially periodic, does not oscillate at the Bragg frequency, consider Figure 6.10(a). Let the reflection coefficient of a wave (at ω) incident from the left on the plane $z = 0$ be r_2, and for a wave incident from the right, r_1. The reflectivity r_2 is given according to (6.4.11) by

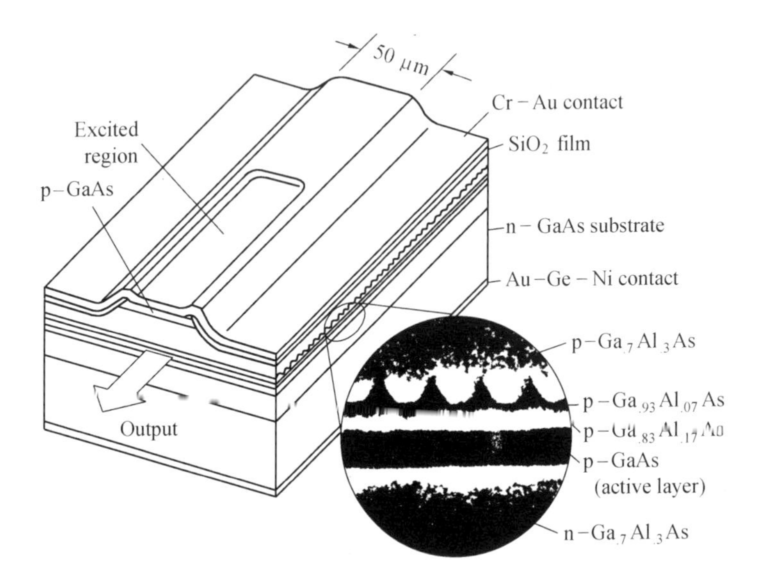

Figure 6.9 A GaAs-GaAlAs *cw* injection laser with a corrugated inferface. The insert shows a scanning electron microscope photograph of the layered structure. The feedback is in third order ($l=3$) and is provided by a corrugation with a period $\Lambda = 3\lambda_g/2 = 0.345$ μm. The thin (0.2 μm) p-$Ga_{83}A_{17}As$ layer provides a potential barrier which confines the injected electrons to the active (p-GaAs) layer. thus increasing the gain. (After Reference [11].)

$$r_1 = \frac{-k\tanh SL_1}{(\gamma - \mathrm{i}\Delta\beta)\tanh SL_1 - S(\cosh SL_1)} \tag{6.4.20}$$

where

$$S = \sqrt{k^2 + (\gamma - \mathrm{i}\Delta\beta)^2}$$

$$\Delta\beta = \frac{\omega}{C}n_{\text{mode}} - \frac{\pi}{\Lambda} = (\omega - \omega_0)\frac{n_{\text{mode}}}{C} \tag{6.4.21}$$

and we used the fact that for an index perturbation of odd symmetry (in z), $\Delta n^2(z) \propto \sin(\eta z)$, the coupling coefficient k given by (3.4.17) is a *real number*,① *i.e.*, $k_{ab} = k_{ba} = k$. The reflectivity r_1, "looking"

① Had we chosen a reference plane other than $z=0$, k_{ab} will not be real, but since $k_{ab} = k^*_{ba}$, all the results remain the same.

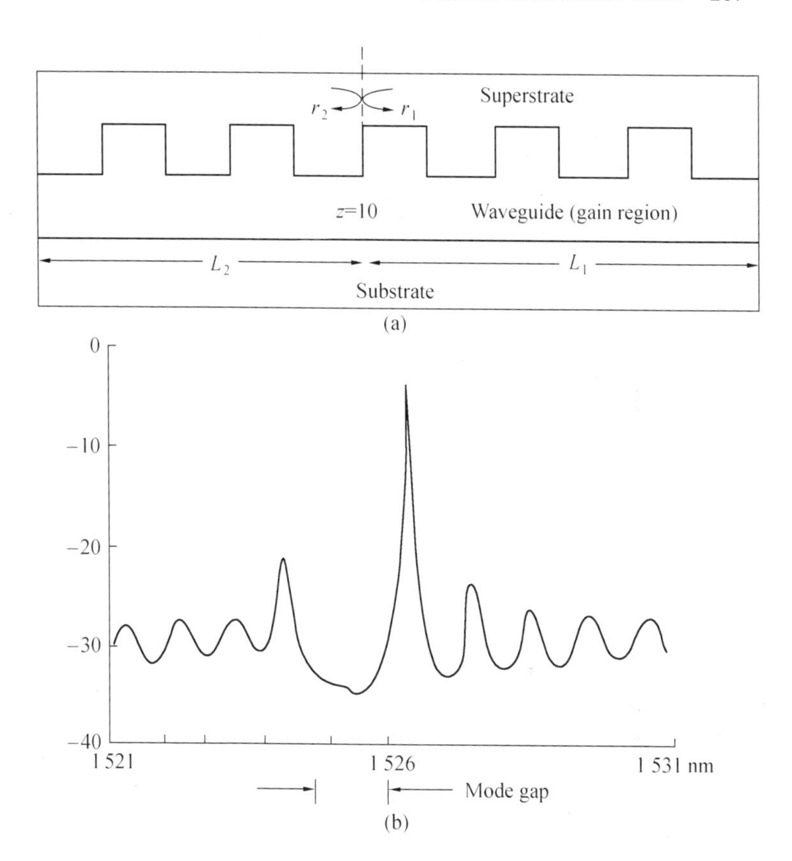

Figure 6.10 A periodic waveguide model used to derive Equation(6.4.12). (a) A periodic (DFB) GalnAsP waveguide laser. (b) The spontaneous emission spectrum below, but near, threshold of a DFB laser showing the mode gap. (c) A DFB laser with a phase shift section. (d) a "quarter wavelength shifted" DFB laser. (e) The spontaneous emission spectrum below threshold of a $\lambda/4$-shifted DFB laster. (Courtesy of P. C. Chen, ORTEL Corporation.)

to the left , is

$$r_1 = \frac{-k\sinh SL_1}{(\gamma - \mathrm{i}\Delta\beta)\sinh SL_1 - S\cosh SL_1}$$

The reason for the difference in sign between r_2 and r_1 is due to the fact that we chose, in(3.4.10), the index perturbation to have odd symmetry. An observer "looking" to the right sees $\Delta n(z) \propto \sin \eta z$, while an observer

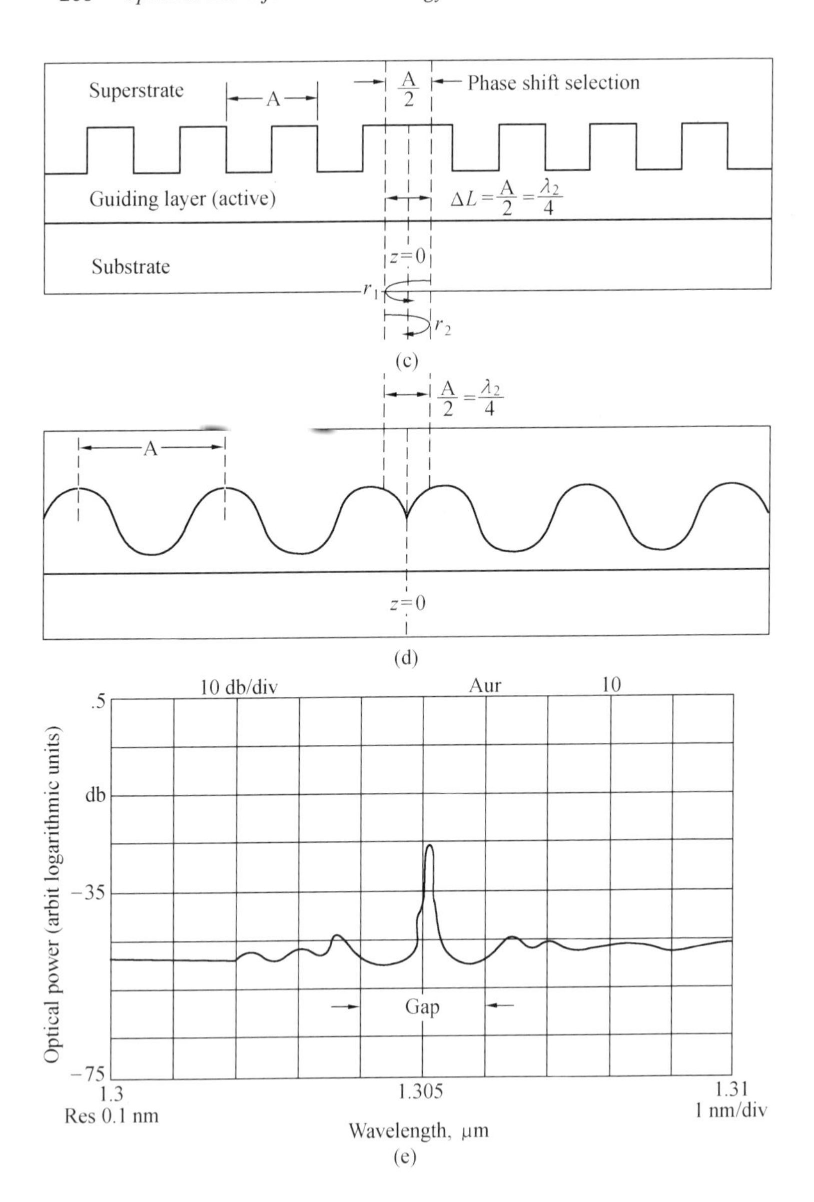

Figure 6.10 (*Continued*)

"looking" to the left will see a perturbation in index $\Delta n(z) \propto -\sin \eta z$. It follows that on resonance ($\Delta\beta = 0$), $r_1 r_2$ = negative number. The oscillation condition for a laser, on the other hand, is[①]

$$r_1(\omega) r_2(\omega) = 1 \tag{6.4.22}$$

It follows that the periodic index DFB laser can not oscillate at the Bragg frequency ω_0 where $\Delta\beta = 0$. Oscillation thus takes place at the symmetrically situated frequencies shown in Figure 6.8. The two oscillation frequencies nearest the Bragg frequency require the lowest gain and are given, according to (6.4.17), by

$$\omega_0 \pm \frac{\pi c}{2 n_{\text{eff}} L} \tag{6.4.23}$$

The threshold gain for oscillation at these two frequencies, is, according to (6.4.19) and Figure 6.8, equal, so that they are in practice equally likely to oscillate. This situation is highly undesirable in practice, since it results in wavelength instabilities and spectral broadening. This is unacceptable, for example in long-haul, high-dat-rate fiber links where the increased spectral width due to multiwavelength oscillation was shown in Chapter 3 to limit the data rate due to pulse broadening by group velocity dispersion.

The existence of two such oscillating wavelengths is shown in the spectrum of Figure 6.10(b) as the two peaks on either side of the "gap."

A widely employed method[12] for forcing the DFB laser to oscillate preferentially at a single midgap frequency is shown in Figure (6.10)(c, d). An extra section of length $\lambda_g/4$ is inserted at the center of the laser (λ_g is the "guide" wavelength). The reflectivities r_1 and r_2 "looking" to the left and right, respectively. from the midplane are now given by their previous values [i. e., those corresponding to Figure 6.10(a)], each multiplied by $\exp[-i(\pi/2)]$ to account for the added propagation delay

① This is just a sophisticated way of saying that at steady state, the oscillation condition is equivalent to demanding that a wave launched, say, to the right, returns after one round trip with the same amplitude and the same phase (modulo $m2\pi$).

in the $\lambda_g/4$ section. At $\omega = \omega_0 (\Delta\beta = 0)$, r_1 becomes $r_1 \exp[-i(\pi/2)]$, and r_2 becomes $r_2 \exp[-i(\pi/2)]$. The product of the reflectivities $r'_1 r'_2$ is now $-r_1 r_2$, which is a *positive* number, so that oscillation at ω_0 is possible. This is illustrated in Figure 6.10(e).

Gain-Coupled Distributed Feedback Lasers(13)

Another type of distributed feedback laser is one where the periodic modulation is not of the index of refraction but of the gain or losses of the medium. To analyze this situation, we remind ourselves that gain or losses can be represented by taking the dielectric constant ϵ of a medium as complex. It is a straightforward matter to show (see Problem 6.6) that ϵ can be expressed as

$$\epsilon = \epsilon_0 n^2 \left(1 + i\frac{2\gamma}{k_0 n}\right)$$

$$k_0 = \frac{2\pi}{\lambda}\omega\sqrt{\mu\epsilon_0} \qquad (6.4.24)$$

and $(\gamma \ll k_0)$ is the exponential gain constant of the field amplitude. In a lossy medium. $\gamma < 0$.

In the case where n and λ are periodic, we can write

$$n(z) = n_0 + n_1 \cos\frac{2\pi}{\Lambda}z$$

$$\gamma(z) = \gamma_0 + \gamma_1 \cos\frac{2\pi}{\Lambda}z \qquad (6.4.25)$$

Limiting ourselves to the case $n_1 \ll n_0$, $\gamma_1 \ll \gamma_0$, we can write

$$\omega^2 \mu\epsilon(z) E = \left[k_0^2 n_0^2 + i2k_0 n_0 \gamma_0 + 4k_0 n_0 \left(\frac{\pi n_1}{\lambda} + i\frac{\gamma_1}{2}\right)\cos\frac{2\pi}{\Lambda}z\right] E \qquad (6.4.26)$$

This last result shows that we can use the coupled mode equations (6.4.5, 6) in the general case of both index and gain modulation. provided we generalize the definition of the coupling constant to

$$k = \frac{\pi n_1}{\lambda} + i\frac{\gamma_1}{2} \qquad (6.4.27)$$

Not surprising: The coupling constant due to gain modulation differs by a factor exp ($i\pi/2$) from that due to index modulation. We can use, for example, the expression for the reflectivity r_2 in (6.4.20) but replace k by ik. This renders the product $r_1 r_2$ at ω_0 a *positive* number, so that laser oscillation can now take place at the exact Bragg frequency (ω_0). If we plot $| r^2 |$ vs, $\Delta\beta$ for this case, we obtain the result shown in Figure 6.11(a). For comparison, we show in Figure 6.11(b) a plot of the reflectivity in the case of index modulation, which shows two modes situated symmetrically about ω_0. The experimental oscillation spectrum of a gain-coupled laser is shown in Figure 6.11(c). It demonstrates the strong supression of higher-order modes.

A cross section of a commercial gain-coupled DFB laser is shown in Figure 6.12. The periodic modulation of the gain is achieved by photolithographic corrugation [13] of an absorbing layer near the active region. The layer is incorporated for this purpose in the epitaxially grown laser structure. Additional layers grown epitaxially result in "burying" the periodic loss layer.

6.5 Vertical Cavity Surface Emitting Semiconductor Lasers(14,15,16)

Vertical cavity surface emitting semiconductor lasers(VCSELs) differ from their more conventional relatives in that the optical beam travels at right angles to the active region instead of in the plane of the active regions. A typical VCSEL structure is illustrated in Figure 6.13. The top and bottom reflectors consist, each, of alter nating layers of semiconductor GaInAlAs with different x and y compositions. The difference in the index of refraction between adjacent layers gives rise to a high reflection(>99 percent) at the vicinity of the Bragg wavelength from each such "stack." The mirror layers are grown epitaxially along with the rest of the laser layers. The laser biasing current flows through the mirrors so that they are

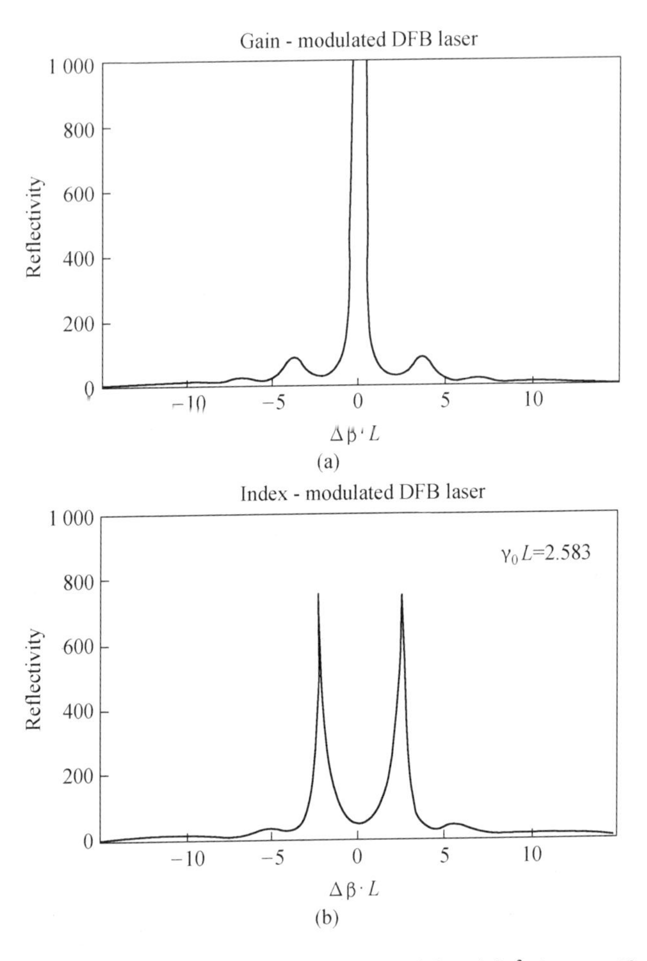

Figure 6.11 (a) A theoretical plot of the reflectivity $|E_r(0)/E_i(0)|^2$ of a waveguide with a gain $\gamma = \gamma_0 + \gamma_1 \sin(2\pi\Lambda)z$. (b) A similar plot for an index modulated waveguide, $\gamma = \gamma_0$, $n = n_0 + n_1 \sin(2\pi/\Lambda)z$. (c) The measured oscillation spectrum of a GaInAsP distributed feedback laser with gain coupling. $\lambda = 1.5427$ μm. A single oscillating mode is present at the Bragg wavelength. Higher-order modes have output powers that are down by a factor of > 45 db (i.e., > 32 000), compared to that of the fundamental mode. (a) and (b) Courtesy of M. McAdam – Caltech. (c) Courtesy of Dr. P. C. Chen. ORTEL Corporation.

highly doped to reduce the series resistance. The gain is provided by a

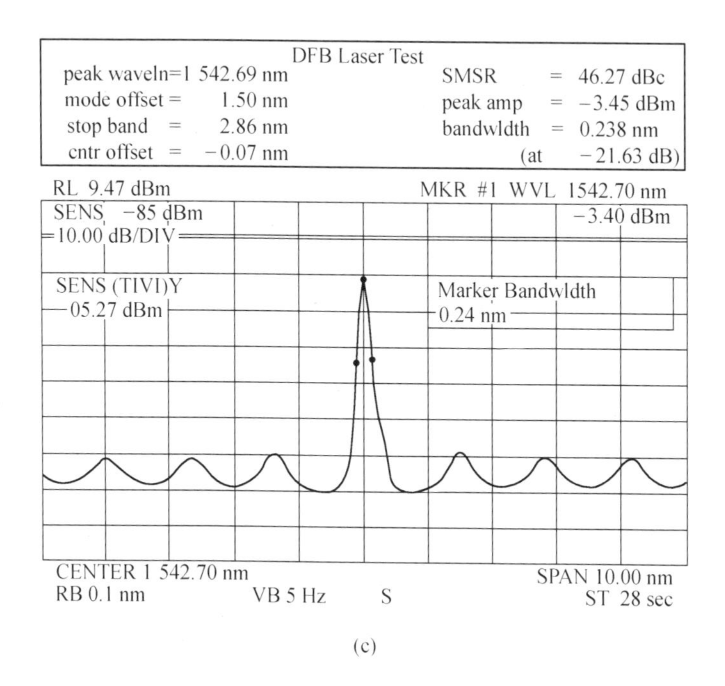

(c)

Figure 6.11 (*Continued*)

small number, typically 1 to 4, of quantum wells that are placed near a maximum of the standing wave pattern to maximize the stimulated emission rate into the oscillation field. The total length of the spacer region, 2 and 3. that straddles the active region is typically $L = \lambda$. where λ is the wavelength in the medium. This translates, near $\lambda = 1\ \mu m$, to $L \approx 0.3\ \mu m$. Typical mode diameters are in the range of 3 to 10 μm. A typical Bragg stack consisting of, say, 15 $\lambda/4$layers is 2 μm thick.

The field distribution inside a vertical cavity laser is shown in Figure 6.14. We note that inside the Bragg mirror the optical wave amplitude undergoes exponential evanescence. This is in agreement with Equation (3.5.4)and Figures 3.7 and 3.8 which describe the evanescent decay of an optical wave inside a periodic medium for optical frequencies within the "forbidden"frequency gap[17].

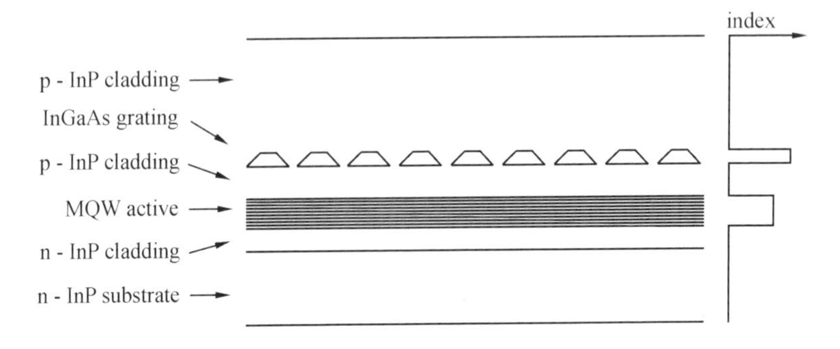

Figure 6.12 A periodic lossy layer, i.e., a layer with an energy gap smaller that $\hbar\omega_{0scill}$, provides the periodic gain coupling in a semiconductor DFB laser. (Courtesy of Dr. P. C. Chen, ORTEL Corporation.)

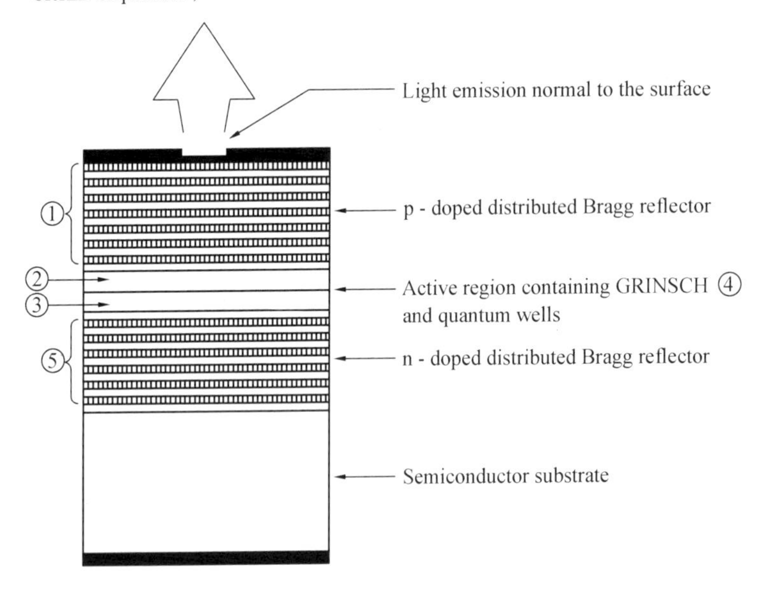

Figure 6.13 Aschematic cross section of a vertical cavity surface emitting semiconductor laser based on the GaInAlAs alloy system.

Since the distance L_z traveled in the amplifying medium is small (approximately 100 Å per quantum well), the gain per pass is very small, and laser oscillation is made possible by the extremely high reflectance (> 99 percent) of the Bragg mirror and the very low losses in regions 2

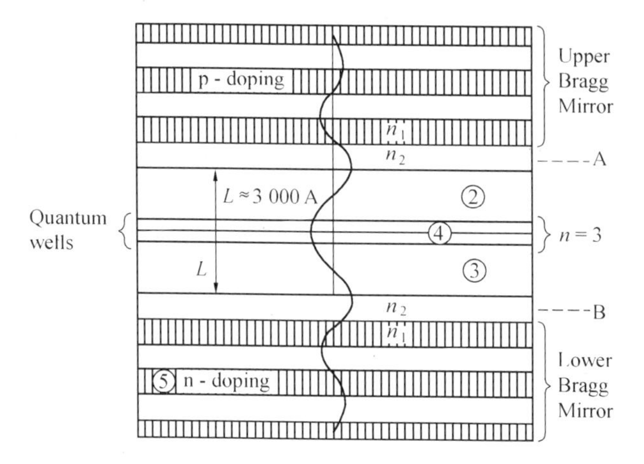

Figure 6.14 The field distribution of the laser mode inside a vertical cavity laser with $L \approx \lambda / n$ with three quantum wells. Notice the evanescent decay of the field envelope inside the Bragg mirrors and the constant amplitude standing wave between the mirrors.

and 3. Figure 6.14 conveys the relative scale of the key layer thicknesses. The Oscillation Condition of a Vertical Cavity Laser.

The Oscillation condition of the VCSEL can be written as

$$r_1(\omega) r_2(\omega) \exp\left[2 \sum_{m=1}^{N} \gamma m(\omega) L_z - i2 \frac{\omega}{C} nL \right] = 1 \quad (6.5.1)$$

which is a statement of the requirement that after one round trip a wave returns to its, arbitrary, starting plane with the same amplitude and, to within an integral multiple of 2π, the same phase. The factor of 2 in the exponent accounts for the fact that the quantum wells are placed at the peak of the standing wave pattern where they are exposed to an intensity that is twice the spatially averaged value. The number of quantum wells is N. In what follows, we will assume that each quantum well contributes equally to the gain so that $\sum_{m=1}^{N} \gamma_m(\omega) L_z = N\gamma(\omega) L_z$, The average index of refraction of the path is n. The Bragg mirrors' (amplitude) reflectances $r_1(\omega)$ and $r_2(\omega)$ refer to their respective input planes A and B in Figure 6.14.

The amplitude condition of (6.5.1) is

$$|\gamma(\omega)|^2 = \exp(-2N\gamma(\omega)L_z) \qquad (6.5.2)$$

Since the optical wave travels at right angles to the plane of the quantum wells, the gain γ is not the modal gain g_{model} of(6.3.13) but the bulk gain γ of a quantum well medium. We note that according to (6.3.9), the product $\gamma(\omega)L_z$ is independent of L_z, (This is strictly true when L_z is sufficiently small so that contributions to the gain from excited states ($l > 1$) in the quantum well are negligible. In practice, this is satisfied at room temperature for $L_z < 70$ Å.) From experimental data of edge emitting quantum well lasers, we determine that for $L_z \approx 70$Å the maximum gain due to the $l = 1$ quantized well level with a fully inverted population is $\gamma(\omega_0) \cong 5 \times 10^3 \text{ cm}^{-1}$. Using this value in(6.5.2), taking $L_z = 70$Å. leads to

$$|\gamma(\omega)|^2 = \exp|-2N \times 5 \times 10^3 \times 7 \times 10^{-7}|$$

for the reflectivity needed for oscillation.

$$N = 1. \ |\gamma(\omega)|^2 = 0.993$$
$$N = 2. \ |\gamma(\omega)|^2 = 0.986$$
$$N = 3. \ |\gamma(\omega)|^2 = 0.979$$
$$N = 4. \ |\gamma(\omega)|^2 = 0.972$$

where N is the number of quantum wells so that reflectivities around $R(\equiv |\gamma(\omega)|^2) = 98$ percent are required of the Bragg reflectors. We will next make a small detour to study these reflectors。

The Bragg Mirror

The analysis of the Bragg mirror is an excellent example of the power of the coupled mode formalism developed in Chapter 3. The periodic perturbation of the index of refraction couples. exactly as in the case of the DFB laser. two waves propagating in opposite directions. The coupling is strongest when the propagation constants $\pm\beta$ of the two coupled waves

$$B(z)\mathscr{E}_y(x,y)\exp[i(\omega t - \beta z)] \qquad \text{(forward wave)}$$

$$A(z)\mathscr{E}_y(x,y)\exp[i(\omega t+\beta z)] \qquad \text{(backward wave)} \tag{6.5.3}$$

obey very nearly the Bragg condition

$$\beta = l\frac{\pi}{\Lambda} \qquad l = 1,2,\ldots \tag{6.5.4}$$

for some integer l. If we retain in Equation(3.4.3) only te two Bragg-coupled waves $A_s^{(-)} \to A$, $A_s^{(+)} \to B$, we obtain

$$\frac{\mathrm{d}A}{\mathrm{d}z} = \frac{\mathrm{i}\omega\kappa_0}{4}B\exp(-\mathrm{i}2\beta z)\int_{-\infty}^{\infty}\Delta n^2(x,y,z)\mathscr{E}_y^2(x,y)\mathrm{d}x\mathrm{d}y \tag{6.5.5}$$

We also assume that the modes A and B are both y-polarized and have a normalized transverse distribution, $\mathscr{E}_y(x,y)$. The index of refraction of the Bragg mirror can be represented by

$$n^2(x,y,z) = \frac{1}{2}(n_1^2+n_2^2) + \frac{1}{2}(n_1^2-n_2^2)f(z)$$

where n_1, n_2 are the indices of refraction of the two alternating layers , and $f(z)$ is a square wave of unity amplitude as shown in Figure 6.14.

$$f(z) = \sum_l a_l \mathrm{e}^{\mathrm{i}l\frac{2\pi}{\Lambda}z},\ a_l = i\,\frac{(\mathrm{e}^{-\mathrm{i}\pi l}-1)}{l\pi}$$

$$\Delta n^2(x,y,z) = \left(\frac{n_1^2-n_2^2}{2}\right)f(z) \tag{6.5.6}$$

Assuming that the Bragg condition (6.5.4) is satisfied by the lth term in the Fourier series expansion of $f(z)$, we can rewrite (6.5.5) as

$$\frac{\mathrm{d}A}{\mathrm{d}z} = \frac{i\omega\kappa_0(n_1^2-n_2^2)a_l}{8}\int_{-\infty}^{\infty}\mathscr{E}^2(x,y)\mathrm{d}x\mathrm{d}yB\exp\left[i\left(l\frac{2\pi}{\Lambda}-2\beta\right)z\right] \tag{6.5.7}$$

when $l = 1$ we have

$$\frac{\mathrm{d}A}{\mathrm{d}z} = kB\exp(i\Delta\beta z)$$

$$\frac{\mathrm{d}B}{\mathrm{d}z} = kA\exp(i\Delta\beta z) \tag{6.5.8}$$

$$k = \frac{\omega\kappa_0}{4\pi}(n_1^2 - n_2^2)\int_{-\infty}^{\infty} \mathscr{E}_y^2(x, y)\mathrm{d}x\mathrm{d}y \approx \frac{2\Delta n}{\lambda}$$

$$\Delta\beta(\omega) = 2\left(\frac{\pi}{\Lambda} - \beta(\omega)\right) \tag{6.5.9}$$

In the second approximate equality of (6.5.9), we assumed $|\Delta n| |n_1 - n_2| \leqslant n_1, n_2$, $\beta \approx \omega\sqrt{\mu\epsilon_0}n$, $n^2 \equiv (1/2)(n_1^2 + n_2^2)$, and used the normalization integral, (3.2.8). Equations (6.5.8) constitute a pair of first-order, linear-coupled differential equations. Their solution requires that we specify two boundary conditions. Our chief interest is in the operation of the Bragg stack as a reflector The incident amplitude $B(0)$ thus becomes one of the given conditions. Since the backward-going wave A is due completely to internal reflections, we take $A(L) = 0$. The solution is thus given by Equations (3.5.2) so that the amplitude reflectance is

$$\gamma(\omega) = \frac{A(0)}{B(0)} = \frac{-ik\sinh(SL)}{-\Delta\beta(\omega)\sinh(SL) + iS\cosh(SL)} \tag{6.5.10}$$

$$S(\omega) = \sqrt{k^2 - \Delta\beta(\omega)^2}$$

where $\omega_0 = \pi c/\Lambda n$ is the Bragg frequency.

To obtain an appreciation for the magnitude of reflectivities that we may expect in a typical Bragg mirror, we will design a Bragg mirror to operate at a center wavelength of $\lambda_0 = 0.875\ \mu$m. The unit cell consists of a pair of epitaxially grown $Ga_{0.8}Al_{0.2}As$ and AlAs layers. The index of refraction difference is as $\Delta n = n_{Ga_{0.8}Al_{0.2}As} - n_{AlAs} = 0.55$. The average index is $n = 3.3$. The peak reflectivity is obtained from (6.5.10) with $\Delta\beta = 0$. Since the thickness of a unit cell is $\lambda_0/2n$, the length of the Bragg mirror with N_m periods is $L = N_m\lambda_0/2n$. The result in the case of $N_m = 15$ is $R(\omega_0) = |r(\omega_0)|^2 = \tanh^2\left(N_m\frac{\Delta n}{n}\right) = \tanh^2\left(\frac{15 \times 0.55}{3.3}\right) = 0.973$. This value is sufficient, according to the discussion following (6.5.2), to satisfy the oscillation conditions in vertical cavity lasers wity more than four inverted ($N \geqslant 4$) quantum wells.

A plot of the reflectivity $|r(\omega)|^2$ based on (6.5.10) and the experimental parameters of the above example is shown in Figure 6.15 (a). An experimental plot of a Bragg mirror with the same parameters is shown is Fig. 6.15(b). The phase shift $\phi(\omega)$ of the complex reflectance $r(\omega) = |r(\omega)| \exp(-i\phi(\omega))$ is shown in Fig. 6.15(c). For a more detailed treatment of Bragg mirrors and light propagation in stratified media, the reader is referred to Reference [17].

The Oscillation Frequencies

The phase part of (6.5.1) is used to obtain an expression for the oscillation frequencies of a surface-emitting Bragg mirror laser. If, for simplicity's sake, we take two identical $r_1(\omega) = r_2(\omega) = r(\omega) \cdot e^{-i\phi(\omega)}$, the phase condition is

$$-\phi(\omega) + \frac{\omega}{c} nL = m\pi \tag{6.5.11}$$

$$m = 1, 2, \ldots$$

Let us denote the two neighboring oscillation frequencies corresponding to m and $m+1$ as ω_m and ω_{m+1}, respectively

$$-\phi(\omega_m) + \frac{\omega_m}{c} nL = m\pi$$

$$-\phi(\omega_{m+1}) + \frac{\omega_{m+1}}{c} nL = (m+1)\pi \tag{6.5.12}$$

so that

$$\left[-\phi(\omega_{m+1}) + \phi(\omega_m) + \frac{\omega_{m+1} - \omega_m}{c} nL\right] = \pi \tag{6.5.13}$$

According to Figure 6.15(c), we can approximate $\phi(\omega)$ in the region of high reflectivity by

$$\phi(\omega) \cong -a(\omega - \omega_0)$$

$$a \approx \frac{\pi n}{2kc} \tag{6.5.14}$$

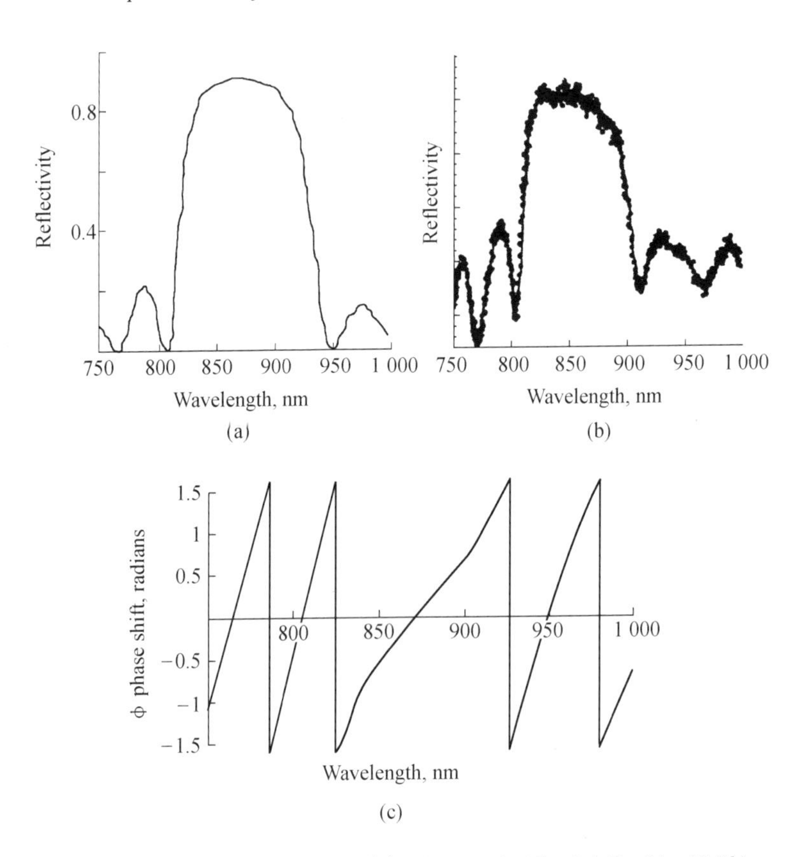

Figure 6.15 Calculated (a) and measured (b) reflectivity of a 15-period $Al_{0.2}GA_{0.8}AS$/AlAs distributed Bragg reflector. The calculated phase shift $\phi(\omega)$ is plotted in (c). (Courtesy of J. Obrien, Caltech.)

The expression for the slope a is obtained by dividing the maximum phase deviation of πin Figure 6.15 (c) by the corresponding (horizontal) frequency interval that, according to Equation (3.5.7), is $(\Delta\omega)_{gap} = 2kc/n$.

which when applied to (6.5.13)results in

$$2\pi\Delta\nu = (\omega_{m+1} - \omega_m) = \frac{\pi C}{n\left(L + \frac{\pi}{2k}\right)} \tag{6.5.15}$$

for the intermode frequency interval.

The effective length of the Bragg mirror resonator is thus not the mirror spacing L but

$$L_{\text{eff}} = L + \frac{\pi}{2k} \tag{6.5.16}$$

The contribution $\pi/2k$ is due to the evanescent penetration of the oscillating laser field into the Bragg mirrors, as illustrated in Figure 6.14. Since two Bragg mirrors are assumed in the analysis, the Bragg penetration distance into a single mirror is $\pi/4k$

> We recall that the field behavior inside the periodic Bragg mirror (at the Bragg frequency ω_0) is given by (13.5.6) as
>
> $$\exp(-i\beta' z) = \exp(-i\frac{\pi}{\Lambda}z)\exp(-kz)$$
>
> which corresponds to an effective penetration distance of $\sim k^{-1}$ to be compared to the value of $\pi/4k$ of (6.5.16)

Numerical example – intermode frequency separation. To obtain an appreciation for the intermode frequency spacing of (6.5.15), we will consider the laser depicted in Figure 6.14. The data for the Bragg mirror is the same as used in the example following (6.5.10). The basic parameters are

$$\lambda = 1\ \mu\text{m} \qquad L = \lambda = 1\ \mu\text{m}$$

$$k = \frac{2\Delta n}{\lambda} = \frac{2 \times 0.55}{1} = 1.1\ \mu\text{m}^{-1}$$

$$L_{\text{eff}} = L + \frac{\pi}{2k} = (1 + 1.427)\mu\text{m} = 2.427\ \mu\text{m}$$

(Note that the penetration depth, 1.427 μm, is larger than the intermirror spacing, L.)

$$\Delta\nu \equiv \frac{\omega_{m+1} - \omega_m}{2\pi} = \frac{c}{2nL_{\text{eff}}} = \frac{3 \times 10^{10}}{2 \times 3.3 \times 2.427 \times 10^{-4}} =$$
$$1.873 \times 10^{13}\ \text{Hz} = 624.3\ \text{cm}^{-1}$$

The high-reflectivity region of the Bragg mirror is given by (see box following (6.5.14))

$$(\Delta\nu)_{\text{Bragg}} \approx \frac{kc}{\pi n} = \frac{1.1\times 10^4(\text{cm}^{-1})\times 3\times 10^{10}}{\pi\times 3.3} = 3.183\times 10^{13}\ \text{Hz}$$

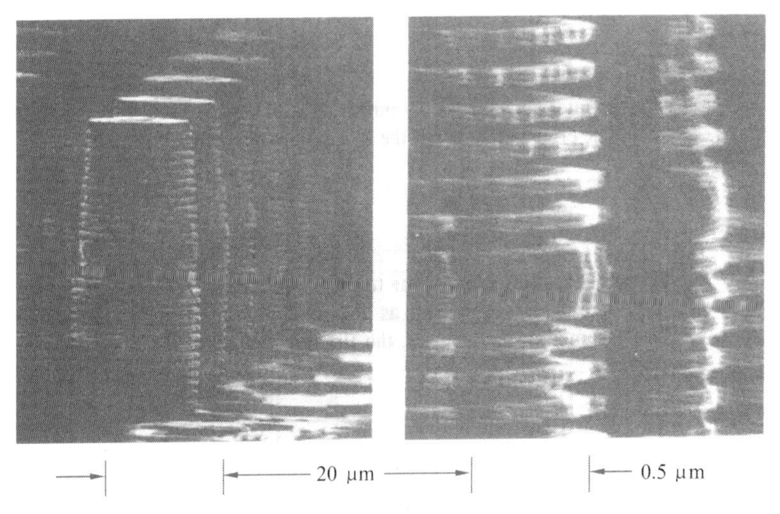

Figure 6.16 A scanning electron micrograph of a two-dimensional array of vertical-cavity. surface-emitting lasers. The multilayered structure of alternating GaAs and AlAs layers is demonstrated by partial preferential etch of the AlAs layers. (Courtesy of Scientific American magazine and of A. Scherer, Caltech.)

This number is comparable to the intermode spacing $\Delta\nu = 1.873\times 10^{13}$ Hz, so that only one mode at a time will experience high reflectivity and will satisfy the oscillation condition. This leads in most cases to a single-mode oscillation. This is to be contrasted with more conventional cleaved-facet-reflectors, edge-emitting semiconductor lasers, where $L \approx 300\ \mu$m and the mode spacing is correspondingly shorter.

We conclude by showing in Figure 6.16 a photograph of a two-dimensional array of surface emitting lasers.

Problems

6.1 Solve the one-dimensional Schrödinger equation (6.2.3) in the case of a simple square potential well where

$$V(z) = -V_0, \frac{L}{2} < z < \frac{L}{2}, V(z) = 0 \text{ elsewhere}$$

6.2 Assume that as we scale the length L of a quantum well laser we maintain the differential quantum efficiency η_{ex} constant by increasing R.

a. Derive the expression relating R (mirror reflectivity) to L.

b. Show that $I_{threshold}$ is proportional to L.

6.3 Show qualitatively that for a given m_c and injection current the maximum gain obtains when $m_v = m_c$.

6.4 Estimate the coupling constant k of the DFB laser whose spontaneous emission spectrum is given in Figure 6.10(e).

6.5

a. Using a computer program, plot the magnitude of the reflection coefficient [see Equation(6.4.11)] of a periodic amplifying waveguide as a function of $\Delta\beta L$. *assuming* kL = 0.4. *Let* γL be the parameter, and generate plots with $\gamma L = 2, 2.9, 3.5$, and 3.8.

b. Plot the equigain contours in the $\gamma L - \Delta\beta L$ plane as in Figure 6.12.

6.6

a. Derive the coupled-mode equations of a DFB laser with a periodic modulation of its losses. The spatial periodic loss can be accounted for according to Maxwell's equations by taking the dielectric constant of the waveguide as $\in'(\mathbf{r}) = \in(\mathbf{r}\left[1 - i\frac{\sigma(\mathbf{r})}{\omega\epsilon(\mathbf{r})}\right]$, with $\sigma(\mathbf{r}) = \sigma_0 + \sigma_1 x \cos\frac{2\pi}{\Lambda}z$ where $\sigma(\mathbf{r})$ is the medium conductivity.

b. Compare the coupling coefficient k in this case to that of index modulation [See(13.4.7).]

c. Estimate the magnitude of k in the case of a loss-modulated waveguide where the effective index is $n_{eff} = 3.5$, $\Lambda = 0.22\mu m$, $\lambda \cong 1.55\ \mu m$. The lossy layer has an absorption coefficient of $\alpha = 300\ cm^{-1}$

and a thickness of 1 000 Å. It is situated at the center of guiding layer. Assume that the waveguide mode is highly confined to the inner layer ($n = 3.51$).

References

1. van der Ziel, J. P., R. Dingle, R. C. Miller, W. Wiegmann, and W. A. Nordland, Jr., "Laser oscillation from quantum states in very thin GaAs-$Al_{0.2}Ga_{0.8}$As multilayer structures," *Appl. Phys. Lett.* 26: 463, 1975.

Dupuis, R. D., P. D. Dapkus, *IEEE J. Quant. Elec.* QE-16: 170, 1980.

2. Dingle R., W. Wiegmann, and C. H. Henry, "Quantum states of confined carriers in very thin Al_x Ga_{1-x} As-GaAs-$Al_x GA_{1-x}$ A_xAs heterostructures," *Phys, Rev, Lett*. 33: 827, 1974. Also see G. Bastard and J. A. Brum, "Electronic states in semiconductor heterostructures," *IEEE, J. Quant. Elec.* QE-22: 1625, 1986.

3. Arakawa. Y., and A. Yariv. "Theory of gain, modulation response and spectral linewidth in AlGaAs quantum-well lasers," *IEEE J. Quantum Elec.* QE-21: 1666, 1985.

4. M. Mittelstein, "Theory and experiments on unstable resonators and quantum well GaAs/GaAlAS lasers," Ph. D. thesis in applied physics. California Institute of Technology, Pasadena, CA, p. 54, 1989.

5. Tsang, W. T., "Extremely low threshold (AlGa) As modified multiquantum well heterostructure laser grown by MBE," *Appl . Phys. Lett.* 39: 786, 1981.

6. Mehuys, D., "Linear, nonlinear and tunable guided wave modes for high power (GaAl) As semiconductor lasers," Ph. D. thesis, California Institute of Technology, Pasadena, CA, June 1989.

7. Derry. p., et al., "Ultra low threshold graded-index separate confinement single quantum well buried heterostructure (Al, Ga) as lasers with high reflectivity coatings," *Appl, Phys. Lett.* 50: 1773, 1987.

8. Eng, L. E., et al., "Sub milliampere threshold current pseudomorphic InGaAs/AlGaAs buried heterostructure quantum well lasers grown by molecular beam epitaxy," *Appl. Phys. Lett.* 55 : Oct. 1989.

9. Kogelnik. h., and c. v. sHANK, "Coupled wave theory of distributed feedback lasers," *J. Appl. Phys.* 43:2328.1972.

10. Nakamura. M., A. Yariv. HW. Yen. S. Somekh, and H. L. Garvin. "Optically pumped GaAs surface laser with corrugation feedback," *Appl. Phys. Lett.* 22:515, 1973.

11. K. Aiki, M. Nakmura, J. Umeda, A. Yariv, A. Katzir, and H. W. Yen, "GaAsGaAlAs distributed feedback laser with separate optical and carrier confinement." *Appl. Phys. Lett.* 27:145, 1975.

12. Haus, H. A., and C. V. Shank, "Antisymmetric taper of distributed feedback lasers," *IEEE. J. Quant. Elec.* QE-12:532, 1976.

13. Nakano. Y., Y. Luo. and K. Tada, "Facet reflection independent, single longitudinal mode oscillation in a GaAlAs/GaAs distributed feedback laser equipped with a gain-coupling mechanism," *Appl. Phys. Lett.* 55:16:1606, 1989.

14. K. Iga. S. Ishikawa, S. Ohkouchi, and T. Nishimura, "Room-temperature pulsed oscillation of GaAlAs/GaAs surface emitting injection laser," *Appl Phys. Lett.* 45:348, 1984.

15. Jewell. J. L., J. P. Harbison. A. Scherer, Y. H. Lee, and L. T. Florez, "Vertical cavity surface emitting lasers," *IEEE J. Quant. Elec.* 27:1332, 1991.

16. Jewell. J. L., J. P. Harbison, and A. Scherer, "Microlasers." *Scientific American*. 86:1991.

17. Yariv. A., and P. Yeh. *Optical Waves in Crystals*, New York; *Wiley Interscience*, 1984,

18. A good general reference to vertical cavity lasers is L. A. Coldren, and S. W. Corzine, *Diode Lasers and Photonic Integrated*

Circuits, New York; Wiley, 1995.

第6章　先进的半导体激光器:量子激光器、分布式反馈激光器及垂直空腔表面发射激光器

New Words and Expressions

semiconductor	*n*. 半导体
quantumn	量子
conventionaladj	惯用的,传统的
heterostructure	*n*. 异质(异晶)结构
profound	*adj*. 深刻的,极度的
performance	*n*. 特征性能
derive from	*v*. 以……得出
decrease	*n*. 减少
threshold	阈值
proportional	*adj*. 成比例的,与……相称,调和
transparent	*adj*. 透明的,可穿透的
injection	*n*. 注入,喷入
sustain	*v*. 维持
scattering	*n*. 漫射,扩散
increment	*n*. 增加
feedback	*n*. 反馈
bulk	*n*. 大容量,大体积
expression	*n*. 表达式,公式
procedure	*n*. 过程,方式
electromagnetic	*adj*. 电磁的
eigen function	*n*. 本征函数
conduction band	*n*. 导带
confined	*adj*. 限制的
potential barrier	*n*. 势垒

simplification	*n*. 简化,理想化
substitute	*v*. 代替
vanish	*v*. 消失,消散
two dimensional	*adj*. 二维的
crystal periodicity	*n*. 晶格周期
valence band	*n*. 价带
diagram	*n*. 图表,简图
dielectric	*adj*. 非传导性的,绝缘的
rectangle	*n*. 长方开,矩形
component	*n*. 分力,分量投影
orientation	*n*. 定向,定位,校正方向
occupancy	*n*. 占有率
staircase	*n*. 阶梯现象
Heariside function	n. 亥维塞函数,阶跃函数
Volumetric	*adj*. 容量的,体积的
Polarization	*n*. 极化作用
transition	*n*. 转变,变化
correspondence	*n*. 一致性
inversion	*n*. 颠倒,倒置
sub-band	*n*. 次能带
collision	*n*. 碰撞,振动
recombination	*n*. 复合,合成
fermin	费米
depict	*n*. 描述,叙述
curve	*n*. 曲线
quasi fermi levels	*n*. 准费米能级
respective	*adj*. 相应的,相关的
increment	*n*. 增量
yield	*n*. 产出,击出

saturate	*v*. 使……饱和
SQW single quantum well	
approximation	*n*. 接近,逼近
oscillator	*n*. 谐振腔
integral	*adj*. 累积的
resonators	*n*. 共振器,共振腔
propagation	*n*. 传递
cross talk	*n*. 窜扰
refraction	*n*. 折射
spatial	*adj*. 空间的
perturbation	*n*. 微扰,干扰
diffraction	*n*. 衍射
predetermined	*adj*. 预定的
chromatic dispersion	*n*. 色散
derivation	*n*. 衍生
hypothetically	*adv*. 假想
propagation	*n*. 传播,增殖
amplitude	*n*. 幅角,幅度
boundary	*n*. 边界
transmittance	*n*. 透光度,透明性
sketch	*n*. 素描,速写
numerical	*adj*. 数值的
threshold	*n*. 阈值
evident	*adj*. 显然的
undesirable	*adj*. 不和需要的,不希望的
corrugation	*n*. 波纹
coefficient	*n*. 率,系数
discontinuity	*n*. 突变
quaternary	*adj*. 四元的

longitudinal	*adj*. 纵向的
manifest	*n*. 清单
discrimination	*n*. 识别率
symmetrically	*adv*. 对称地
velocity	*n*. 速度,周转率
spectrum	*n*. 光谱
insert	*v*. 镶嵌,插入
straightforward	*adj*. 简单的,直接的
exponential	*adj*. 指标,指数的
modulation	*n*. 调制,调幅
vertical cavity surface emitting semiconductor lasers(VCSELs)	垂直空腔表面发射半导体激光器
reflector	*n*. 反射镜
index of refraction	*n*. 折射率
Bragg wavelength	*n*. 布拉格波长
biasing current	*n*. 偏置电流
oscillation	*n*. 振荡
cladding	*n*. 包层
grating	*n*. 栅格
lossy	*n*. 有损耗的
Bragg reflector	*n*. 布拉格反射镜
Semiconductor substrate	*n*. 半导体基片
Evanescent	*n*. 衰逝
Envelope	*n*. 包络线
Staddle	*n*. 支撑架
Exponential	*adj*. 指数的
Arbitrary	*adj*. 随机的
Gain	*n*. 增益
Edge-emitting	*n*. 边发射

Reflectivity *n*. 反射比

Eetour *n*. 迂回

Perturbation *n*. 干扰

Formalism *n*. 形式

Mode *n*. 模式

First-order Linear-coupled Differential equation
一阶线性耦合微分方程

Backward-going wave *n*. 反向波

approcciation *n*. 增值、评价

epitaxially *adv*. 外延地

stratified *adj*. 分层的

denote *v*. 表示

intermode frequency interval 模间频率间隔

resonator *n*. 共振腔、共振器

evanescent *adj*. 渐消失的

penetration *n*. 穿透率

parameter *n*. 参数、参量

etch *n*. 蚀刻

cleaved *adj*. 夹层的、裂开的

facet *n*. 小平面、小刻面

Schrodinger equation 薛定谔方程

quantum *n*. 定量、量子

spectrum *n*. 光谱

magnitude *n*. 大小、数量

NOTES

1. The main advantage to derive from the thinning of active region is almost too obvious to state-a decrease in the threshold current that is nearly proportional to the thinning.

活性区厚度减小的主要优点几乎是显而易见的,这就是阈值电流的减小,因为阈值电流与厚度几乎是成比例的。

2. The student without a quantum mechanical background but with a good electromagnetic preparation can simply think if the electron using the de Broglie picture as ware obeying not Maxwell's equations but the Schrodinger equation.

没有量子物理背景但具备良好的电磁学知识的同学可以简单地利用 de Broglie 思想考虑电子作为德布罗意波,不满足 Maxwell 方程而是满足 Schrodinger 方程

3. In a real semiconducting quantum well, the height ΔE_c of the confining well (see Figure 5.10) is finite, which causes the number of confined states, i. e., states with exponential decay in the z direction outside the well to be finite

在实际半导体井中,这就限制了的电子状态数,即:井外 Z 方向指数衰减的状态数也是受限的。

4. An expression similar to (6.15) but with mc→mv applies to the valence band. In the reasoning leading to (6.15), we considered only one transverse u(z) quantum state with a fixed 1 quantum number (see Equation 6.15).

除了右价带质量由 mc 变化为 mv 外,公式和(6.15)式是相似的,在推导(6.15)式的理由中,我们只考虑 u(z)一种量子状态,即有固定的 1 个量子数目(6.15 式)。

5. The first two steps of the staircase density of states are shown in figure 6.2. In the figure we plotted the volumetric density of states of the quantum well medium $\rho_{\theta\omega}/L_z$ so that we can compare it to the density of states in a cWonventional semiconductor medium.

状态密度阶跃现象的最初两阶段显示在图 6.2 中,在图中,我们能够测出量子阱有质得体密度,因此,我们能够将它与传统的半导体介质的体密度相比较。

6. As the current is increased and with it the density of electrons (holes), in the conduction (valence) band. The quasi Fermi levels (E_{F_c}, E_{F_v}) move deeper into their respective band.

当电流增加时,在价电层中电子的密度也随着增加,准费米能组进一步进入它们各自的能带中。

7. We note that equal increments of current will yield larger increments of gain in the SQE case, that, at low currents.

我们注意到在 SQE 条件下,低电流的相同电流增量将会产生更大的增益。

8. This enables the laser designer, through a choice of Λ to force the laser to oscillate at any predetermined wavelength, provided that the amplifying medium is capable of providing sufficient gain at that wavelength.

这能够使激光器设计者通过选择空间频率 Λ 来决定激光器在任何预定波长振荡,并且配备能够在预定波长产生足够增益的放大介质。

9. We can thus use the solution (3.2) of Equations (3.1) to write down directly the solutions for the complex amplitudes $E_i(z)$ and $E_r(z)$ of the incident and reflected waves, respectively, inside the amplifying periodic waveguide.

因此,我们能够使用方程(3.1)的解(3.2)直接写出分析在那个扩大周期的波导管之内的入射和反射波的复振幅 $E_i(z)$和 $E_r(z)$。

10. This achieved by interrupting the crystal growth at the appropriate stage and wet-chemical etching a corrugation into topmost layer by using an interferometrically produced photoresist mask.

这些通过干扰晶体在某一阶段的增长和通过干扰仪产生的干扰条纹而得到的对最上层所产生的湿化学侵蚀。

11. This is unacceptable, for example, in long-haul, high-data-rate fiber links where the increased spectral width due to multiwavelength oscillation was shown in Chapter 3 to limit the data rate due to pulse broadening by group velocity dispersion.

这是不能接受的,例如,在长距离高数据传输率的时候,由于在第三章所述的多波长振荡这些都会增大光谱宽度,从而由于群速度色散所造成的脉冲展宽而限制数据传输效率。

12. Vertical cavity surface emitting semiconductor lasers (VCSELs) differ from their more conventional relatives in that the optical beam travels at right angles to the active region instead of in the plane of the active regions.

垂直空腔表面发射半导体激光器不同于常见的半导体激光器之处在于,其腔中的光束与激活区成直角,而不是处在激活区之中。

13. The laser biasing current flows through the mirrors so that they are highly doped to reduce the series resistance.

为使激光偏置电流能够透过镜子,要求高度 掺杂来减少串联电阻。

14. This is in agreement with Equation (3.4) and Figures 3.7and 8.8, which describe the evanescent decay of an optical wave inside a periodic medium for optical frequencies within the "forbidden" frequency gap.

这与方程(3.4)和图 3.7、8.8 相符合。方程(3.4)和图 3.7、3.8描述了处在禁带频率中的周期媒质中以光波频率传输的光波的衰逝。

15. To obtain an appreciation for the magnitude of reflectivities that we may expect in a typical Bragg mirror to operate at a center wavelength of

$$\lambda_0 = 0.875.$$

为了获得反射比大小的增值,我们可以用一个工作中心波长是0.875微米的典型布拉格镜。

16. This value is sufficient, according to the discussion following (6.2), to satisfy the oscillation conditions in vertical cavity lasers with more than four inverted quantum wells.

根据(6.2)的讨论,这个值是满足条件的。为了在垂直腔激光器里满足振动条件,必须有四个以上的反向量子阱。

PART THREE OPTOELECTRONIC PROCESS METHODS

7

Detection of Optical Radiation

7.1 Introduction

The detection of optical radiation is often accomplished by converting the radiant energy into an electric signal whose intensity is measured by conventional techniques. Some of the physical mechanisms that may be involved in this conversion include

1. The generation of mobile charge carriers in sold-state photoconductive detectors

2. Changing through absorption the temperature of thermocouples, thus causing a change in the junction voltage

3. The release by the photoelectric effect of free electrons from photoemissive surfaces

In this chapter we consider in some detail the operation of four of the most important detectors:

1. The photomultiplier

2. The photoconductive detector

3. The photodiode

4. The avalanche photodiode

The limiting sensitivity of each is discussed and compared to the theoretical limit. We will find that by use of the heterodyne mode of detection the theoretical limit of sensitivity may be approached.

7.2 Optically Induced Transition Rates

A common feature of all the optical detection schemes discussed in this chapter is that the electric signal is proportional to the rate at which electrons are excited by the optical field. This excitation involves a transition of the electron from some initial bound state, say a , to a final state(or a group of states) b in which it is free to move and contribute to the current flow. For example, in an n-type photoconductive detector, state a corresponds to electrons in the filled valence band or localized donor impurity atoms, while state b corresponds to electrons in the conduction band. The two levels involved are shown schematically in Figure 7.1. Aphoton of energy hv is absorbed in the process of exciting an electron from a "bound" state a to a"free" state b in which the electron can contribute to the current flow.

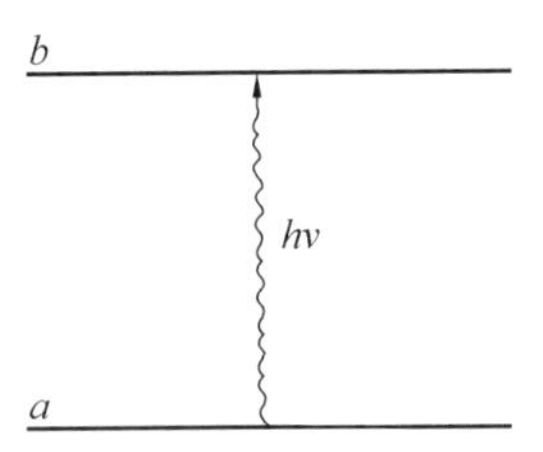

Figure 7.1 Most high-speed optical detectors depend on absorption of photons of energy hv accompanied by a simultaneous transition of an electron(or hole) from a quantum state of low mobility (a) to one of higher mobility (b).

An important point to understand before proceeding with the analysis of different detection schemes is the manner of relating the transition rate per electron from state a to b to the intensity of the optical field. This rate is derived by quantum mechanical considerations.① In our case it can be stated in the following form: Given a nearly sinusoidal optical field②

$$e(t) = \frac{1}{2}[E(t)e^{i\omega_0 t} + E*(t)e^{-i\omega_0 t}] \equiv \mathrm{Re}[V(t)] \qquad (7.2.1)$$

where $V(t) = E(t)\exp(i\omega_0 t)$, ③ the transition rate per electron induced by this field is proportional to $V(t)V*(t)$. Denoting the transition rate as $W_{a\to b}$, we have

$$W_{a\to b} \propto V(t)V*(t) \qquad (7.2.2)$$

We can easily show that $V(t)V*(t)$ is equal to twice the average value of $e^2(t)$, where the averaging is performed over a few optical periods.

To illustrate the power of this seemingly simple result, consider the problem of determining the transition rate due to a field

$$e(t) = E_0\cos(\omega_0 t + \phi_0) + E_1\cos(\omega_1 t + \phi_1) \qquad (7.2.3)$$

taking E_0 and E_1 real and $\omega_1 - \omega_0 \equiv \omega \leqslant \omega_0$. We can rewrite (7.2.3) as

$$e(t) = \mathrm{Re}(E_0 e^{i(\omega_0 t + \phi_0)} + E_1 e^{i(\omega_1 t + \phi_1)}) =$$

$$\mathrm{Re}[(E_0 e^{i\phi_0} + E_1 e^{i(\omega_t + \phi_1)})e^{i\omega_0 t}] \qquad (7.2.4)$$

and, using (7.2.1), identify $V(t)$ as

$$W_{a\to b} \propto (E_0 e^{i\phi_0} + E_1 e^{i(\omega t + \phi_1)})(E_0 e^{-i\phi_0} + E_1 e^{-i(\omega t + \phi_1)}) =$$

$$E_0^2 + E_1^2 + 2E_0E_1\cos(\omega t + \phi_1 - \phi_0) \qquad (7.2.5)$$

This shows that the transition rate has, in addition to a constant term $E_0^2 + E_1^2$, a component oscillating at the difference frequency ω with a phase equal to the difference of the two original phases. This coherent "beating" effect forms the basis of the heterodyne detection scheme, which is discussed in detail in Section 7.4.

---

① More specifically, from first order time-dependent perturbation theory; see, for example, Reference[1].

② By "nearly sinusoidal" we mean a field where $E(t)$ occupies a bandwidth that is small compared to ω_0. Under these conditions the variation of the amplitude $E(t)$ during a few optical periods can be neglected.

③ $V(t)$ is referred to as the "analytic signal" of $e(t)$. See Problem 7.1.

7.3 Photomultiplier

The photomultiplier, one of the most common optical detectors, is used to measure radiation in the near ultraviolet, visible, and near infrared regions of the spectrum. Because of its inherent high current amplification and low noise, the photomultiplier is one of the most sensitive instruments devised by man and under optimal operation—which involves long integration time, cooling of the photocathode, and pulseheight discrimination—has been used to detect power levels as low as about 10^{-19} watt[2].

A schematic diagram of a conventional photomultiplier is shown in Figure 7.2. It consists of a photocathode (C) and a series of electrodes, called dynodes, that are labeled 1 through 8. The dynodes are kept at progressively higher potentials with respect to the cathode, with a typical potential difference between adjacent dynodes of 100 volts. The last electrode (A), the anode, is used to collect the electrons. The whole assembly is contained within a vacuum envelope in order to reduce the possibility of electronic collisions with gas molecules.

The photocathode is the most crucial part of the photomultiplier, since it converts the incident optical radiation to electronic current and thus determines the wavelength-response characteristics of the detector and, as will be seen, its limiting sensitivity. The photocathode consists of materials with low surface work functions. Compounds involving Ag-O-Cs and Sb-Cs are often used; see References[2,3]. These compounds possess work functions as low as 1.5 eV, as compared to 4.5 eV in typical metals. As can be seen in Figure 7.3, this makes it possible to detect photons with longer wavelengths. It follows from the figure that the low-frequency detection limit corresponds to $h\nu = \phi$. At present the lowest-work-function materials make possible photoemission at wavelengths as long as 1 – 1.1 μm.

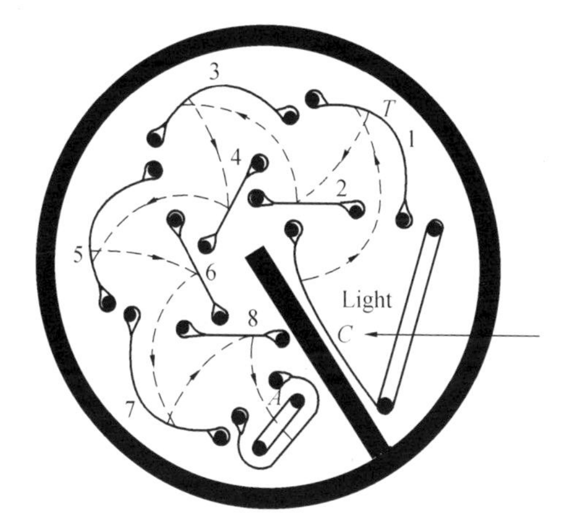

Figure 7.2 Photocathode and focusing dynode configuration of a typical commercial photomultiplier. C = cathode; 1 - 8 = secondary-emission emission dynodes; A = collecting anode. (After Reference[3].)

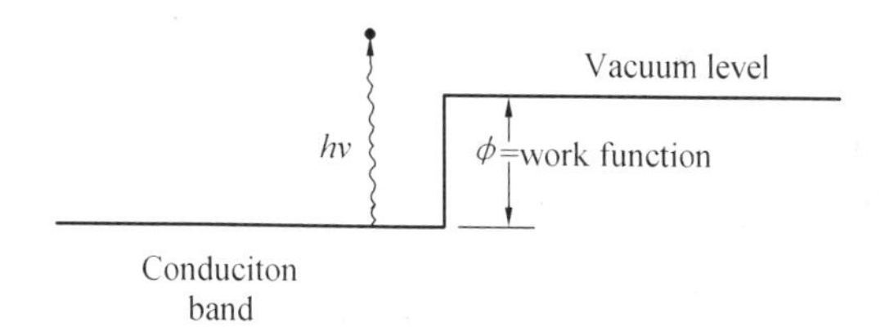

Figure 7.3 Photomultiplier photocathode. The vacuum level corresponds to the energy of an electron at rest at infinite distance from the cathode. The work function ϕ is the minimum energy required to lift an electron from the metal into the vacuum level, so only photons with $h\nu > \phi$ can be detected.

Spectral response curves of a number of commercial photocathodes are shown in Figure 7.4. The quantum efficiency (or quantum yield as it is often called) is defined as the number of electrons released per incident photon.

The electrons that are emitted from the photocathode are focused electrostatically and accelerated toward the first dynode, arriving with a

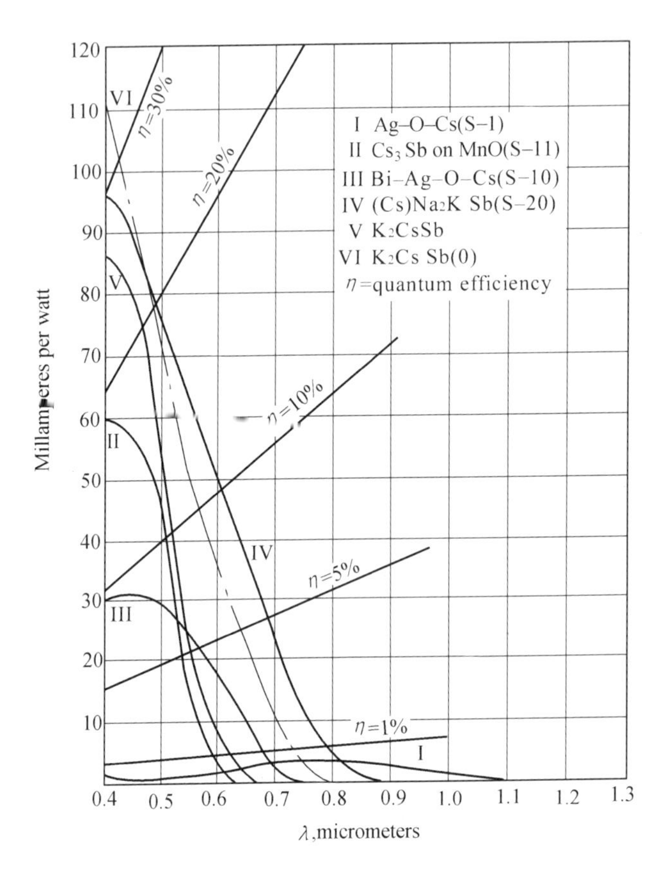

Figure 7.4 Photoresponse versus wavelength characteristics and quantum efficiency of a number of commercial photocathodes. (After Reference[3], p.228.)

kinetic energy of, typically, about 100 eV. Secondary emission from dynode surfaces causes a multiplication of the initial current. This process repeats itself at each dynode until the initial current emitted by the photocathode is amplified by a very large factor. If the average secondary emission multiplication at each dynode is δ (that is, δ secondary electrons for each incident one) and the number of dynodes is N, the total current multiplication between the cathode and anode is

$$G = \delta^N$$

which, for typical values[①] of $\delta = 5$ and $N = 9$, gives $G \simeq 2 \times 10^6$.

7.4 Noise Mechanisms in Photomultipliers

The random fluctuations observed in the photomultiplier output are due to

1. Cathode shot noise, given according to (10.4.9) by

$$\overline{(i_{N_1}^2)} = G^2 2e(\bar{i}_c + i_d)\Delta\nu \tag{7.4.1}$$

where $\bar{i}_c$ is the average current emitted by the photocathode due to the signal power that is incident on it. The current i_d is the so-called dark current, which is due to random thermal excitation of electrons from the surface as well as to excitation by cosmic rays and radioactive bombardment.

2. Dynode shot noise, which is the shot noise due to the random nature of the secondary emission process at the dynodes. Since current originating at a dynode does not exercise the full gain of the tube, the contribution of all the dynodes to the total shot noise output is smaller by a factor of $\sim \delta^{-1}$ than that of the cathode; since $\delta \simeq 5$ it amounts to a small correction and will be ignored in the following.

3. Johnson noise, which is the thermal noise associated with the output resistance R connected across the anode. Its magnitude is given by (10.5.9) as

$$\overline{(i_{N_2}^2)} = \frac{4kT\Delta\nu}{R} \tag{7.4.2}$$

Minimum Detectable Power in Photomultipliers—Video Detection

Photomultipliers are used primarily in one of two ways. In the first, the optical wave to be detected is modulated at some low frequency ω_m before impinging on the photocathode. The signal consists then, of an

① The value of δ depends on the voltage V between dynodes, and values of $\delta \simeq 10$ can be obtained (for $V \simeq 400$ volts). In commercial tubes, values of $\delta \simeq 5$, achievable with $V \simeq 100$ volts, are commonly used.

output current oscillating at ω_m, which, as will be shown below, has an amplitude proportional to the optical intensity, This mode of operation is known as *video*, or straight, detection.

In the second mode of operation, the signal to be detected, whose optical frequency is ω_s, is combined at the photocathode with a much stronger optical wave of frequency $\omega_s + \omega$. The output signal is then a current at the offset frequency ω. This scheme, known as *heterodyne* detection, will be considered in detail in Section 7.4.

The optical signal in the case of video detection may be taken as

$$e_s(t) = E_s(1 + m\cos\omega_m t)\cos\omega_s t =$$

$$\mathrm{Re}[E_s(1 + m\cos\omega_m t)\mathrm{e}^{\mathrm{i}\omega_s t}] \qquad (7.4.3)$$

where the factor $(1 + m\cos\omega_m t)$ represents amplitude modulation of the carrier.① The photocathode current is given, according to (7.2.2), by

$$i_c(t) \propto [(E_s(1 + m\cos\omega_m t)]^2 =$$

$$E_s^2\left[\left(1 + \frac{m^2}{2}\right) + 2m\cos\omega_m t + \frac{m^2}{2}\cos 2\omega_m t\right] \qquad (7.4.4)$$

To determine the proportionality constant involved in (7.4.4), consider the case of $m = 0$. The average photocathode current due to the signal is then②

$$\overline{i_c} = \frac{Pe\eta}{h\nu_s} \qquad (7.4.5)$$

where $v_s = \omega_s/2\pi$, P is the average optical power and η (the quantum efficiency) is the average number of electrons emitted from the photocathode per incident photon. This number depends on the photon frequency, the photocathode surface, and in practice (see Figure 7.4) is found to approach 0.3 Using (7.4.5), we rewrite (7.4.4) as

① The amplitude modulation can be due to the information carried by the optical wave or, as an example, to chopping before detection.

② $P/h\nu_s$ is the rate of photon incidence on the photocathode; thus, if it takes $1/\eta$ photons to generate one electron, the average current is given ty (7.4.5).

$$\overline{i_c}(t) = \frac{Pe\eta}{h\nu_s}\left[\left(1 + \frac{m^2}{2}\right) + 2m\cos\omega_m t + \frac{m^2}{2}\cos 2\omega_m t\right] \quad (7.4.6)$$

The signal output current at ω_m is

$$i_s = \frac{GPe\eta}{h\nu_s}(2m)\cos\omega_m t \quad (7.4.7)$$

If the output of the detector is limited by filtering to a bandwidth $\Delta\nu$ centered on ω_m, it contains a shot-noise current, which, according to (7.4.1), has a mean-squared amplitude

$$\overline{(i_{N_1}^2)} = 2G^2 e(\overline{i_c} + i_d)\Delta\nu \quad (7.4.8)$$

where $\overline{i_c}$ is the average signal current and i_d is the dark current.

The noise and signal equivalent circuit is shown in Figure 7.5, where for the sake of definiteness we took the modulation index $m = 1$. R represents the output load of the photomultiplier. T_e is chosen so that the term $4kT_e\Delta\nu/R$ accounts for the thermal noise of R as well as for the noise generated by the amplifier that follows the photomultiplier.

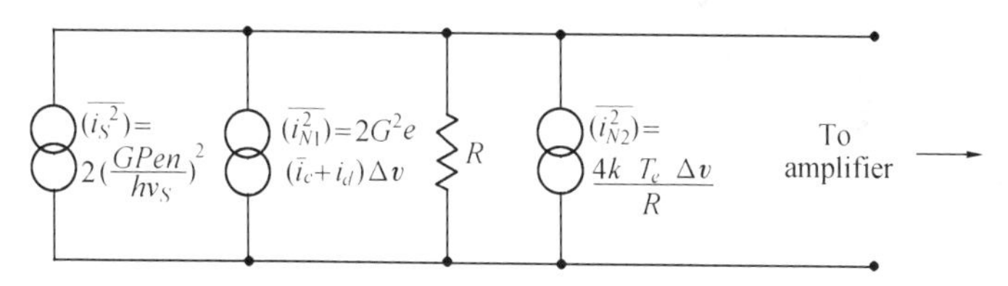

Figure 7.5 Equivalent circuit of a photomultiplier.

The signal-noise power ratio at the output is thus

$$\frac{S}{N} = \frac{\overline{i_s^2}}{\overline{(i_{N_1}^2)} + \overline{(i_{N_2}^2)}} =$$

$$\frac{2(Pe\eta/h\nu_s)^2 G^2}{2G^2 e(\overline{i_c} + i_d)\Delta\nu + (4kT_e\Delta\nu/R)} \quad (7.4.9)$$

Due to the large current gain ($G \simeq 10^6$), the first term in the denominator of (7.4.9), which represents amplified cathode shot noise, is much larger than the thermal and amplifier noise term $4kT_e\Delta\nu/R$.

Neglecting the term $4kT_e\Delta\nu/R$. Neglecting the term $4kT_e\Delta\nu/R$, assuming $i_d \geqslant \bar{i}_c$, and setting $S/N = 1$, we can solve for the minimum detectable optical power as

$$P_{\min} = \frac{h\nu_s(i_d\Delta\nu)^{1/2}}{\eta e^{1/2}} \tag{7.4.10}$$

Numerical Example: Sensitivity of Photomultiplier

Consider a typical case of detecting an optical signal under the following conditions:

$$\nu_s = 6\times 10^{14}\text{Hz}(\lambda = 0.5\ \mu\text{m})$$

$$\eta = 10\ \text{percent}$$

$$\Delta\nu = 1\text{Hz}$$

$i_d = 10^{-15}$ampere(a typical value of the dark photocathode current)

Substitution in (7.4.10) gives

$$P_{\min} = 3\times 10^{-16}\text{watt}$$

The corresponding cathode singal current is $\bar{i}_c \sim 2.4\times 10^{-17}$ ampere, so the assumption $i_d \geqslant \bar{i}_c$ is justified.

Signal-Limited Shot Noise

If one could, somehow, eliminate the Johnson noise and the dark current altogether, so that the only contribution to the average photocathode current is $\bar{i}_c$, which is due to the optical signal, then , using (7.4.5)and(7.4.9)to solve self-consistently for $P_{\min}$

$$P_{\min} \simeq \frac{h\nu_s\Delta\nu}{\eta} \tag{7.4.11}$$

This corresponds to the quantum limit of optical detection. Its significance will be discussed in the next section. The practical achievement of this limit in video detection is nearly impossible since it depends on near total suppression of the dark current and other extraneous noise sources such as background radiation reaching the photocathode and causing shot noise.

The quantum detection limit(7.4.11)can, however, be achieved in the heterodyne mode of optical detection. This is discussed in the next section.

7.5 Heterodyne Detection With Photomultipliers

In the heterodyne mode of optical detection, the signal to be detected $E_s \cos \omega_s t$ is combined with a second optical field, referred to as the local-oscillator field, $E_L \cos(\omega_s + \omega)t$, shifted in frequency by ω ($\omega \ll \omega_s$). The total field incident on the photocathode is therefore given by

$$e(t) = Re[E_L e^{i(\omega_s + \omega)t} + E_s e^{i\omega_s t}] \equiv \mathrm{Re}[V(t)] \qquad (7.5.1)$$

The local-oscillator field originates usually at a laser at the receiving end, so that it can be made very large compared to the signal to be detected. In the following we will assume that

$$E_L \gg E_s \qquad (7.5.2)$$

A schematic diagram of a heterodyne detection scheme is shown in Figure 7.6. The current emitted by the photocathode is given, according to (7.2.2) and (7.5.1), by

$$i_c(t) \propto V(t)V*(t) = E_L^2 + E_s^2 + 2E_L E_s \cos \omega t$$

which, using (7.5.2) can be written as

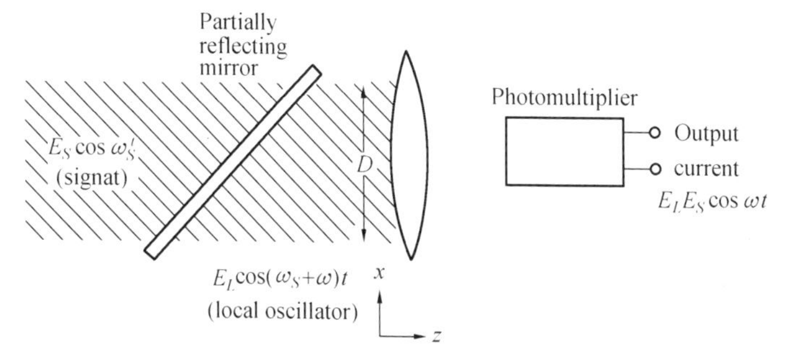

Figure 7.6 Schematic diagram of a heterodyne detector using a photomultiplier.

$$i_c(t) = aE_L^2\left(1 + \frac{2E_s}{E_L}\cos\omega t\right) = aE_L^2\left(1 + 2\sqrt{\frac{P_s}{P_L}}\cos\omega t\right) \qquad (7.5.3)$$

where P_s and P_L are the signal and local-oscillator powers. respectively. The proportionality constant a in (7.5.3) can be determined as in (7.5.6) by requiring that when $E_s = 0$ the direct current be related to the local-oscillator power P_L by $\bar{i}_c = P_L \eta e / h\nu_L$, ① so taking $\nu \approx \nu_L \approx \nu_s$

$$i_c(t) = \frac{P_L e\eta}{h\nu}\left(1 + 2\sqrt{\frac{P_s}{P_L}}\cos\omega t\right) \tag{7.5.4}$$

The total cathode shot noise is thus

$$\overline{(i_{N_1}^2)} = 2e\left(i_d + \frac{P_L e\eta}{h\nu}\right)\Delta\nu \tag{7.5.5}$$

where i_d is the average dark current while $P_L e\eta / h\nu$ is the dc cathode current due to the strong local-oscillator field. The shot-noise current is amplified by G, resulting in an output noise

$$\overline{(i_N^2)}_{\text{anode}} = G^2 2e\left(i_d + \frac{P_L e\eta}{h\nu}\right)\Delta\nu \tag{7.5.6}$$

The mean-square signal current at the output is, according to (7.5.4).

$$\overline{(i_s^2)}_{\text{anode}} = 2G^2\left(\frac{P_s}{P_L}\right)\left(\frac{P_L e\eta}{h\nu}\right)^2 \tag{7.5.7}$$

The signal – to-noise power ratio at the output is given by

$$\frac{S}{N} = \frac{2G^2(P_s P_L)(e\eta/h\nu)^2}{[G^2 2e(i_d + P_L e\eta/h\nu) + 4kT_e/R]\Delta\nu} \tag{7.5.8}$$

where, as in (7.4.9), the last term in the denominator represents the Johnson (thermal) noise generated in the output load, plus the effective input noise of the amplifier following the photomultiplier. The big advantage of the heterodyne detection scheme is now apparent. By increasing P_L the S/N ratio increases until the denominator is dominated by the term $G^2 2eP_L e\eta/h\nu$. This corresponds to the point at which the *shot noise produced by the local oscillator current dwarfs all the other noise contributions*. When this state of affairs prevails, we have, according to

① This is just a statement of the fact that each incident photon has a probability η of releasing an electron.

(7.5.8)

$$\frac{S}{N} \simeq \frac{P_s}{h\nu\Delta\nu/\eta} \tag{7.5.9}$$

which corresponds to the quantum-limited detection limit. The minimum detectable signal—that is, the signal input power leading to an output signal-to-noise ratio of 1—is thus

$$(P_s)_{\min} = \frac{h\nu\Delta\nu}{\eta} \tag{7.5.10}$$

This power corresponds for $\eta = 1$ to a flux at a rate of one photon per $(\Delta\nu)^{-1}$ seconds—that is , one photon per resolution time of the system.①

Numerical Example: Minimum Detectable Power with a Heterodyne System

It is interesting to compare the minimum detectable power for the heterodyne system as given by (7.5.10) with that calculated in the example of Section 7.3 for the video system. Using the same data,

$$\nu = 6 \times 10^{14} \text{ Hz} (\lambda = 0.5 \ \mu\text{m})$$

$$\eta = 10 \text{ percent}$$

$$\Delta\nu = 1 \text{ Hz}$$

we obtain

$$(P_s)_{\min} \simeq 4 \times 10^{-18} \text{ watt}$$

to be compared with $P_{\min} \simeq 3 \times 10^{-16}$ watt in the video case.

Limiting Sensitivity as a Result of the Particle Nature of Light

The quantum limit to optical detection sensitivity is given by (7.5.10) as

$$(P_s)_{\min} = \frac{h\nu\Delta\nu}{\eta} \tag{7.5.11}$$

This limit was shown to be due to the shot noise of the photoemitted current. We may alternatively attribute this noise to the granularity—that

① A detection system that is limited in bandwidth to $\Delta\nu$ can not resolve events in time that are separated by less than $\sim (\Delta\nu)^{-1}$ second. Thus $(\Delta\nu)^{-1}$ is the resolution time of the system.

is, the particle nature—of light, according to which the minimum energy increment of an electromagnetic wave at frequency ν is $h\nu$. The average power P of an optical wave can be written as

$$P = \overline{N}h\nu \tag{7.5.12}$$

where $\overline{N}$ is the average number of photons arriving at the photocathode per second. Next assume a hypothetical noiseless photomultiplier in which *exactly* one electron is produced for each η^{-1} incident photons. The measurement of P is performed by counting the number of electrons produced during an observation period T and then averaging the result over a large number of similar observations.

The average number of electrons emitted per obervation period T is

$$\overline{N}_e = \overline{N}T\eta \tag{7.5.13}$$

If the photons arrive in a perfectly random manner, then the number of photons arriving during the fixed observation period obeys poissonian statistics.① Since in our ideal example, the electrons that are emitted mimic the arriving photons, they obey the same statistical distribution law. This leads to a fluctuation

$$\overline{(\Delta N_e)^2} \equiv \overline{(N_e - \overline{N}_e)^2} = \overline{N}_e = \overline{N}T\eta$$

Defining the minimum detectable number of quanta as that for which the rms fluctuation in the number of emitted photoelectrons equals the average value, we get

$$(\overline{N}_{\min}T\eta)^{1/2} = \overline{N}_{\min}T\eta$$

---

① This follows from the assumption that the photon arrival is perfectly random. so the probability of having N photons arriving in a given time interval is given by the Poisson law

$$p(N) = \frac{\overline{(N)}^N e^{-\overline{N}}}{N!}$$

The mean-square fluctuation is given by

$$\overline{(\Delta N)^2} = \sum_{N=0}^{\infty} p(N)(N - \overline{N})^2 = \overline{N}$$

where

$$\overline{N} = \sum_{0}^{\infty} Np(N)$$

is the average N.

or

$$\overline{N}_{\min} = \frac{1}{T\eta} \tag{7.5.14}$$

If we convert the last result to power by multiplying it by $h\nu$ and recall that $T^{-1} \simeq \Delta\nu$, where $\Delta\nu$ is the bandwidth of the system, we get

$$(P_s)_{\min} = \frac{h\nu\Delta\nu}{\eta} \tag{7.5.15}$$

in agreement with (7.5.10).

The above discussion points to the fact that the noise (fluctuation) in the photo current can be blamed on the physical process that introduces the randomness. In the case of Poissonian photon arrival statistics (as is the case with ordinary lasers) and perfect photon emission ($\eta = 1$), the fluctuations are due to the photons. The opposite, hypothetical, case of no photon fluctuations but random photoemission ($\eta < 1$) corresponds to pure shot noise. The electrical measurement of noise power will yield the same result in either case and can not distinguish between them.

7.6 Photoconductive Detectors

The operation of photoconductive detectors is illustrated in Figure 7.7. A semiconductor crystal is connected in series with a resistance R and a supply voltage V. The optical field to be detected is incident on and absorbed in the crystal, thereby exciting electrons into the conduction band (or, in p-type semiconductors, holes into the valence band.) Such excitation results in a lowering of the resistance R_d of the semiconductor crystal and hence in an increase in the voltage drop across R, which, for $\Delta R_d / R_d \ll 1$, is proportional to the incident optical intensity.

To be specific, we show the energy levels involved in one of the more popular semiconductive detectors—Mercury-doped germanium [7]. Mercury atoms enter germanium as acceptors with an ionization energy of 0.09 eV. It follows that it takes a photon energy of at least 0.09 eV (that

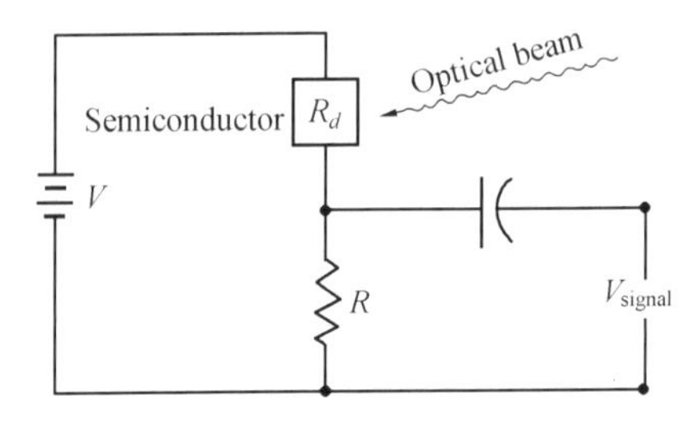

Figure 7.7 Typical biasing circuit of a photoconductive detector.

is ,a photon with a wavelength shorter than 14 μm) to lift an electron from the top of the valence band and have it trapped by the Hg(acceptor) atom. Usually the germanium crystal contains a smaller density N_D of donor atoms, which at low temperatures find it energetically profitable to lose their valence electrons to one of the far more numerous Hg acceptor atoms, thereby becoming positively ionized and ionizing (negatively) an equal number of acceptors.

Since the acceptor density $N_A \gg N_D$, most of the acceptor atoms remain neutrally charged.

An incident photon is absorbed and lifts an electron from the valence band onto an acceptor atom, as shown in process A in Figure 7.8. The electronic deficiency(that is ,the hole) thus created is acted upon by the electric field, and its drift along the field direction gives rise to the signal current. The contribution of a given hole to the current ends when an electron drops from an ionized acceptor level back into the valence band, thus eliminating the hole as in B. This process is referred to as electron-hole recombination or trapping of a hole by an ionized acceptor atom.

By choosing impurities with lower ionization energies, even lower-energy photons can be detected, and, indeed, photoconductive detectors commonly operate at wavelengths up to $\lambda = 50$ μm. Cu, as an example, enters into Ge as an acceptor with an ionization energy of 0.04 eV, which would correspond to long-wavelength detection cutoff of $\lambda \simeq 32$ μm. The

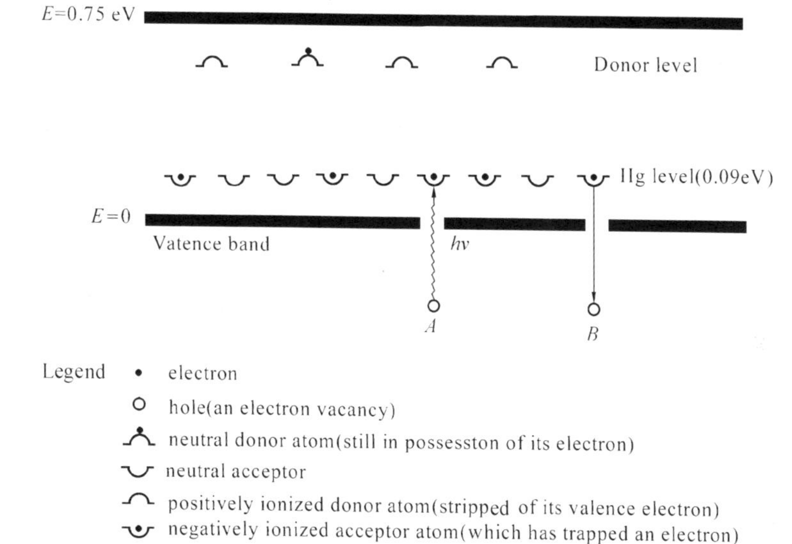

Figure 7.8 Donor and acceptor impurity levels involved in photoconductive semiconductors.

response of a number of commercial photoconductive detectors is shown in Figure 7.9.

It is clear from this discussion that the main advantage of photoconductors compared to photomultipliers is their ability to detect long-wavelength radiation. since the creation of mobile carriers does not involve overcoming the large surface potential barrier. On the debit side we find the lack of current multiplication and the need to cool the semiconductor so that photoexcitation of carriers will not be masked by thermal excitation.

Consider an optical beam, of power P and frequency ν, that is incident on a photoconductive detector. Taking the probability for excitation of a carrier by an incident photon—the so-called quantum efficiency—as η. the carrier generation rate is $G = P\eta/h\nu$. If the carriers last on the average τ_0 seconds before recombining, the average number of carriers N_c is found by equating the generation rate to the recombination

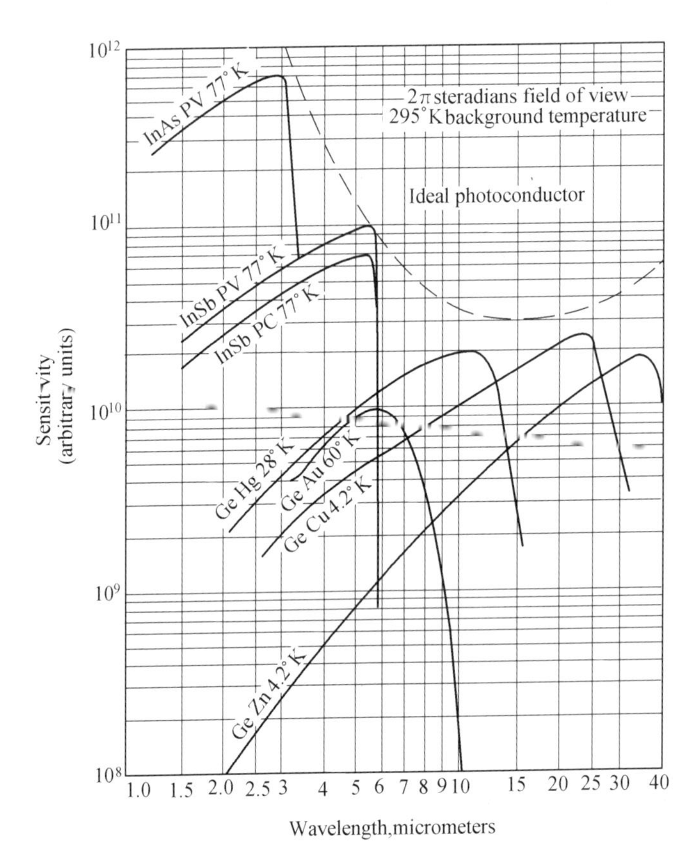

Figure 7.9 Relative sensitivity of a number of commercial photoconductors. (Courtesy Santa Barbara Research Corp.)

rate (N_c/τ_0), so

$$N_c = G\tau_0 = \frac{P\eta\tau_0}{h\nu} \tag{7.6.1}$$

Each one of these carriers drifts under the electric field influence① at a velocity $\bar{v}$ giving rise, according to (7.4.1), to a current in the external circuit of $i_c = ev/d$. where d is the length (between electrodes)

① The drift velocity is equal to μE. where μ is the mobility and E is the electric field.

of the semiconductor crystal. The total current is thus the product of i_e and the number of carriers present, or, using (7.6.1),

$$\bar{i} = N_c i_e = \frac{P\eta\tau_0 e \bar{v}}{h\nu d} = \frac{e\eta}{h\nu}\left(\frac{\tau_0}{\tau_d}\right)P \tag{7.6.2}$$

where $\tau_d = d\sqrt{v}$ is the drift time for a carrier across the length d. The factor (τ_0/τ_d) is thus the fraction of the crystal length drifted by the average excited carrier before recombining.

Equation (7.6.2) describes the response of a photoconductive detector to a constant optical flux. Our main interest, however, is in the heterodyne mode of photoconductive detection, which, as has been shown in Section 7.4, allows detection sensitivities approaching the quantum limit. In order to determine the limiting sensitivity of photoconductive detectors, we need first to understand the noise contribution in these devices.

Generation Recombination Noise in Photoconductive Detectors

The principal noise mechanism in cooled photoconductive detectors reflects the randomness inherent in current flow. Even if the incident optical flux were constant in time, the generation of individual carriers by the flux would constitute a random process. This is exactly the type of randomness involved in photoemission, and we may expect, likewise, that the resulting noise will be shot noise. This is almost true except for the fact that in a photoconductive detector a photoexcited carrier lasts τ seconds① (its recombination lifetime) before being captured by an ionized impurity. The contribution of the carrier to the charge flow in the external circuit is thus $e(\tau/\tau_d)$, as is evident from inspection of (7.6.2). Since the lifetime τ is not a constant, but must be described statistically, another element of randomness is introduced into the current flow.

Consider a carrier excited by a photon absorption and lasting τ seconds. Its contribution to the external current is, according to (10.4.1).

---

① The parameter τ_0 appearing in (7.6.2) is the value of τ averaged over a large number of carriers.

$$i_e(t) = \begin{cases} \frac{e\bar{v}}{d} & 0 \le t \le \tau \\ 0 & \text{otherwise} \end{cases} \tag{7.6.3}$$

which has a Fourier transform

$$I_e(\omega, \tau) = \frac{e\bar{v}}{2\pi d}\int_0^\tau e^{-i\omega t}dt = \frac{-ie\bar{v}}{2\pi\omega d}[1 - e^{-i\omega\tau}] \tag{7.6.4}$$

so that

$$|I_e(\omega, \tau)|^2 = \frac{e^2\bar{v}^2}{4\pi^2\omega^2 d^2}[2 - e^{-i\omega\tau} - e^{i\omega\tau}] \tag{7.6.5}$$

According to (10.3.10) we need to average $|I_e(\omega, \tau)|^2$ over τ. This is done in a manner similar to the procedure used in Section 10.5 Taking the probability function① $g(\tau) = \tau_0^{-1}\exp(-\tau/\tau_0)$. We average (7.6.5) over all the possible values of τ according to

$$\overline{|I_e(\omega)|^2} = \int_0^\infty |I_e(\omega, \tau)|^2 g(\tau)d\tau = \frac{2e^2\bar{v}^2\tau_0^2}{4\pi^2 d^2(1 + \omega^2\tau_0^2)} \tag{7.6.6}$$

The spectral density function of the current fluctuations is obtained using Carson's theorem (10.3.10) as

$$S(v) = 2\bar{N}\frac{2e^2(\tau_0^2/\tau_d^2)}{1 + \omega^2\tau_0^2} \tag{7.6.7}$$

where we used $\tau_d = d\sqrt{v}$ and where $\bar{N}$, the average number of carriers generated per second, can be expressed in terms of the average current $\bar{I}$ by use of the relation②

$$\bar{I} = \bar{N}\frac{\tau_0}{\tau_d}e \tag{7.6.8}$$

Leading to

---

① $(\tau)\,d\tau$ is the probability that a carrier lasts between τ and $\tau + d\tau$ seconds before recombining.

② This relation follows from the fact that the average charge per carrier flowing through the external circuit is $e(\tau_0/\tau_d)$, which, when multiplied by the generation rate $\bar{N}$, gives the current.

$$S(v)=\frac{4e\bar{I}(\tau_0/\tau_d)}{1+4\pi^2 v^2\tau_0^2}$$

Therefore, the mean-square current representing the noise power in a frequency interval v to $v+\Delta v$ is

$$\overline{i_N^2}\equiv S(v)\Delta v=\frac{4e\bar{I}(\tau_0/\tau_d)\Delta v}{1+4\pi^2 v^2\tau_0^2} \tag{7.6.9}$$

which is the basic result for generation-recombination noise.

Numerical Example: Generation Recombination Noise in Hg Doped Germanium Photoconductive Detector

To better appreciate the kind of numbers involved in the expression for $\overline{i_N^2}$ we may consider a typical mercury-doped germanium detector operating at 20 K with the following characteristics:

$d=10^{-1}$ cm

$\tau_0=10^{-9}$s

V(across the length d) = 10 volts$\Rightarrow E=10^2$ V/cm

$\mu=3\times10^4$ cm^2/V − s

The drift velocity is $\bar{v}=\mu E=3\times10^6$ cm/s and $\tau_d=d\sqrt{v}\simeq3.3\times10^{-8}$ second, and therefore $\tau_0/\tau_d=3\times10^{-2}$. Thus, on the average, a carrier traverses only 3 percent of the length($d=1$ mm) of the sample before recombining. Comparing(7.6.9) to the shot-noise result(10.4.9), we find that for a given average current $\bar{I}$ the generation recombination noise is reduced from the shot-noise value by a factor

$$\frac{\overline{(i_N^2)}_{\text{generation-recombination}}}{\overline{(i_N^2)}_{\text{shot noise}}}\underset{\omega\tau_0\ll1}{=}2\left(\frac{\tau_0}{\tau_d}\right) \tag{7.6.10}$$

which. in the foregoing example. has a value of about 1/15 Unfortunately, as will be shown subsequently, the reduced noise is accompanied by a reduction by a factor of (τ_0/τ_d) in the magnitude of the signal power. which wipes out the advantage of the lower noise.

Heterodyne Detection in Photoconductors

The situation here is similar to that described by Figure 7.6 in connection with heterodyne detection using photomultipliers. The signal field

$$e_s(t) = E_s \cos\omega_s t$$

is combined with a strong local-oscillator field

$$e_L(t) = E_L \cos(\omega + \omega_s)t \qquad E_L \gg E_s$$

so the total field incident on the photoconductor is

$$e(t) = Re(E_s e^{i\omega_s t} + E_L e^{i(\omega_s + \omega)t}) \equiv Re[V(t)] \quad (7.6.11)$$

The rate at which carriers are generated is taken, following (7.2.2), as $aV(t)V*(t)$ Where a is a constant to be determined . The equation describing the number of excited carriers N_c is thus

$$\frac{dN}{dt} = aVV* - \frac{N_c}{\tau_0} \quad (7.6.12)$$

where τ_0 is the average carrier lifetime, so N_c/τ_0 corresponds to the carrier's decay rate. We assume a solution for $N_c(t)$ that consists of the sum of dc and a sinusoidal component in the form of

$$N_c(t) = N_0 + (N_1 e^{i\omega t} + c.c.) \quad (7.6.13)$$

where c.c. stands for "complex conjugate."

Substitution in (7.6.12) gives

$$N_c(t) = a\tau_0(E_s^2 + E_L^2) + a\tau_0\left(\frac{E_s E_L e^{i\omega t}}{1 + i\omega\tau_0} + c.c.\right) \quad (7.6.14)$$

where we took E_s and E_L as real. The current through the sample is given by the number of carriers per unit length N_c/d times e $\bar{v}$, where $\bar{v}$ is the drift velocity

$$i(t) \frac{N_c(t) e \bar{v}}{d} \quad (7.6.15)$$

which, using (7.6.14), gives

$$i(t) = \frac{e \bar{v} a\tau_0}{d}\left(E_s^2 + E_L^2 + \frac{2E_s E_L \cos(\omega t - \phi)}{\sqrt{1 + \omega^2\tau_0^2}}\right) \quad (7.6.16)$$

where $\phi = \tan^{-1}(\omega\tau_0)$.

The current is thus seen to contain a signal component that oscillates at ω and is proportional to E_s. The constant a in (7.6.16) can be determined by requiring that, when $P_s = 0$, the expression for the direct current predicted by (7.6.16) agree with (7.6.2). This condition is satisfied if we rewrite(7.6.16) as

$$i(t) = \frac{e\eta}{h\nu}\left(\frac{\tau_0}{\tau_d}\right)\left[P_s + P_L + \frac{2\sqrt{P_sP_L}}{\sqrt{1+\omega^2\tau_0^2}}\cos(\omega t - \phi)\right] \tag{7.6.17}$$

where P_s and P_L refer, respectively, to the incident-signal and local-oscillator powers and $\nu = \nu_s = \omega_s/2\pi$ and η, the quantum efficiency, is the number of carriers excited per incident photon. The signal current is thus

$$i_s(t) = \frac{2e\eta}{h\nu}\left(\frac{\tau_0}{\tau_d}\right)\frac{\sqrt{P_sP_L}}{\sqrt{1+\omega^2\tau_0^2}}\cos(\omega t - \phi) \tag{7.6.18}$$

while the dc(average) current is

$$\overline{I} = \frac{e\eta}{h\nu}\left(\frac{\tau_0}{\tau_d}\right)(P_s + P_L) \tag{7.6.19}$$

Since the average current $\overline{I}$ appearing in the expression (7.6.9) for the generation recombination noise is given in this case by

$$\overline{I} = \left(\frac{e\eta}{h\nu}\right)\left(\frac{\tau_0}{\tau_d}\right)P_L \qquad P_L \gg P_s$$

we can, by increasing P_L, increase the noise power $\overline{i_N^2}$ and at the same time, according to (7.6.18), the signal i_s^2 until the generation recombination noise (7.6.9) is by far the largest contribution to the total output noise. When this condition is satisfied, the signal-to-noise ratio can be written, using (7.6.9), (7.6.18), and (7.6.19) and taking $P_L \gg P_s$, as

$$\frac{S}{N}=\frac{\overline{i_s^2}}{\overline{i_N^2}}=\frac{2(e\eta\tau_0/h\upsilon\tau_d)^2P_sP_L/(1+\omega^2\tau_0^2)}{4e^2\eta(\tau_0/\tau_d)^2P_L\Delta\upsilon/(1+\omega^2\tau_0^2)h\upsilon}=\frac{P_s\eta}{2h\upsilon\Delta\upsilon} \tag{7.6.20}$$

The minimum detectable signal—that which leads to a signal-to-noise ratio of unity—is found by setting the left side of (7.6.20) equal to unity and solving for P_s. It is

$$(P_s)_{\min}=\frac{2h\upsilon\Delta\upsilon}{\eta} \tag{7.6.21}$$

which, for the same η, is twice that of the photomultiplier heterodyne detection as given by (7.6.10). In practice, however, η in photoconductive detectors can approach unity, whereas in the best photomultipliers $\eta\simeq 30$ percent.

Numerical Example: Minimum Detectable Power of a Heterodyne Receiver Using a Photoconductor at 10.6 μm

Assume the following :

$$\lambda = 10.6\ \mu\text{m}$$

$$\Delta\upsilon = 1\ \text{HZ}$$

$$\eta\simeq 1$$

Substitution in (7.6.21) gives a minimum detectable power of

$$(P_s)_{\min}\simeq 10^{-19}\ \text{watt}$$

Experiments ([8, 9]). have demonstrated that the theoretical signal-to-noise ratio as given by (7.6.20) can be realized quite closely in practice; see Figure 7.10.

7.7 The *p-n* Junction

Before embarking on a description of the *p-n* diode detector. we need to understand the operation of the semiconductor *p-n* junction. Consider the junction illustrated in Figure 7.11. It consists of an abrupt transition from a donor-doped (that is, n – type) region of a semiconductor, where the

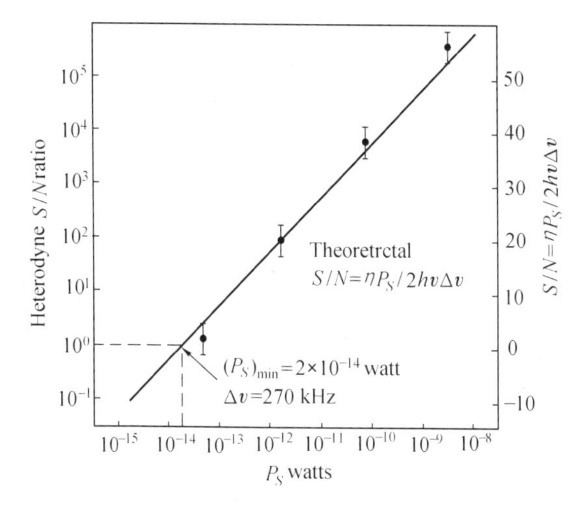

Figure 7. 10 Signal-to-noise ratio of heterodyne signal to Ge: Cu detector at a heterodyne frequency of 70 MHz. Data points represent observed values. (After Reference[8].)

charge carriers are predominantly electrons, to an acceptor-doped (p-type) region, where the carriers are holes. The doping profile—that is, the density of excess donor (in the n region) atoms or acceptor atoms (in the p region)—is shown in Figure 7. 11 (a). This abrupt transition results usually from diffusing suitable impurity atoms into a substrate of a semiconductor with the opposite type of conductivity. In our slightly idealized abrupt junction we assume that the n region ($x > 0$) has a constant (net) donor density N_D and the p region ($x < 0$) has a constant acceptor density N_A.

The energy-band diagram at zero applied bias is shown in Firure 7.11(b). The top (or bottom) curve can be taken to represent the potential energy of an electron as a function of position x, so the minimum energy needed to take an electron from the n to the p side of the junction is eV_d. Taking the separations of the Fermi level from the respective band edges as ϕ_n and ϕ_P as shown, we have

$$eV_d = E_g - (\phi_n + \phi_P)$$

V_d is referred to as the "built-in" junction potential.

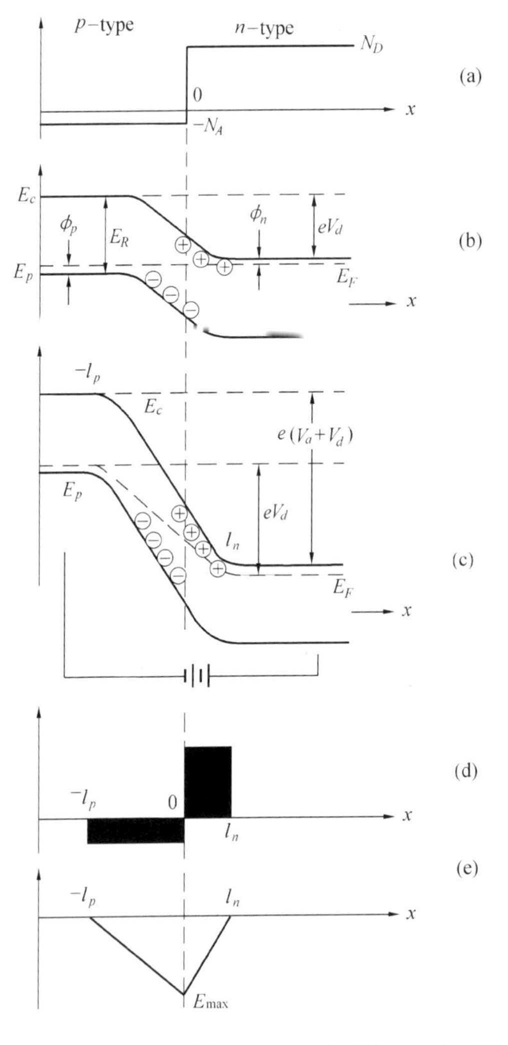

Figure 7.11 The abrupt *p-n* junction. (a) Impurity profile. (b) Energy-band diagram with zero applied bias. (c) Energy-band diagram with reverse applied bias. (d) Net charge density in the depletion layer. (e) The electric field. The circles in (b) and (c) represent ionized impurity atoms in the depletion layer.

Figure 7.11(c) shows the energy band diagram in the junction with an applied reverse bias of magnitude V_a. This leads to a separation of eV_a between the Fermi levels in the p and n regions and causes the potential barrier across the junction to increase from eV_d to $e(V_d + V_a)$. The change of potential between the p and n regions is due to a sweeping of the mobile charge carriers from the region $-l_p < x < l_n$. giving rise to a charge double layer of stationary (ionized) impurity atoms, as shown in Figure 7.11(d).

In the analytical treatment of the problem we assume that in the depletion layer($-l_p < x < l_n$)the excess impurity atoms are fully ionized and thus, using $\nabla \cdot \mathrm{E} = \rho/\in$ and $\mathrm{E} = -\nabla V$. where V is the potential. we have

$$\frac{\mathrm{d}^2 V}{\mathrm{d}x^2} = \frac{eN_A}{\in} \qquad \text{for } -l_p < x < 0 \qquad (7.7.1)$$

and

$$\frac{\mathrm{d}^2 V}{\mathrm{d}x^2} = -\frac{eN_D}{\in} \qquad 0 < x < l_n \qquad (7.7.2)$$

where the charge of the electron is $-e$ and the permittivity is $\in$. The boundary conditions are

$$E = -\frac{\mathrm{d}V}{\mathrm{d}x} = 0 \text{ at } x = -l_p \text{ and} x = +l_n \qquad (7.7.3)$$

$$V \text{ and } \frac{\mathrm{d}V}{\mathrm{d}x} \text{ are continuous at } x = 0 \qquad (7.7.4)$$

$$V(l_n) - V(-l_p) = V_d + V_a \qquad (7.7.5)$$

The solutions of (7.7.1)and(7.7.2)conforming with the arbitrary choice of $V(0) = 0$ are

$$V = \frac{e}{2\in} N_A(x^2 + 2l_p x) \qquad \text{for } -l_p < x < 0 \qquad (7.7.6)$$

$$V = -\frac{e}{2\in} N_D(x^2 - 2l_n x) \qquad 0 < x < l_n \qquad (7.7.7)$$

which, using(7.7.4), gives

$$N_A l_p = N_D l_n \tag{7.7.8}$$

so the double layer contains an equal amount of positive and negative charge.

Condition (7.7.5) gives

$$V_d + V_a = \frac{e}{2\epsilon}(N_D l_n^2 + N_A l_p^2) \tag{7.7.9}$$

which, together with (7.7.8) leads to

$$l_p = (V_d + V_a)^{1/2}\left(\frac{2\epsilon}{e}\right)^{1/2}\left(\frac{N_D}{N_A(N_A + N_D)}\right)^{1/2} \tag{7.7.10}$$

$$l_n = (V_d + V_a)^{1/2}\left(\frac{2\epsilon}{e}\right)^{1/2}\left(\frac{N_A}{N_A(N_A + N_D)}\right)^{1/2} \tag{7.7.11}$$

and, therefore, as before,

$$\frac{l_p}{l_n} = \frac{N_D}{N_A} \tag{7.7.12}$$

Differentiation of (7.7.6) and (7.7.7) yields

$$E = -\frac{e}{\epsilon}N_A(x + l_p) \qquad \text{for } -l_p < x < 0$$

$$E = -\frac{e}{\epsilon}N_D(l_n - x) \qquad 0 < x < l_n \tag{7.7.13}$$

The field distribution of (7.7.13) is shown in Figure 7.11(e). The maximum field occurs at $x = 0$ and is given by

$$E_{\max} = -2(V_d + V_a)^{1/2}\left(\frac{e}{2\epsilon}\right)^{1/2}\left(\frac{N_D N_A}{N_A + N_D}\right)^{1/2} =$$

$$-\frac{2(V_d + V_a)}{l_p + l_n} \tag{7.7.14}$$

The presence of a charge $Q = -eN_A l_p$ per unit junction area on the p side and an equal and opposite charge on the n side leads to a junction capacitance. The reason is that l_p and l_n depend, according to (7.7.10) and (7.7.11), on the applied voltage V_a, so a change in voltage leads to change in the charge $eN_A l_p = eN_D l_n$ and hence to a differential *capacitance per unit area*,[①] *given by*

$$\frac{C_d}{\text{area}} \equiv \frac{dQ}{dV_a} = eN_A \frac{dl_p}{dV_a} = \left(\frac{\in e}{2}\right)^{1/2} \left(\frac{N_A N_D}{N_A + N_D}\right)^{1/2} \left(\frac{1}{V_a + V_d}\right) \quad (7.7.15)$$

which, using (7.7.10) and (7.7.11), can be shown to be equal to

$$\frac{C_d}{\text{area}} = \frac{\in}{l_p + l_n} \quad (7.7.16)$$

as appropriate to a parallel-plate capacitance of separation $l = l_p + l_n$. The equivalent circuit of a *p-n* junction is shown in Figure 7.12. The capacitance C_d was discussed above. The diode shunt resistance R_d in back-biased junctions is usually very large($> 10^6$ ohms) compared to the load impedance R_L and can be neglected. The resistance R_s represents ohmic losses in the bulk p and n regions adjacent to the junction.

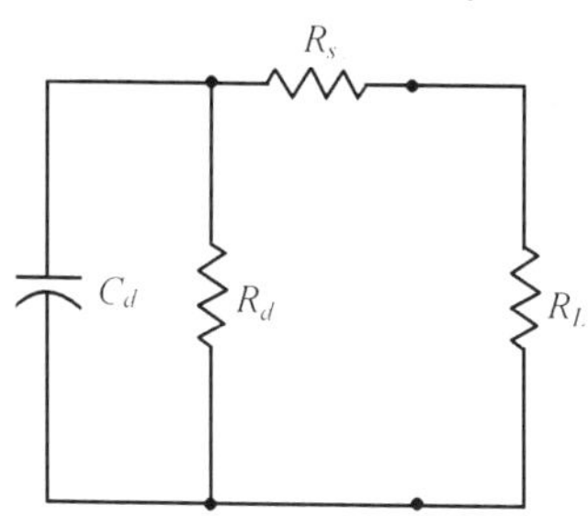

Figure 7.12 Equivalent circuit of a *p-n* junction. In typical back-biased diodes . $R_d \gg R_s$ and R_L, and $R_L \gg R_s$. so the resistance across the junction can be taken as equal to the load resistance R_L.

7.8 Semiconductor Photodiodes

Semiconductor *p-n* junctions are used widely for optical detection: see References [10 - 12]. In this role they are referred to as junction

① The capacitance is defined by $C = Q/V_a$, whereas differential capacitance $C_d = dQ/dV_a$ is the capacitance "seen" by a small ac voltage when the applied bias is V_a.

photodiodes. The main physical mechanisms involved in junction photodetection are illustrated in Figure 7.13. At A. an incoming photon is absorbed in the p side creating a hole and a free electron. If this takes place within a diffusion length (the distance in which an excess minority concentration is reduced to e^{-1} of its peak value. or in physical terms, the average distance a minority carrier traverses before recombining with a carrier of the opposite type) of the depletion layer, the electron will, with high probability, reach the layer boundary and will drift under the field influence across it. An electron traversing the junction contributes a charge e to the current flow in the external circuit, as described in Section 10.1. If the photon is absorbed near the n side of the depletion layer, as shown at C, the resulting hole will diffuse to the junction and then drift across it again, giving rise to a flow of charge e in the external load. The photon may also be absorbed in the depletion layer as at B, in which case both the hole and electron that are created drift (in opposite directions) under the field until they reach the p and n sides, respectively. Since in this case each carrier traverses a distance that is less than the full junction width, the contribution of this process to charge flow in the external circuit is , according to (10.4.1) and (10.4.7), e. In practice this last process is the most desirable, since each absorption gives rise to a charge e, and delayed current response caused by finite diffusion time is avoided. As a result, photodiodes often use a p-i-n structure in which an intrinsic high resistivity (i) layer is sandwiched between the p and n regions. The potential drop occurs mostly across this layer, which can be made long enough to ensure that most of the incident photons are absorbed within it. Typical construction of a p-i-n photodiode is shown in Figure 7.14.

It is clear from Figure 7.13 that a photodiode is capable of detecting only radiation with photon energy $hv > E_g$, where E_g is the energy gap of the semiconductor. If, on the other hand, $hv \gg E_g$, the absorption, which in a semiconductor increases strongly with frequency , will take place entirely near the input face (in the n region of Figure 7.14) and the minority carriers generated by absorbed photons will recombine with majority carriers before diffusing to the depletion layer. This event does not contribute to the current flow and, as far as the signal is concerned, is wasted. This is why the photoresponse of diodes drops off when $h\nu > E_g$. Typical frequency response curves of photodiodes are shown in Figure 7.15. The number of carriers flowing in the external circuit per incident photon, the so-called quantum efficiency, is seen to approach 50 percent in Ge.

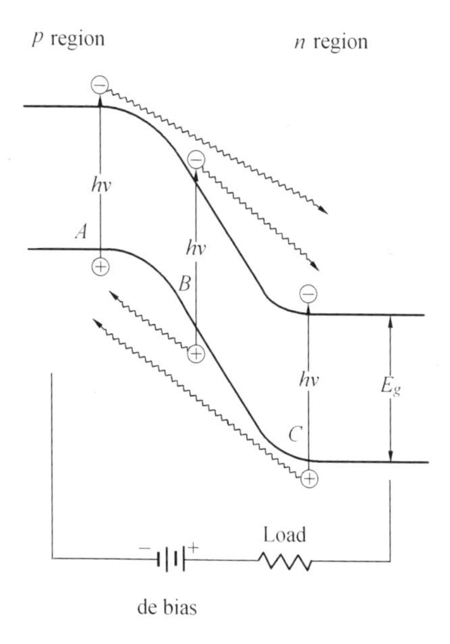

Figure 7.13 The three types of electron-hole pair creation by absorbed photons that contribute to current flow in a p-n photodiode.

Frequency Response of Photodiodes

One of the major considerations in optical detectors is their frequency response—that is, the ability to respond to variations in the incident intensity such as those caused by high-frequency modulation. The three

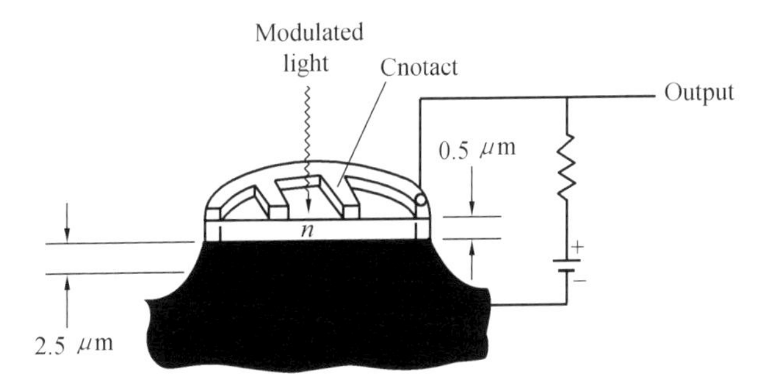

Figure 7.14 A *p-i-n* photodiode. (After Reference[13].)

main mechanisms limiting the frequency response in photodiodes are;

1. The finite diffusion time of carriers produced in the p and n regions. This factor was described in the last section, and its effect can be minimized by a proper choice of the length of the depletion layer.

2. The shunting effect of the signal current by the junction capacitance C_d shown in Figure 7.12. This places an upper limit of

$$\omega_m \simeq \frac{1}{R_e C_d} \tag{7.8.1}$$

on the intensity modulation Frequency where R_e is the equivalent resistance in parallel with the capacitance C_d.

3. The finite transit time of the carriers drifting across the depletion layer.

To analyze first the limitation due to transit time, we assume the slightly idealized case in which the carriers are generated in a *single* plane, say point A in Figure 7.13, and then drift the full width of the depletion layer at a constant velocity v. For high enough electric fields, the drift velocity of carriers in semiconductors tends to saturate, so the constant velocity assumption is not very far from reality even for a nonuniform field distribution, such as that shown in Figure 7.11(e), provided the field exceeds its saturation value over most of the depletion layer length. The saturation of the whole velocity in germanium, as an

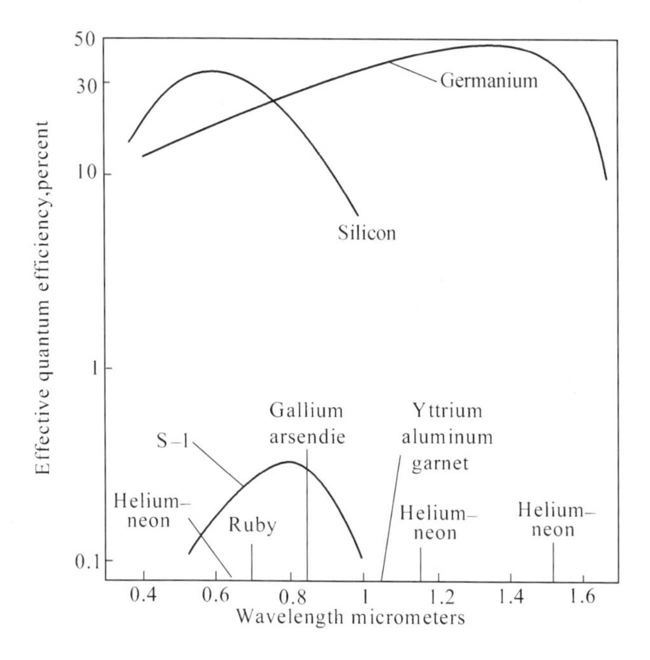

Figure 7.15 Quantum efficiencies for silicon and germanium photodiodes compared with the efficiency of the S-1 photodiode used in a photomultiplier tube. Emission wavelengths for various lasers are also indicated. (After Reference[13].)

example, is illustrated by the data of Figure 7.16.

The incident optical field is taken as

$$e(t) = E_s(1 + m\cos\omega_m t)\cos\omega t \equiv \mathrm{Re}[V(t)] \tag{7.8.2}$$

where

$$V(t) \equiv E_s(1 + m\cos\omega_m t)e^{i\omega t} \tag{7.8.3}$$

Thus, the amplitude is modulated at a frequency $\omega_m/2\pi$. Following the discussion of Section 7.1 we take the generation rate $G(t)$; that is, the number of carriers generated per second as proportional to the average of $e^2(t)$ over a time long compared to the optical period $2\pi/\omega$. This average is equal to $\frac{1}{2}V(t)V^*_{(t)}$. so the generation rate is taken as

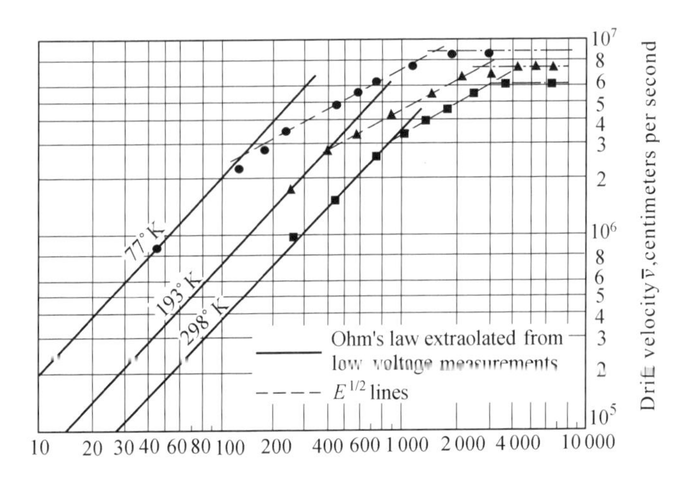

Figure 7.16 Experimental data showing the saturation of the drift velocity of holes in germanium at high electric fields. (After Reference[14].)

$$G(t) = aE_s^2\left[\left(1 + \frac{m^2}{2}\right) + 2m\cos\omega_m t + \frac{m^2}{2}\cos 2\omega_m t\right] \quad (7.8.4)$$

where a is a proportionality constant to be determined. Dropping the term involving $\cos 2\omega_m t$ and using complex notation, we rewrite $G(t)$ as

$$G(t) = aE_s^2\left[1 + \frac{m^2}{2} + 2m\mathrm{e}^{\mathrm{i}\omega_m t}\right] \quad (7.8.5)$$

A single carrier drifting at a velocity $\bar{v}$ contributes, according to (10.4.1), an instantaneous current

$$i = \frac{e\bar{v}}{\mathrm{d}} \quad (7.8.6)$$

to the external circuit. where d is the width of the depletion layer. The current due to carriers generated between t' and $t' + \mathrm{d}t'$ is $(e\bar{v}/\mathrm{d})G(t')\mathrm{d}t'$ but, since each carrier spends a time $\tau_d = \mathrm{d}\sqrt{v}$ in transit. the instantaneous current at time t is the sum of contributions of carriers generated between t and $t - \tau_d$

$$i(t) = \frac{e\bar{v}}{d}\int_{t-\tau_d}^{t} G(t')\mathrm{d}t' = \frac{e\bar{v}aE_s^2}{d}\int_{t-\tau_d}^{t}\left(1 + \frac{m^2}{2} + 2m\mathrm{e}^{\mathrm{i}\omega_m t'}\right)\mathrm{d}t'$$

and, after integration.

$$i(t) = \left(1 + \frac{m^2}{2}\right)eaE_s^2 + 2meaE_s^2\left(\frac{1 - \mathrm{e}^{-\mathrm{i}\omega_m\tau_d}}{i\omega_m\tau_d}\right)\mathrm{e}^{\mathrm{i}\omega_m t} \qquad (7.8.7)$$

The factor $(1 - \mathrm{e}^{-\mathrm{i}\omega_m\tau_d})/i\omega_m\tau_d$ represents the phase lag as well as the reduction in signal current due to the finite drift time τ_d. If the drift time is short compared to the modulation period, so $\omega_m\tau_d \leqslant 1$, it has its maximum value of unity, and the signal is maximum. This factor is plotted in Figure 7.17 as a function of the transit phase angle $\omega_m\tau_d$. We can determine the value of the constant a in (7.8.7) by requiring that (7.8.7) agree with the experimental observation according to which in the absence of modulation, $m = 0$, each incident photon will create η carriers. Thus the dc (average) current is

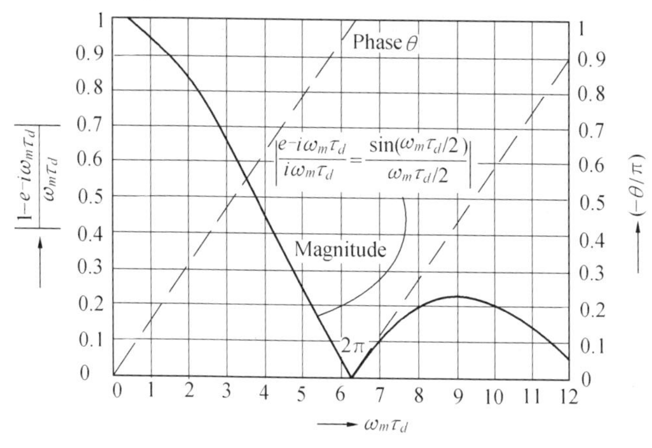

Figure 7.17 Phase and magnitude of the transit-time reduction factor $(1 - \mathrm{e}^{-\mathrm{i}\omega_m\tau_d})/\omega_m\tau_d$.

$$\bar{I} = \frac{Pe\eta}{hv} \qquad (7.8.8)$$

where P is the optical (signal) power when $m = 0$. Using (7.8.8). we can rewrite (7.8.7) as

$$i(t)=\frac{Pe\eta}{h\nu}\left(1+\frac{m^2}{2}\right)+\frac{Pe\eta}{h\nu}2m\left(\frac{1-e^{-i\omega_m\tau_d}}{i\omega_m\tau_d}\right)e^{-i\omega_m\tau_d} \qquad (7.8.9)$$

To evaluate the effect of the other limiting factors on the modulation frequency response of a photodiode, we refer to the diode equivalent ac circuit in Figure 7.18. Here R_d is the diode incremental (ac) resistance, C_d the junction capacitance, R_s represents the contact and series resistance, L_p the parasitic inductance associated mostly with the contact leads, and C_p the parasitic capacitance due to the contact leads and the contact pads.

Recent advances [20 22, 24] have resulted in metal – GaAs (Schottky) diodes with frequency response extending up to 10^{11} Hz. Figure 7.19 shows a schematic diagram of such a diode. This high-frequency limit was achieved by using a very small ares (5 μm × 5 μm) that minimizes C_d, by using extremely short contact leads to reduce R_s and L_p, by fabricating the diode on semi-insulating GaAs substrate [20] to reduce C_p, and by using a thin (0.3 μm) n – GaAs drift region to reduce the transit time. The resulting measured frequency response is shown in Figure 7.20. The measurement of the frequency response up to 100 GHz is by itself a considerable achievement. This was accomplished by first obtaining the impulse response of the photodiode by exciting it with picosecond pulses (which, for the range of frequencies of interest, may be considered as delta functions) from a mode-locked laser [21]。The diode response, which is only a few picoseconds long. is measured by a new electrooptic sampling technique [23,24]. The frequency response, as plotted in Figure 7.20, is obtained by taking the Fourier transform of the measured impulse response.

The laser transition can be pumped by radiation at $\lambda \sim 0.98$ μm or $\lambda \sim 1.49$ μm as shown. This pumping field is usually obtained from semiconductor lasers and is coupled into the amplifying fiber whose length is typically between a few meters and a few tens of meters. A schematic

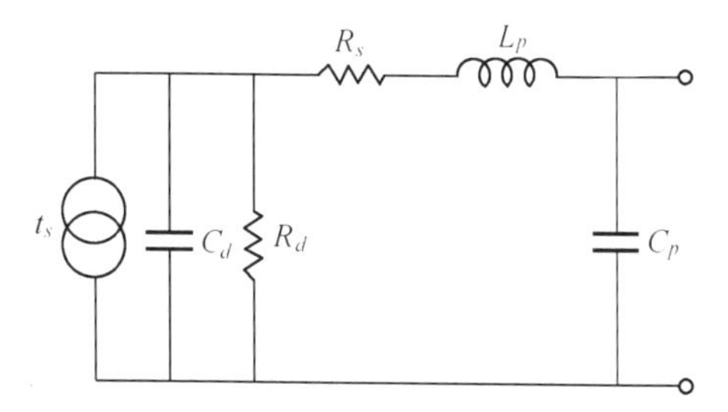

Figure 7.18 The equivalent high-frequency circuit of a semiconductor photodiode.

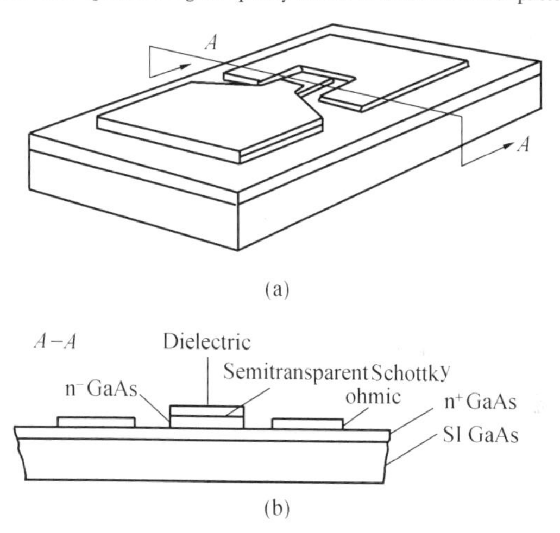

Figure 7.19 (a) Planar GaAs Schottky photodiode. (b) Cross section along A – A. The n^- GaAs layer (0.3 μm thick) and the n^+ GaAs(0.4 μm thick) are grown by liquid-phase epitaxy on semi-insulating GaAs. The semitransparent Schottky consists of 100Å of Pt (After Reference [22].)

diagram of the amplifier configuration is shown in Figure 11.31 (b). The fiber amplifier section can be spliced smoothly into the fiber. A plot of the gain vs. signal wavelength is shown in Figure 7.21 (a).

The main effect of the optical amplifier on the SNR of the detected signal is to add, upon detection, a noise current component, at frequencies near that of the signal current. This noise is due to beating between the amplified (optical) spontaneous emission (ASE) power of the amplifier

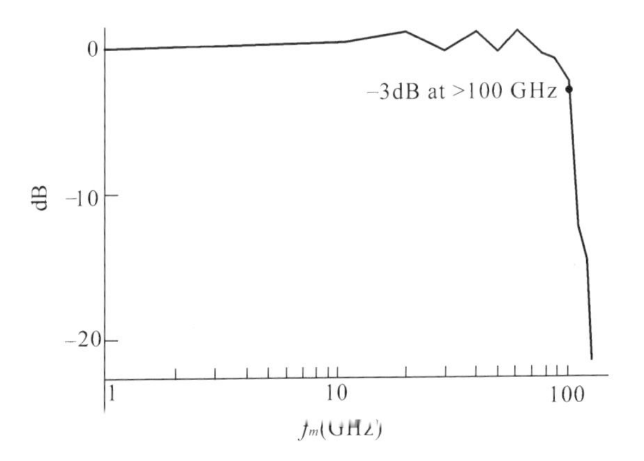

Figure 7.20 The modulation frequency response of the Schottky photodiode shown in Figure 7.19. (After Reference [22].)

and the signal optical field.

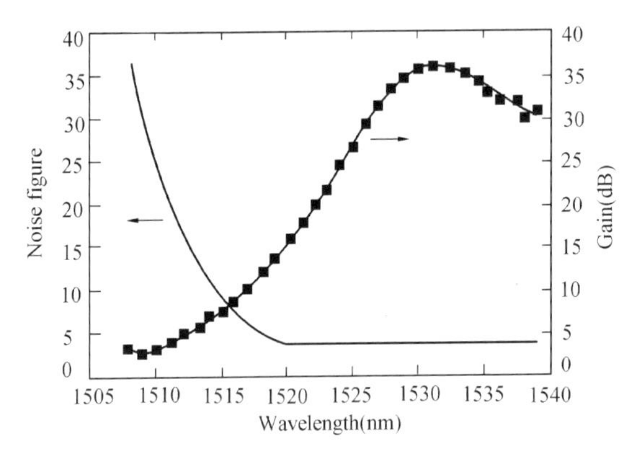

Figure 7.21 (a) Noise factor and gain spectrum of the silica E_r^{3+} fiber amplifier for a constant pump power of 34.2 mW at 0.98 μm. (After Reference[35].)

Figure 7.21 (b) shows the two spectral windows of the amplified output spontaneous emission power that beat (at the detector) with the optical signal field S_0 at ω_0 to generate an output noise current at some arbitrary frequency ω_m, This mechanism thus gives rise to a spectral continuum of RF noise current extending from dc to approximately $\Delta\omega_{gain}$, the width of the (amplified) spontaneous emission spectrum. To estimate this current we first need to obtain an expression for the optical spontaneous emission power at the output of an optical amplifer. This topic is the subject of Appendix C. The main result, Equation C-8, is that the (amplified) spontaneous emission power in a *single* mode within a spectral bandwidth Δv_{opt} at the output of an optical amplifier is [36]

$$F_0 = \mu hv \Delta v_{opt} (G - 1) \tag{7.8.10}$$

where $G = \exp(\gamma l)$ is the power gain of the optical amplifier with a distributed gain γ and length l and

$$\mu = \frac{N_2}{N_2 - N_1 \frac{g_2}{g_1}} \tag{7.8.11}$$

is the atomic inversion factor of the transition. It accounts for the larger value of N_2, and hence larger spontaneous emission power, in atomic

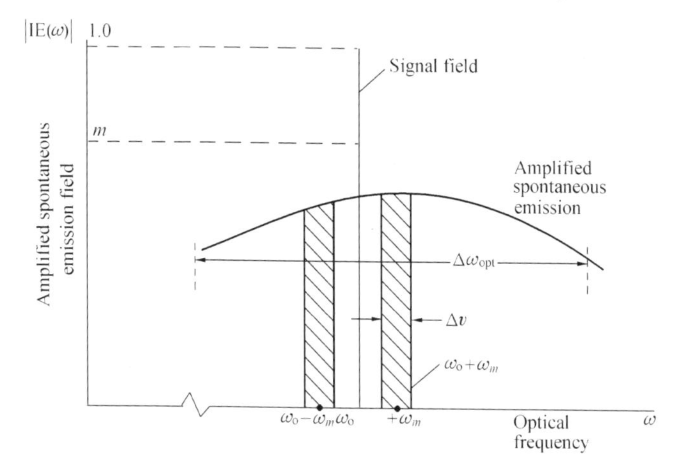

Figure 7.21 (b) At the output of the laser, the amplified spontaneous emission fields near $\omega_0 + \omega_m$ and $\omega_0 - \omega_m$ will. each, beat (mix) at the detector with the amplified optical signal field at ω_0 to yield radio frequency (RF) currents with frequencies near ω_m. These currents, which occupy a spectral width of $\Delta\omega_{opt}$, can not be separated from those due to intentional (signal) modulation of the intensity at ω_m and thus constitute RF noise.

(amplifier) systems in which $N_1 \neq 0$. [1]

If we denote the output optical power at ω_2 as S and that of the spontaneous emission at ω_1 as F, then the beat current component with a frequency $\omega_m = \omega_2 - \omega_1$ is (see Equation 7.8.4)

$$i = \frac{Se\eta}{hv}\left[1 + \frac{F}{S} + 2\sqrt{\frac{F}{S}}\cos(\omega_m t + \Phi_{\text{ASE}} - \Phi_s)\right] \quad (7.8.12)$$

where ϕ_{ASE} and ϕ_S are the phases of the ASE field and the signal optical field, respectively. The mean-squared beat current is then

$$\overline{(i^2)}_{\text{ASE-singnal}} = 2FS\left(\frac{e\eta}{hv}\right)^2 \quad (7.8.13)$$

which, using(7.8.10) and putting $\Delta v_{\text{opt}} = \Delta\omega_{\text{opt}}/2\pi = 2\Delta f_{sig}$, yields [2]

$$\overline{(i^2)}_{\text{ASE-signal}} = \frac{4e^2\eta^2 S(G-1)\mu\Delta f_{\text{sig}}}{hv} \quad (7.8.14)$$

In the remainder we will drop the subscript "sig" and use Δf only.

Consider an optical in-line amplifier as shown in Figure 7.22. The input signal power is S_0, and it enters the amplifier in a *single* transverse (usually the fundamental) fiber mode. The amplified output signal is GS_0, while F_0, as given by (7.8.10), represents the (optical) amplified spontaneous emission power at the output, which is generated within the amplifier in a band Δv. If we were to detect the signal at the input to the amplifier, the main noise contribution would, in an ideal case, i.e., a noiseless receiver, be that of the signal shot noise so that the signal-to-noise power ratio (SNR) at the input to the amplifier is

$$SNR_{in} = \frac{\left(\frac{S_0 e}{hv}\right)^2}{2e^2\frac{S_0}{hv}\Delta f} = \frac{S_0}{2hv\Delta f} \quad (7.8.15)$$

---

[1] In a laser the gain per pass is given by $G = \exp[a(N_2 - N_1)L_{\text{amp}}]$ where L_{amp} is the length and a is a constant depending on the atoms. Alarge N_1 thus causes a larger N_2 for a given gain. The SE power is proportional to N_2.

[2] Two ASE frequency bands, each with a width Δv_{sig}, one above and one below the signal frequency contribute incoherently to the beat power so that the effective $\Delta v_{\text{opt}} = 2\Delta f_{\text{sig}}$.

where we assume 100% detection efficiency $\eta = 1$ for simplicity.

The detected signal "power"① at the output is

$$\overline{(i^2)}_{\text{out}} = \left(\frac{GS_0 e}{hv} \right)^2 \tag{7.8.16}$$

while the noise power is that of the ASE-signal noise (7.8.14) and the shot noise

$$\overline{(i_{\text{shot}}^2)}_{\text{out}} = \frac{2e^2 GS_0}{hv} \Delta f \tag{7.8.17}$$

The noise current component that is due to beating of ASE frequencies with themselves is proportional to F_0^2 and can be made to be negligible compared to the ASE-signal current if the signal power $S(z)$ is not allowed to drop too far and/or by optical filtering. We have neglected for similar reasons the shot noise due to the ASE. The (S/N) ratio at the output of the amplifier is thus

$$SNR_{\text{out}} = \frac{\left(\dfrac{GS_0 e}{hv} \right)^2}{\dfrac{2e^2 GS_0}{hv} \Delta f + \dfrac{4e^2 G(G-1) S_0 \mu \Delta f}{hv}} \tag{7.8.18}$$

where we assumed a 100 percent detector quantum efficiency. For large gain $G \gg 1$, the second term in the denominator of (7.8.18), dominates, and

$$SNR_{\text{out}} \approx \frac{S_0}{4\mu hv \Delta f} \tag{7.8.19}$$

The ratio of the input (SNR) to the output value is thus

$$\frac{SNR_{\text{in}}}{SNR_{\text{out}}} \approx 2\mu$$

① The "power" everywhere is taken as the mean square of the current. Since our final results involve only (signal-to-noise) power ratios, this procedure is justified.

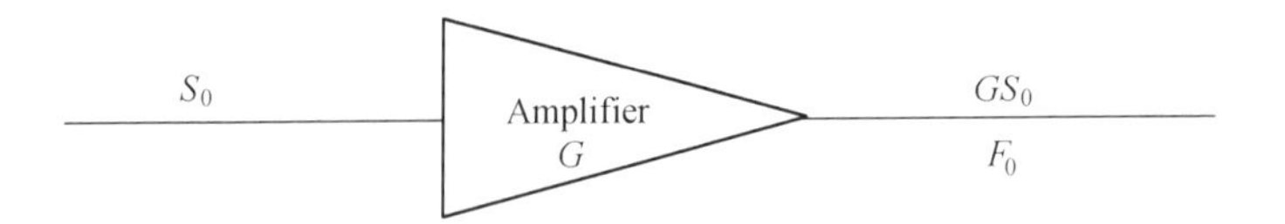

Figure 7.23 An optical amplifier with a power gain G and an input signal power S_0. F_0 is the total power of the amplified spontaneous emission (ASE) at the output of the amplifier in the appropriate bandwidth Δv.

which in an ideal, four-level ($N_1 = 0, \mu = 1$) amplifier is equal to 2. The single high-gain optical amplifier will thus degrade the *SNR* of the detected output by a factor of 2(3 db). We recall that this degradation is tolerated only in order to save the signal from the , far worse, fate of succumbing, in its attenuated state, to the noise of the receiver. An experimental verification of the 3 db limit is shown in Figure 7.21(a).

In a very long (100 km) fiber link, we will need to amplify the signal a number of times We will consequently develop in what follows a formalism for treating systematically cascades of amplifiers.

A generalization of the expression (7.8.18) for the *SNR* of the detected signal at an arbitrary point z along the link is to write

$$SNR(z) = \frac{\left[\frac{eS(Z)}{hv}\right]^2}{\frac{2e^2 S(z)\Delta f}{hv} + \frac{4e^2 F(z) S(z)}{(hv)^2} + \frac{4kT_e \Delta f}{R}} \tag{7.8.20}$$

where the last term in the denominator represents the mean-squared thermal noise current of the receiver (at point z) whose effective noise temperature is T_e. R is the output impedance of the detector including the receiver's input impedance. Equation (7.8.20) neglects, again, the shot noise due to the ASE, the ASE-ASE beat noise, and intensity fluctuation noise of the source laser. If the signal power $S(z)$ can be maintained above a certain level by repeated amplification, we can neglect the receiver noise term. Under these realistic circumstances, the *SNR* expression (7.8.20) becomes

$$SNR(z) = \frac{S^2(z)}{2S(z)h\nu\Delta f + 4S(z)F(z)} \tag{7.8.21}$$

$S(z)$ is the signal power at z, while $F(z)$ is the total ASE power at z originating in *all* the preceding amplifiers $(z' < z)$.

Let us next consider the realistic scenario of a long fiber with amplifiers employed serially at fixed and equal intervals (z_0), as illustrated in Figure 7.24.

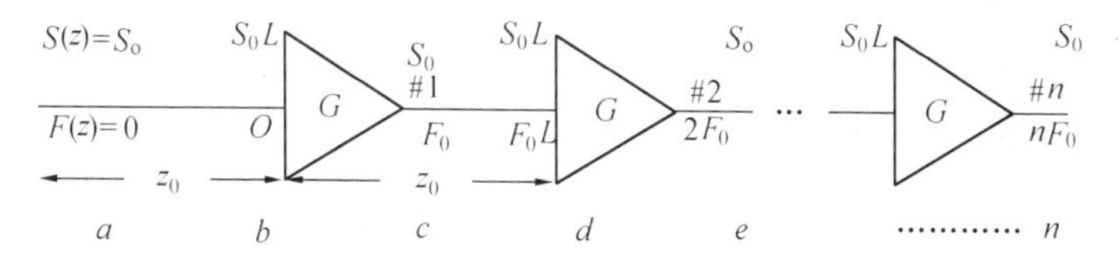

Figure 7.24 Afiber link with periodic amplification. The spontaneous emission power F(z) and the signal power S(z) at the amplifiers' input and output planes are indicated.

The signal power level $S(z)$ at the fiber input and at the output of each amplifier is S_0. The signal is attenuated by a factor of $L \equiv \exp(-\alpha z_0)$ in the distance z_0 between amplifiers and is boosted back up by the gain $G = L^{-1} = e^{\alpha z 0}$ at each amplifier to the initial level S_0. The spontaneous emission power $F(z)$ is attenuated by a factor L between two neighboring amplifiers and increases by an increment of F_0 at the output of each amplifier. We employ Equation (7.8.10) to calculate the SNR of the detected current at the output of the nth amplifier. Assuming $G \geqslant 1$, the result is

$$\text{SNR}_n = \frac{S_0}{2h\nu\Delta f[1 + 2n\mu(e^{\alpha z_0} - 1)]} \tag{7.8.22}$$

where, because of the high short noise and ASE levels, we neglected the thermal receiver noise. When $\exp(\alpha z_0) = G \gg 1$, we find a z^{-1} (more exactly an n^{-1}) dependence of the SNR rather than the $\exp(-\alpha z)$ dependence of a fiber without amplification in which the main noise mechanism is shot noise. The physical reason for this difference is that the repeated amplification keeps the signal level high as well as the level of the signal-ASE beat noise. The latter is kept well above the signal shot nois. A fixed amount of beat noise power is thus added at each stage leading to the inverse distance dependence of the SNR.

Equation(7.8.22) suggests that the SNR at z can be improved by reducing z_0, i. e., by using smaller intervals between the amplifiers which, of course, entails reducing the gain $G = \exp(\alpha z_0)$ of each. Let us take the limit of Equation (7.8.22) as $z_0 \to 0$, i. e., the separation between amplifiers tends to zero. In this limit the whole length of the fiber acts as a distributed amplifier with a gain constant $g = \alpha$, just enough to maintain the signal at a constant value. Since $S(z)$ is a constant, we need only evaluate the ASE optical power $F(z)$ in order to obtain, using (7.8.21), an expression for the SNR at z. To find how much noise power is added by the amplifying fiber, we consider a differential length dz. It may be viewed as a discrete amplifier with a gain of $\exp(g dz)$ so that its contribution to $F(z)$ is given by (7.8.10) as

$$dF = (e^{g(dz)} - 1)\mu h\nu \Delta f \tag{7.8.23}$$

or

$$\frac{dF}{dz} = g\mu h\nu \Delta f, \quad F(z) = g\mu h\nu \Delta f z \tag{7.8.24}$$

where, since no spontaneous emission is present at the input, we used $F(0) = 0$. Using (7.8.24) in (7.8.20) and taking $S(z) = S_0$, $g = \alpha$ results in

$$SNR(z) = \frac{S_0}{2[1 + 2\mu\alpha z]h\nu\Delta f} \tag{7.8.25}$$

We can also obtain (7.8.25) as the limit of (7.8.22) when $z_0 \to 0$. It is interesting to compare the (ideal) distributed amplifier to the discrete amplifier case of Equation(7.8.22)

$$(SNR)(z) = \frac{S_0}{2[1 + 2(z/z_0)\mu(e^{\alpha z 0} - 1)]h\nu\Delta f} \tag{7.8.26}$$

where we used $G = \exp(\alpha z_0)$ and $n = z/z_0$.

Figure 7.25 shows plots of the ideal continuous amplification case described by Equation (7.8.25) as well as two cases of discrete amplifier cascades [Equation (7.8.22)]. The advantage of continuous amplification compared to, say, amplification every α^{-1} is seen to be less than 2 db so that the latter may be taken as a practical optimum configuration. In a low-loss optical fiber, say with $\alpha = 0.2$ dB/km, the distance between amplifiers that are placed every α^{-1} km would be 21.7 km, Figure 7.26 shows the *SNR* of the detected signal along a realistic link for the case of (a) continuous amplification; (b) discrete amplifiers spaced by $z_0 = \alpha_0^{-1}$; and (c) for the case of no amplification at all. The launched power is $P_0 = 5$ mW, $\lambda = 1.55$ μm, $\Delta f = 10^9$Hz, and $\alpha = 0.2$ dB/km. Curve (b) is to be read only at multiples of $z = \alpha^{-1} =$ 21.7 km, which are the output planes of the optical amplifiers. Curve(c) assumes detection with a receiver with $T_e = 725$ K($F = 4$ dB) and an input impedance of 1 000 Ω.

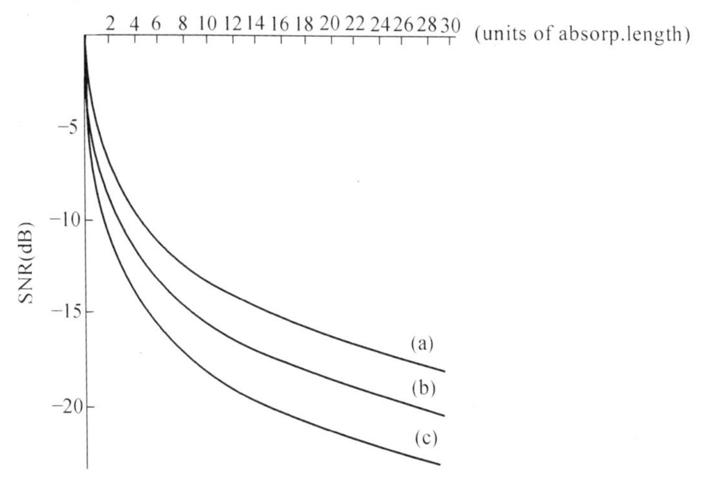

Figure 7.25 A universal plot of the degradation of the SNR compared to the initial ($z = 0$) value in the cases of (a) continuous amplification ($g = \alpha$)($\mu = 1$); (B) periodic amplification every $z_0 = \alpha^{-1}$($z' = 1, 2, 3...$), ($\mu = 1$), (curve is to be read only at $z' = 1, 2, 3...$); and . (c)periodic amplification every $z_0 = 2\alpha^{-1}$($z' = 1, 2, 3...$), ($\mu = 1$), (curve is to be read only at $z' = 2, 4, 6...$).

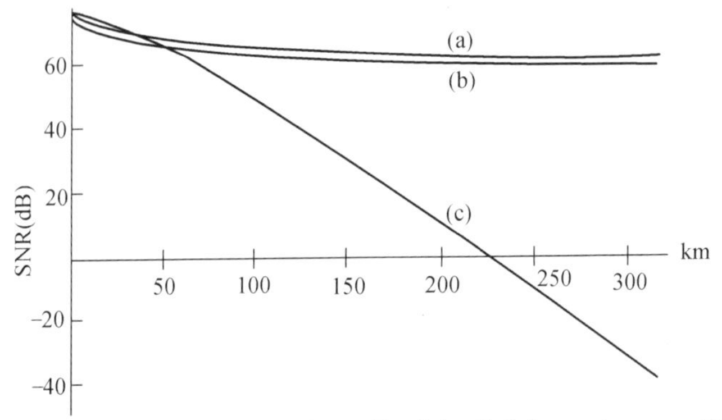

Figure 7.26 SNR of detected signal in a fiber link with (a) a continuous amplifier $g = \alpha$, ($\mu = 1$); (b) discrete amplifiers employed every absorption length $\alpha^{-1} = 21.7$ km (0.2 dB/km fiber loss), ($\mu = 1$) (curve is to be read only at multiples of 21.7 km); and (c) no optical amplification and detection with a receiver with a noise figure of 4 dB. The power launched into the fiber is 5 mW, the fiber loss is 0.2 dB/km, $\lambda = 1.55$ μm, the detection banwidth is $\Delta f = 10^9$, and the detector load impedance is 1 000 ohms.

We note that if, for example, we need to maintain a SNR exceeding 50 db, we must use a fiber link shorter than 100 km if no amplifier is used, but if laser amplifiers are used every, say, $z_0 = \alpha^{-1}$ ($= 21.7$ km), fiber length in excess of 1 000 km can be employed.

Serious consideration has also been devoted to the use of semiconductor (SC) laser amplifiers [31]. These are identical in their construction to semiconductor laser oscilators, which are discussed in Chapter 5, except that the facets are coated with. antireflection layers to reduce optical feedback and thus prevent oscillation from taking place. The main advantage is the possibility of very large gains (> 20 dB) in a short ($< \sim 400$ μm) semiconductor chip. The main disadvantages of the SC amplifier compared to the fiber amplifier is the presence of residual reflections and the resulting need for optical isolators. The presence of even minute reflection ($R < 10^{-5}$) can give rise to instabilities and excess noise in the source laser oscillator. Impressive results, however, have been demonstrated [36].

The above discussion centers on the use of optical amplification in long distance transmission of data. A second class of applications, no less important, is that of distribution systems with a very large number of subscribers. The use of optical amplifiers makes it possible to maintain the power arriving at a subscriber's premises at sufficiently high levels so as not to be degraded by the receiver noise. The number of subscribers that can thus be served by a single laser can be increased by anywhere from 1 to 3 orders of magnitude. This topic is the subject of Problem 7.13.

Problems

7.1 Show that the total output shot-noise power in a photomultiplier including that originating in the dynodes is given by

$$\overline{(i_N^2)} = G^2 2e(\bar{i}_c + i_d)\Delta f \frac{1 - \delta^{-N}}{1 - \delta^{-1}}$$

where δ is the secondary-emission multiplication factor and N is the number of stages.

7.2 Calculate the minimum power that can be detected by a photoconductor in the presence of a strong optical background power P_B.

Answer:

$$(P_s)_{\min} = 2\left(\frac{P_B h\nu \Delta f}{\eta}\right)^{1/2}$$

7.3 Derive the expression for the minimum detectable power using a photoconductor in the video mode(that is , no local-oscillator power) and assuming that the main noise contribution is the generation-recombination noise. The optical field is given by $e(t) = E(1 + \cos \omega_m t)\cos \omega t$, and the signal is taken as the component of the photocurrent at ω_m.

7.4 Derive the minimum detectable power of a Ge:Hg detector with characteristics similar to those described in Section 7.7 when the average current is due mostly to blackbody radiation incident on thephotocathode. Assume $T = 295$ K, an acceptance solid angle $\Omega = \pi$ and a photocathode area of 1 mm^2. Assume that the quantum yield η for blackbody radiation at $\lambda < 14$ μm is unity and that for $\lambda > 14$ μm. $\eta = 0$[*Hint*: Find the flux of photons with wavelengths 14 μm > λ > 0 using blackbody radiation formulas or, more easily, tables or a blackbody "slide rule."]

7.5 Find the minimum detectable power in Problem 7.4 when the input field of view is at $T = 4.2$ K.

7.6 Derive Equations(7.7.15) and (7.7.16).

7.7 Show that the transit time reduction factor$(1 - e^{-i\omega_m \tau_d})/i\omega_m \tau_d$ in Equation (7.8.7) can be written as

$$\alpha - i\beta$$

where

$$\alpha = \frac{\sin \omega_m \tau_d}{\omega_m \tau_d} \qquad \beta = \frac{\sin \omega_m \tau_d}{\omega_m \tau_d}$$

Plot α and β as functions of $\omega_m \tau_d$.

7.8 Derive the minimum detectable optical power for a photodiode operated in the heterodyne mode. (*Answer*: $P_{\min} = h\nu\Delta\nu/\eta$)

7.9 Discuss the limiting sensitivity of an avalanche photodiode in which the noise increases as M^2. Compare it with that of a photomultiplier. What is the minimum detectable power in the limit of $M \gg 1$, and of zero background radiation and no dark current ?

7.10 Derive an expression for the magnitude of the output current in a heterodyne detection scheme as a function of the angle θ between the signal and local-oscillator propagation directions. Taking the aperture diameter(see Figure 7.6) as D, show that if the output is to remain near its maximum ($\theta = 0°$) value, θ should not exceed λ/D. [*Hint*: You may replace the lens in Figure 7.6 by the photoemissive surface.] Show that instead of Equation (7.5.4) the current from an element $\mathrm{d}x\ \mathrm{d}y$ of the detector is

$$\mathrm{d}i(x, t) = \frac{P_L e\eta}{h\nu(\pi D^2/4)}\left[1 + 2\sqrt{\frac{P_s}{P_L}}\cos(\omega t + kx\sin\theta)\right]\mathrm{d}x\mathrm{d}y$$

The propagation directions lie in the $z - x$ plane. The contribution of $\mathrm{d}x$ $\mathrm{d}y$ to the (complex) signal current is thus

$$dI_s(x,t)=\frac{2\sqrt{P_sP_L}}{h\nu(\pi D^2/4)}e^{ikx\sin\theta}dx\,dy$$

7.11 show that for a Poisson distribution (footnote 9) $\overline{(\Delta N)^2}=\overline{N}$.

7.12 Calculate the smallest temperature increment that can be measured by an infrared detector "looking" at an object at $T=350$ K with a background temperature of $T=300$ K. The detector has a $D_\lambda^*=10^{11}$ cm $(\mathrm{Hz})^{1/2}/\mathrm{W}$ and responds to $\Delta\lambda\sim 0.1\ \lambda$ centered on $\lambda=10\ \mu$m. The output circuit bandwidth is $\Delta f=10^3$ Hz.

7.13 Assume a fiber distribution network fed by a single semiconductor laser at $\lambda=1.55\ \mu$m with a power output $P_0=10$ mW. The power is divided into N branches, amplified by an optical fiber amplifier(in each branch) and then divided again into M branches.

Determine the maximum number of "subscribers" NM that can be serviced by the system assuming: $\Delta f=10^9$ Hz; R (receiver input impedance) is 10^3 ohms, $T_e=1\ 000$ K; and a minimum SNR at the subscriber of 42 db. The maximum power level at the output of the amplifiers is 10 mW.

References

1. Yariv, a., *Quantum Electronics*, 3d ed. New York: Wiley, 1988, p.54.

2. Engstrom, R. W., "Multiplier phototube characteristics: Application to low light levels," *J. Opt. Soc. Am.* 37:420, 1947.

3. Sommer, A. H., *Photo-Emissive Materials*. New York: Wiley, 1968.

4. Forrester, A. T., "Photoelectric mixing as a spectroscopic tool," *J. Opt. Soc. Am.* 51:253, 1961.

5. Siegman, A. E., S. E. Harris and B. J. McMurtry, "Optical heterodyning and optical demodulation at microwave frequencies." In *Optical Masers*, J. Fox, ed. New York: Wiley, 1963, p.511.

6. Mandel, L., "Heterodyne detection of a weak light beam," *J.*

Opt. Soc. Am. 56:1200,1966

7. Chapman, R. A., and W. G. Hutchinson, "Excitation spectra and photoionization of neutral mercury centers in germanium," *Phys. Rev.* 157:615,1967.

8. Teich, M. C., "Infrared heterodyne detection," *Proc. IEEE* 56:37,1968.

9. Buczek, C., and G. Picus, "Heterodyne performance of mercury doped germanium," *Appl. Phys. Lett.* 11:125,1967.

10. Lucovsky, G., M. E. Lasser, and R. B. Emmons, "Coherent light detection in solid-state photodiodes," *Proc. IEEE* 51:166,1963.

11. Riesz, R. P., "High speed semiconductor photodiodes," *Rev. Sci. Instr.* 33:994,1962.

12. Anderson, L. K., and B. J. McMurtry, "High speed photodetectors," *Appl. Opt.* 5:1573,1966.

13. D'Asaro, L. A., and L. K. Anderson, "At the end of the laser beam, a more sensitive photodiode," *Electroniscs*, May 30,1966, p.94.

14. Shockley, W., "Hot electrons in germanium and Ohm's law," *Bell Syst. Tech. J.* 30:990,1951.

15. McKay, K. G., and K. B. McAfee, "Electron multiplication in silicon and germanium," *Phys. Rev.* 91:1079,1953.

16. McKay, K. G., "Avalanche breakdown in silicon," *Phys. Rev.* 94:877,1954.

17. McIntyre, R., "Multiplication noise in uniform avalanche diodes," *IEEE Trans. Elect. Devices* ED-13:164,1966.

18. Lindley, W. T., R. J. Phelan, C. M. Wolfe, andA. J. Foyt, "GaAs Schottky barrier avalanche photodiodes," *Appl. Phys. Lett.* 14:197,1969.

19. Nahory, R. E., M. A. Pollack, E. D. Beebe, and J. C. DeWinter, "Continuous operation of a 1.0 μm wavelength $GaAs_{1-x}Sb_x/Al_yGa_{1-y}$-$As_{1-x}Sb_x$ double-heterostructure injection laser," *Appl. Phys. Lett.* 28:19,1976.

20. B-Chaim, N., K. Y. Lau, I. Ury, and A. Yariv, "High speed GaAlAs/GaAs photodiode on a semi-insulating GaAs substrate," *Appl. Phys. Lett.* 43:261, 1983.

21. Wang, S. Y., D. M. Bloom, and D. M. Collins, "20-GHz bandwidth GaAs photodiode," *Appl. Phys. Lett.* 42:190, 1983.

22. Wang, S. Y., and D. M. Bloom, "100 GHz bandwidth planar GaAs Schottky photodiode," *Elec. Lett.* 19:554, 1983.

23. Valdmanis, J. A., G. Mourou, and C. W. Gabel, "Picosecond electro-optic sampling system," *Appl. Phys. Lett.* 41:211, 1982.

24. Kolner, B. H., D. M. Bloom, and P. S. Cross, "Characterization of high speed GaAs photodiodes using a 100-GHz electro-optic sampling system," 1983 Conference on Lasers and Electro-optics, paper ThGl.

25. Kinch, M. A. and A. Yariv, "Performance limitations of GaAs/GaAlAs infrared superlattices," *Appl. Phys. Lett.* 55:2093, 1990.

26. R. E. Buregess, "Fluctuations in the number of electrons and holes in semiconductors." *Proc. Phys. Soc. B* 68:661, 1955.

27. Van Der Ziel, a., *Fluctuation Phenomena in Semiconductors* CH.4, New York: Academic Press, 1959.

28. Long, D. "On generation-recombination noise in infrared detector materials," *Infrared Phys.* 7:167, *Pergamon Press*, 1967.

29. L. C. Chiu, J. S. Smith, S. Margalit, A. Yariv, and A. Y. Cho, "Application of internal photoemission fromquantum well heterojunction superlattices to infrared detectors" *Infrared Phys.* 23(2):93, 1983.

30. B. F. Levine, C. G. Bethea, G. Hasnain, J. Walker, and R. J. Malek, "High detectivity $D* = 10^{10}$ cm $\sqrt{Hz}/W$ GaAs/AlGaAs multiquantum well $\lambda = 8.3\mu m$ Infrared Detector," *Appl. Phys. Lett.* 53: 2196. 1988.

31. Simon J. C., "Semiconductor laser amplifier for single mode optical fiber communications," *J. Opt. Commun.* 4:51, 1983.

32. Mears, R. J., L. Reekie, I. M. Jauncey, and D. N. Payne, "Low

noise erbiumdoped fiber amplifier operating at 1.54 mm," *Elec. Lett*. 23: 1026, 1987.

33. Hagimoto, K., et al. "A 212 Km non-repeatered transmission experiment at 1.8 Gb/s using LD pumped Er^{3+}-doped fiber amplifiers in an Im/diect-detection repeater system," inProceedings of the Optical Fiber Conference, Houston, TX, postdeadline Paper PD15, 1989.

34. Olshansky, R., "Noise figure for Er-doped optical fibre amplifiers," *Elec. Letts*. 24: 1363, 1988.

35. Payne, David N., "Tutorial session abstracts," Optical Fiber Communication (OFC 1990) Conference, San Francisco 1990. Published by Opt. Soc. of Am., Washington, D.C.

36. See, for example, Eisenstein, G., U. Koren, G. Raybon, T. L. Koch, M. Wiesenfeld, M. Wegener, R. S. Tucker, and B. I. Miller, "Large-signal and small-signal gain characteristics of 1.5 mm quantum welloptical amplifiers," *Appl. Phys. Lett*. 56: 201, 1990.

Supplementary Reterence

37. Boyd, R. W. *Radiometry and the Detection of Optical Radiation*. New York: Wiley, 1983.

第7章　光辐射的探测

New Words and Expressions

charge carrier	*n*. 载流子
photoconductive detector	*n*. 光电导探测器
thermo couple	*n*. 温差电偶
photomultiplier	*n*. 光电倍增器
photodiode	*n*. 光电二极管
avalanche photodiode	*n*. 雪崩光电二极管
valence band	*n*. 价电子带
sinusoidal	*adj*. 正弦的
shot noise	*n*. 发射噪声
impinge	*v*. 撞击

work function	*n*. 功函
kinetic	*adj*. (运)动的,运动(学)的
ultraviolet	*adj*. 紫外线的
dynode	*n*. 倍增极
band width	*n*. 带宽
denominator	*n*. 分母
substitution	*n*. 替换
extraneous	*adj*. 外来的
proportionality constant	*n*. 比例常数
signal-to-noise power ratio	*n*. 信噪比
incident photon	*n*. 入射光子
photoelectron	*n*. 光电子
photo current	*n*. 光生电流
granularity	*n*. 间隔尺寸　粒度
incident optical intensity	*n*. 入射光强
ionization	*n*. 电离,离子化
mercury-doped germanium	*n*. 掺杂水银的锗
potential barrier	*n*. 势垒
probability function	*n*. 概率函数
complex conjugate	*n*.复共轭
potential energy	*n*. 势能
conductivity	*n*. 传导率
depletion layer	*n*. 耗尽层
permittivity	*n*. 介电常数
substrate	*n*. 底层
complex notation	*n*. 复数符号
parasitic inductance	*n*. 寄生电感
pilosecond	*n*. 皮秒
majority carrier	*n*. 多数载流子
intrinsic high resistivity	*n*. 固有的高电阻

frequency response	*n*. 频率响应
Laser transition	*n*. 激光跃迁
pumping field	*n*. 泵浦场
continuum	*n*. 连续统一体
in-line amplifier	*n*. 同轴放大器
single transverse fiber mode	*n*. 单个横向(光纤)模
succumb	*vi*. 屈服
attenuate	*v*. 削弱
impendence	*n*. 阻抗
discrete	*adj*. 离散的
low-loss optical fiber	*n*. 低损耗光纤
facet	*n*. 小平面
antireflection layer	*n*. 抗反射膜
residual	*n*. 残余量

NOTES

1. This excitation involves a transition of the electron from some initial bound state, say a to a final state(or a group of states)b in which it is free to more and contribute to the current flow.

这种激活包含了电子从某些被称为 a 态的原始束缚状态,到一种(或多种)最终状态 b 的转换,在 b 态中多灵敏电子自由运动引起电流流动。

2. The dynodes are kept at progressively higher potentials with respect to the cathode, with a typical potential difference between adjacent dynodes of 100 volts.

相对与于阴极,倍增管的各倍增极的电位逐次递增,两相邻倍增极之间的典型电位差值为 100 伏。

3. The current i_d the so-called dark current, which is due to random thermal excitation of electrons from the surface as well as to excitation by cosmic rays and radioactive bombardment.

电流 i_d 即所谓的暗电流,它一部分由表面电子的随机热激发产生,一部分由宇宙射线和放射性轰击激发产生。

4. If the output of the detector is limited by filtering to a band width Δv centered on W_m, it contains a shot-noise current , which, according to (1.1), has a mean-squared amplitude.

如果经滤波使得探测器的输出限制在以 WM 为中心,带宽为 Δv 的范围,就包含了发射噪声电流,据(1.1)式,该噪声电流具有如下的均方根振幅。

5. Defining the minimum detectable number of quanta as that for which the rms fluctuation in the number of emitted photoelectrons equals the average value.

当所发射光电子数量波动值的均方根与平均值相等时,将其定义为最小探测量子数。

6. Even if the incident optical flux were constant in time, the generation of individual carriers by the flux would constitute a random process.

即使入射光通量在某一时段是恒定的,其产生单独载流子的过程仍然构成了一个随机过程。

7. This abrupt transition results usually from diffusing suitable impurity atoms into a substrate of a semiconductor with the opposite type of conductivity.

这种突变通常是由合适的杂质原子扩散进入到具有相反电导率的半导体的底层而形成的。

8. For high enough electric fields, the drift velocity of carriers in semiconductors tends to saturate, so the constant velocity assumption is not very far from reality even for a nonuniform field distribution, such as that shown in figure 1.11(e), provided the field exceeds its saturation value over moat of the depletion layer length.

对于足够强的电场,半导体中载流子的漂移电压趋向饱和,因此,只要电场超过其饱和值(该值高于绝大多数耗尽层宽度),恒定的电压消耗和实际情况相关就不是很大,甚至对于一个不均匀的电场分布,也是如此,如图 1.11(e)所示。

8

A Classical Treatment of Quantum Optics, Quantum Noise, and Squeezing

8.1 Introduction

Some of the most important and elegant phenomena related to optical waves and their detection can only be explained using the extension of the formalism of quantum mechanics to optics, i.e., "quantum optics." Important topics that fit this category involve amplitude and phase noise (fluctuations), the statistics of photo-generated electrons, and the new field of nonlinear squeezing. These areas are just too important to forego. Somewhat to my surprise, I found that by asking the student to accept just *one* result from quantum mechanics, it is possible to treat all the above-mentioned phenomena classically and obtain results that agree with those of quantum optics.

8.2 The Quantum Uncertainty Goes Classical

One of the better-known uncertainties of quantum mechanics relates to the simultaneous measurement of the position(x) and momentum(p) of a particle and decrees that the product of the uncertainties Δp and Δx

must obey [1]

$$\Delta p \Delta x \geqslant \hbar/2$$

where $\hbar \equiv h/2\pi$ and $h = 6.623\ 77 \times 10^{-34}$ joule-sec is the Planck's constant. These fundamental uncertainties extend to optical measurements such as measurements of the amplitudes and phases of optical fields. Their proper study involves the elegant formalism of quantum optics [1,2,3]. Since we have foresworn quantum mechanics in this book, we cannot approach this subject from first principles. We can, however, appreciate many of the consequences and even obtain numerically correct results for the important scenarios by accepting just *one* basic consequence of quantum mechanics: that of the *uncertainty principle*.

The Uncertainty Principle

Let us represent the classical monochromatic electric field of some mode oscillating in a resonator as

$$e(t) = |E|\cos(\omega t + \beta) = Re(E\ \exp[i\omega t)] \qquad (8.2.1)$$

where

$$E = |E|\exp\ (i\beta) = E_1 + iE_2 \qquad (8.2.2)$$

is the complex phasor representing the field. It is shown in Figure 8.1(a) as a vector in the complex **E** plane of length $|E|$ and projections E_1 and E_2 along the real and imaginary axes. respectively.

According to quantum mechanics [1.2]. the complex amplitude E in (8.2.2) *cannot be* specified exactly. This uncertainty is represented in Figure 8.1(b) by means of the "uncertainty circle." The most probable position of the tip of the phasor E, on measurement, will be found near the center of the circle. The field phasor corresponding to the center of this circle is denoted as $\langle E \rangle$, the "expectation value" of E. The expectation value corresponds to the quantum mechanical ensemble average, that is, to the average of a large number of independent field determinations (measurements) under *identical* conditions. This

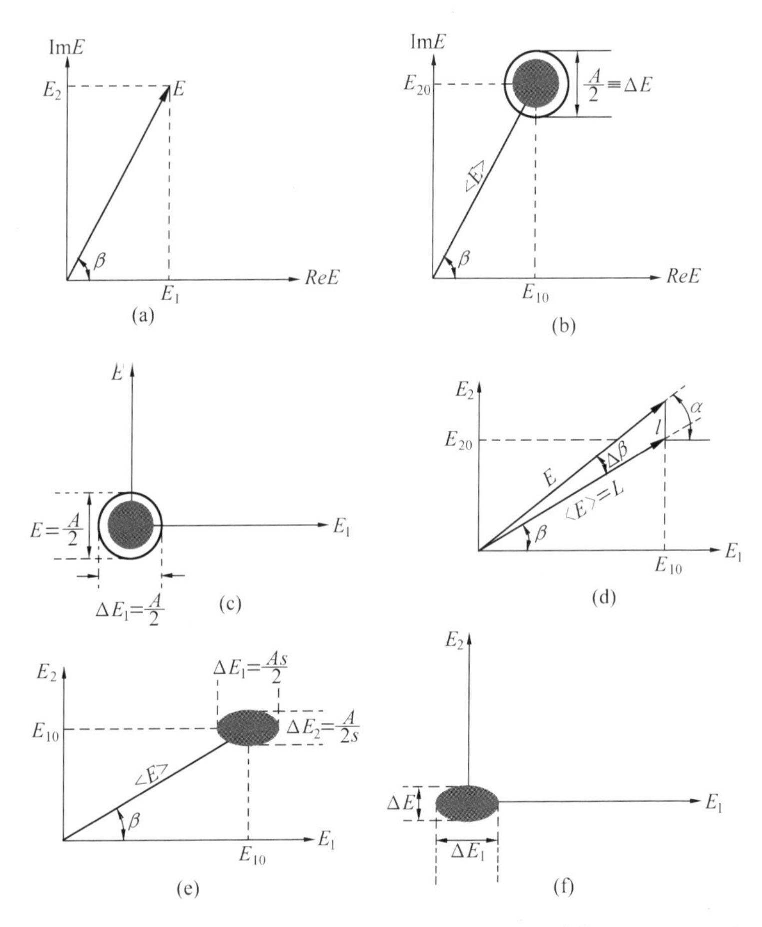

Figure 8.1 (a) A classical phasor representation of the optical field. (b) A representation of a coherent field that is consistent with quantum optics. In this special case, $\Delta E_1 = \Delta E_2$. (c) The electromagnetic field representation of an unexcited vacuum($n = 0$) optical mode. (d) An equivalent representation of the field with a random phasor added, vectorially, to the tip of the classical phasor. (e) A "squeezed" field. (f) A squeezed vacuum ($n = 0$) mode. The squeezing factor is $s > 1$.

expectation value obeys in all respects the same (Maxwell's) equations as its classical counterpart. There is even a theorem in quantum mechanics,

Ehrenfest's theorem [2], to prove it.

Repeated measurements of the projections of E, $E_1(t)$, and $E_2(t)$ will yield different results, and the results for $E(t) = E_1(t) + iE_2(t)$ will tend to cluster about the center of the circle, which is a graphical way to describe the most probable region in which the tip of $E(t)$ will fall. The uncertainty that results from this inherent quantum-imposed spread in the values of $E(t)$ can be thought of as *quantum noise*. We shall devote the rest of this chapter to a consideration of the consequences of this noise in optical measurements.

A classical *approximation* of the quantum physics is to write the basic monochromatic field of an electromagnetic mode as

$$e(t) = Re[E(t)\exp(i\omega t)] = E_1(t)\cos\omega t - E_2(t)\sin\omega t \tag{8.2.2a}$$

where the complex amplitude $E(t)$ is

$$E(t) = E_1(t) + iE_2(t)$$

and

$$\begin{aligned} E_1(t) &= \langle E_1(t)\rangle + \Delta E_1(t) = E_{10} + \Delta E_1(t) \\ E_2(t) &= \langle E_2(t)\rangle + \Delta E_2(t) = E_{20} + \Delta E_2(t) \end{aligned} \tag{8.2.3}$$

where $\langle E_{1.2}(t)\rangle \equiv E_{10.20}$ is the expectation value of $E_{1.2}(t)$. $\Delta E_1(t)$ and $\Delta E_2(t)$ represent the fundamental quantum mechanical uncertainties. They have zero mean $\langle \Delta E_{1.2}(t)\rangle = 0$, and are uncorrelated $\langle \Delta E_1 \Delta E_2\rangle = 0$. $\langle\ \rangle$ everywhere indicates ensemble (or temporal) averaging.

Before proceeding with a description of these fluctuations, we will, as is the practice in quantum optics, find it useful to relate the *mean* electric field of a mode to the mean number n of optical photons (quanta) in the mode. Taking the mode volume as V, the dielectric constant of the medium as ϵ. and using Equation (1.3.22) yields

$$\frac{\epsilon}{2}|\langle E\rangle|^2 V = n\hbar\omega = \text{Field Energy within volume } V$$

so that the mean field amplitude is expressible as

$$|\langle E\rangle| = \left(\frac{2\hbar\omega}{\in V}\right)^{1/2}\sqrt{n} \equiv A\sqrt{n}$$

$$A \equiv \left(\frac{2\hbar\omega}{\in V}\right)^{1/2} \tag{8.2.4}$$

If we define the measure of the fundamental quantum mechanical uncertainties as

$$\Delta E_1 = \langle (E_1 - E_{10})^2\rangle^{1/2} = \langle (\Delta E_1(t))^2\rangle^{1/2}$$

with a similar expression in which $1 \rightarrow 2$ for ΔE_2, then according to quantum mechanics [2]

$$\Delta E_1 \Delta E_2 \geqslant \frac{A^2}{4} \tag{8.2.5}$$

This is the only quantum mechanical result that we will use.

The output field of most laser oscillators is in the so-called *coherent state* [1, 2] in which the uncertainty is divided equally between two quadrature components E_1 and E_2:

$$\Delta E_1 = \Delta E_2 = \frac{A}{2} \tag{8.2.6}$$

or using the normalized dimensionless field, $x \equiv \frac{E}{A}$, we can rewrite (8.2.5) as

$$\Delta x_1, \Delta x_2 \geqslant 1/4 \tag{8.2.7}$$

and for the coherent state field

$$\Delta x_1 = \Delta x_2 = 1/2 \tag{8.2.8}$$

A profound consequence of the uncertainty relation Equation (8.2.5), is that it applies even to a mode that, classically, is *not excited*, that is, $n = 0$. This so-called *vacuum state*, which corresponds to (8.2.3) with $E_{10} = E_{20} = 0$, is illustrated by Figure 8.1(c).

Electromagnetic fields in which the uncertainties of the two quadrature components are unequal, i.e., $\Delta x_1 \neq \Delta x_2$, are called *squeezed states*. Such states have been produced recently using nonlinear optical techniques [4,5]. A squeezed electromagnetic field is illustrated in Figure 8.1(e). A squeezed vacuum field is shown in Figure 8.1(f). Instead of representing the field in terms of the quadrature amplitudes E_1 and E_2, we can use the description of Figure 8.1(d) in which a random phasor

$$\mathscr{l}(t) = |\mathscr{l}| e^{i\alpha} = \Delta E_1(t) + i\Delta E_2(t) \qquad (8.2.9)$$

is added vectorially to the average phasor $\langle E \rangle$.

The phase angle α of this fluctuation phasor measured from $\langle E \rangle$ is uniformly distributed (is equally likely to occur) between 0 and 2π while

$$\langle |\mathscr{l}|^2 \rangle = 2\langle \Delta E_1^2(t) \rangle = \frac{A^2}{2} \qquad (8.2.10)$$

Some of the consequences of the uncertainty relation (8.2.5) are explored in what follows.

The Energy of an Electromagnetic Mode

This energy is given classically by

$$\mathscr{E} = \frac{1}{2} \in V(EE^*) = \frac{1}{2} \in V[(L + |\mathscr{l}|\cos\alpha)^2 + |\mathscr{l}|^2 \sin^2\alpha] \qquad (8.2.11)$$

and choosing, without loss of generality, the direction of $\langle E \rangle$ as the real axis so that

$$E = L + |\mathscr{l}|\cos\alpha + i|\mathscr{l}|\sin\alpha \qquad L \equiv \langle E \rangle$$

$$\langle \mathscr{E} \rangle = \frac{\in V}{2} \langle (L^2 + 2L|\mathscr{l}|\cos\alpha) + |\mathscr{l}|^2\cos^2\alpha + |\mathscr{l}|^2\sin^2\alpha \rangle =$$

$$\frac{\in V}{2}(L^2 + \langle |\mathscr{l}|^2 \rangle) = \frac{\in V}{2}\left(A^2 n + \frac{1}{2}A^2\right) =$$

$$\hbar\omega\left(n + \frac{1}{2}\right)$$

$$\langle \Delta E_1(0,t) \rangle = \langle \Delta E_2(0,t) \rangle = 0 \qquad (8.2.12)$$

where we used the definition of A in (8.2.4) and $\langle \sin^2\alpha \rangle = \langle \cos^2\alpha \rangle = \frac{1}{2}$, as well as the fact that $\langle \cos\alpha \rangle = 0$, since α is distributed uniformly between 0 and 2π. It follows that in the case when the classical field is zero, e., $\langle E\rangle$ and n, according to (8.2.4), are zero, the mode energy is $\hbar\,\omega/2$. This is the so-called *zero point vibration energy of the mode*. It is one of the main consequences of quantum mechanics and it does *not* have a classical counterpart.

Uncertainty in Energy

The uncertainty of the mode energy $\mathscr{E}$ can be defined by

$$\langle(\Delta\mathscr{E})^2\rangle = \langle\mathscr{E} - \langle\mathscr{E}\rangle\rangle^2 = \langle\mathscr{E}^2\rangle - \langle\mathscr{E}\rangle^2$$

Using (8.2.11) for $\mathscr{E}$, taking $\langle\cos^n\alpha\rangle = 0$ for n odd, and neglecting terms 0 $(\Delta\ell/L^4)$, we obtain

$$\Delta\mathscr{E} \equiv \langle\Delta\mathscr{E}^2\rangle^{1/2} = \hbar\,\omega\sqrt{n}$$

or in terms of the number of photons, N, in the resonator

$$\Delta N = \frac{\Delta\mathscr{E}}{\hbar\,\omega} = \sqrt{n} \equiv \sqrt{\langle(N)\rangle} \qquad (\text{recall } n \equiv \langle N\rangle)$$

$$(\Delta N)^2 = \langle N\rangle = n \qquad (8.2.13)$$

where $\Delta N \equiv \langle(N-n)^2\rangle^{1/2}$.

The relation(8.2.13) between the mean square uncertainty in the number of photons N and the average number $\langle N\rangle$ *applies to a* **Poissonian** *probability distribution of the photon number* **N** [2,3]

$$p(N) = \frac{\langle N\rangle^N e^{-\langle N\rangle}}{N!} \qquad (8.2.14)$$

for the number of photons N in the resonator.

Phase Uncertainty

The uncertainty $\Delta\beta$ in the value of the field phase is obtainable from Figure 8.1(d)

$$(\Delta\beta)^2 \equiv \langle(\Delta\beta(t)^2\rangle = \left[\left(\frac{|\mathscr{A}\sin\alpha}{L}\right)^2\right] =$$

$$\frac{\langle|\mathscr{A}^2\rangle\langle\sin^2\alpha\rangle}{L^2} = \frac{\frac{1}{2}A^2\frac{1}{2}}{A^2 n} = \frac{1}{4n}$$

where we assumed $|\ell \ll L$ and used the fact that $\ell(t)$ and $\alpha(t)$ are not correlated.

Using (8.2.13) we obtain

$$\Delta N \Delta \beta = \frac{1}{2} \tag{8.2.15}$$

where $\Delta \beta \equiv \langle \Delta \beta(t)^2 \rangle^{1/2}$. This is a most important result and states the fundamental quantum mechanical limit on the simultaneous measurement of the phase (β) and excitation level N of an electromagnetic field.

Fluctuation of Photoelectron Number

If an optical wave is incident on a perfect photodetector whose area is A_{rea}, then for each incident and absorbed photon, ideally, one photoelectron is emitted. The resulting current is thus

$$i = \frac{ec \in |E|^2 A_{rea}}{2\hbar\omega} \tag{8.2.16}$$

where e is the absolute value of the electronic charge, and the power incident on the detector is $c \in |E|^2 A_{rea}$. If we use the dimensionless field $x \equiv E/A$ (A is defined in Eq. 8.2.4), the expression for the current becomes

$$\begin{aligned} i &= \frac{ecA_{rea}}{V}[x_1^2 + x_2^2] \cong \\ &\frac{ecA_{rea}}{V}(x_{10}^2 + x_{20}^2 + 2x_{10}\Delta x_1 + 2x_{20}\Delta x_2) = \\ &i_0 + \Delta i \end{aligned} \tag{8.2.17}$$

where $x_1 = x_{10} + \Delta x_1$, $x_2 = x_{20} + \Delta x_2$,

$$i_0 = \frac{ecA_{rea}}{V}(x_{10}^2 + x_{20}^2) = \frac{eC}{L_R} n \tag{8.2.18}$$

$$\Delta i = 2\frac{eC}{L_R}(x_{10}\Delta x_1 + x_{20}\Delta x_2) \tag{8.2.19}$$

where we used $x_{10}^2 + x_{20}^2 = |\langle E \rangle|^2 / A^2 = n$:

$$L_R = \frac{V}{A_{\text{rea}}} = \text{length of quantization volume used in Eq.} \qquad (8.2.4)$$

The total number of photoelectrons emitted during a time interval $\tau = L_R / C$ (corresponding to a bandwidth $B = 1/\tau$) is

$$N_e = \frac{i\tau}{e} = \frac{[i_0 + \Delta i(t)]\tau}{e}$$

$$\langle N_e \rangle = \frac{i_0 \tau}{e} = n$$

$$\langle \Delta N_e^2 \rangle = \frac{\langle (\Delta i)^2 \rangle r^2}{e^2} = 4\langle (x_{10}\Delta x_1 + x_{20}\Delta x_2)^2 \rangle =$$

$$4[x_{10}^2 \langle (\Delta x_1)^2 \rangle + x_{20}^2 \langle (\Delta x_2)^2 \rangle] =$$

$$4(x_{10}^2 + x_{20}^2)\langle (\Delta x_1)^2 \rangle = n$$

where we used $\langle \Delta x_1 \Delta x_2 \rangle = 0$, Equation(8.2.18), and $\langle \Delta i \rangle = 0$ as well as

$$\langle (\Delta x_1)^2 \rangle = \langle (\Delta x_2)^2 \rangle \equiv \langle (\Delta x)^2 \rangle = \frac{1}{4}$$

It follows that

$$\langle \Delta N_e^2 \rangle = \langle N_e \rangle \qquad (8.2.20)$$

so that the photoelectrons number N_e obeys Poisson's statistics. This was also shown in(10.10.18) to be true for the photons. Poissonian statistics apply to the case where each event (electron emission in this case) is completely random and independent so that there exist no correlations between individual emission events. This is exactly the scenario shown in Sections 10. 3 and 10. 4, which leads to shot noise in the current spectrum

$$S_i(\upsilon) = \frac{i_N^2(\upsilon)}{\Delta \upsilon} 2ei_0 \qquad (8.2.21)$$

where $S_i(\upsilon)$ is the spectral density of the photocurrent at the radio frequency υ. We have thus *demonstrated that the electronic shot noise in the photocurrent can be attributed to the quantum field fluctuations*.

Minimum Detectable Optical Power Increment

Most of the methods used to measure the power of an electromagnetic wave employ detectors that convert absorbed optical power to a proportional output current. In a perfect detector, which releases one electron into the external circuit for each absorbed photon, we have

$$i = \frac{Pe}{\hbar \omega} \tag{8.2.22}$$

where P is the optical power to be measured. Let our measurement of the current consist of accumulating it for T seconds. which results in a total number of collected electrons (or holes)

$$N_e(T) = \frac{iT}{e} = \frac{PT}{\hbar \omega} \Rightarrow P = \frac{\hbar \omega}{T} N_e(T)$$

The mean-squared uncertainty in the power measurement is thus given by

$$\langle (\Delta P)^2 \rangle = \left(\frac{\hbar \omega}{T}\right)^2 \langle (\Delta N_e)^2 \rangle = \left(\frac{\hbar \omega}{T}\right)^2 \quad \langle N_e(T) \rangle$$

where, in the last equality, we used (8.2.20). The average number of collected electrons during the observation interval T is given by

$$\langle N_e(T) \rangle = \frac{\langle P \rangle}{\hbar \omega} T$$

so that

$$\langle (\Delta P)^2 \rangle = \frac{\hbar \omega}{T} \langle P \rangle$$

Defining, arbitrarily, the minimum detectable power as that power at which the root-mean-squared fluctuation is equal to the average, i.e.,

$$\langle P \rangle_{\min} = \langle (\Delta P)^2 \rangle^{1/2}$$

we obtain

$$P_{\min} = \hbar \omega B \tag{8.2.23}$$

$B = 1/T$ is the bandwidth of the current integrating system. One often refers to the quantity $\hbar \omega B$ as the minimum detectable power. It is treated in some detail in Section 11.4.

8.3 Squeezing of Optical Fields

It is possible to take a coherent optical field, such as the output of a laser, and reduce the fluctuation of one of its quadrature components, say, ΔE_1, at the expense of ΔE_2, or vice versa. The resulting uncertainty diagram becomes elliptical, while the product $\Delta E_1 \Delta E_2$ retains its initial value of $A^2/4$. This is referred to as *squeezing* [4, 5]. Squeezing is accomplished, usually, by a nonlinear optical operation on the field. One of the most common methods of achieving squeezing employs degenerate optical parametric amplification, which is discussed in Section 8.6. In this, degenerate, case, the "pump" frequency is twice that of the "signal," $\omega_{pump} = 2\omega_{signal}$. To demonstrate how squeezing is accomplished in this situation, we will need, first, to revisit the topic of parametric amplification.

Optical parametric amplification is described by Equations (8.7.2). In the case of a degenerate phase-matched parametric amplifier, we have

$$\omega_1 = \omega_2 \equiv \omega \qquad \omega_3 = 2\omega$$

If we designate the complex amplitude of the field at ω as E, the amplifier equations become

$$\frac{dE}{dz} = -i\frac{g}{2}E^* \tag{8.3.1}$$

$$g = \frac{\omega d}{n_0}\sqrt{\mu/\epsilon_0}\,E_3 \tag{8.3.2}$$

where n_0 is the index of refraction.

The coupling constant g is complex, since it is proportional to the complex pump amplitude E_3. Without loss of generality, we can take the pump field at $z=0$ as

$$E_3(t) = -|E_3|\sin 2\omega t = \frac{i|E_3|}{2}(e^{i2\omega t} - e^{-i2\omega t}) \tag{8.3.3}$$

This choice determines the time reference. In this case, $E_3 = +i|E_3|$, the coupling constant g is imaginary, and (8.3.1) assumes the form

$$\frac{\mathrm{d}E}{\mathrm{d}z} = \frac{1}{2}|g|E^*$$

$$|g| = \frac{\omega d}{n_0}\sqrt{\mu/\epsilon_0}|E_3| \tag{8.3.4}$$

It is convenient to express the "signal" field at ω as in Equation (8.1.2) in terms of its quadrature amplitudes, E_1 and E_2

$$E = (E_1 + iE_2) \quad E_1 = \frac{1}{2}(E + E^*) \quad E_2 = \frac{-i}{2}(E - E^*) \tag{8.3.5}$$

so that the time dependent("signal") field is given by

$$\begin{aligned} e(t) &= Re[E\exp(i\omega t)] = Re[(E_1 + iE_2)\exp(i\omega t)] \\ &= E_1 \cos \omega t - E_2 \sin \omega t \end{aligned} \tag{8.3.6}$$

If we substitute the first of Equations (8.3.5) in (8.3.4), we obtain

$$\begin{aligned} \frac{\mathrm{d}E_1}{\mathrm{d}z} &= \frac{|g|}{2}E_1 \\ \frac{\mathrm{d}E_2}{\mathrm{d}z} &= -\frac{|g|}{2}E_2 \end{aligned} \tag{8.3.7}$$

so that at output of the parametric amplifier $z = L$

$$\begin{aligned} E_1(L) &= E_1(0)\exp\left(\frac{|g|}{2}L\right) = E_1(0)s \\ E_2(L) &= E_2(0)\exp\left(-\frac{|g|}{2}L\right) = \frac{E_1(0)}{s} \end{aligned} \tag{8.3.8}$$

where the squeezing factor s is defined, in accordance with Figure 8.1 (e), by

$$s = \exp(|g|L/2)$$

Degenerate parametric amplification is thus seen to lead to amplification of one quadrature component (E_1) and to the attenuation of the other component (E_2). The choice of which component (E_1) and to the attenuation of the other component (E_2). The choice of which component is amplified is determined by the phase of the pump E_3. This is illustrated by Figure 8.2. If we now express the field amplitudes as in

(8.1.3), including their quasi quantum mechanical fluctuations, the last two equations become

$$E_1(L,t) = (E_{10}(0) + \Delta E_1(0,t))\exp\left(\frac{|g|L}{2}\right)$$

$$E_2(L,t) = (E_{20}(0) + \Delta E_2(0,t))\exp\left(-\frac{|g|L}{2}\right) \quad (8.3.9)$$

The mean fields E_{10} and E_{20} *as well as the fluctuations* $\Delta E_1(t)$ and $\Delta E_2(t)$ are thus found to be amplified (attenuated) by the nonlinear parametric interaction, The output $(z = L)$ fluctuations are

$$\Delta E_1(L) = \langle \Delta E_1(L,t)^2\rangle^{1/2} = \Delta E_1(0)\exp\left(\frac{|g|L}{2}\right) = \frac{A}{2}\exp\left(\frac{|g|L}{2}\right) \quad (8.3.10)$$

$$\Delta E_2(L) = \Delta E_2(0)\ \exp\left(-\frac{|g|L}{2}\right) = \frac{A}{2}\exp\left(-\frac{|g|L}{2}\right)$$

The uncertainty product

$$\Delta E_1(L)\Delta E_2(L) = \Delta E_1(0)\Delta E_2(0) = \frac{A^2}{4} \quad (8.3.11)$$

remains unchanged, although the uncertainty area is now elliptical rather than circular. This is illustrated in Figure 8.3(a) for the case of $\exp\left(\frac{|g|L}{2}\right) = 2$.

The case of a parametric amplifier with no input is particularly interesting. Classically we expect no output. Quantum mechanically, however, there exists an input field, the so-called *vacuum field*, represented by the origin-centered circle of Figure 8.1(c). This field can be represented classically by Figure 8.1(a) with $\langle E_1\rangle = \langle E_2\rangle = 0$.

$$e_{in}(t) = Re\{[\Delta E_1(0,t) + i\Delta E_2(0,t)]\exp(i\omega t)\}$$
$$= \Delta E_1(0,t)\cos\omega t - \Delta E_2(0,t)\sin\omega t \quad (8.3.12)$$
$$\langle \Delta E_1(0,t)\rangle = \langle \Delta E_2(0,t)\rangle = 0$$
$$\langle \Delta E_1(0,t)^2\rangle = \langle(\Delta E_2(0,t)^2)\rangle = \frac{A^2}{4}$$

The resulting output is given by (8.3.9) with $E_{10}(0) = E_{20}(0) = 0$ and

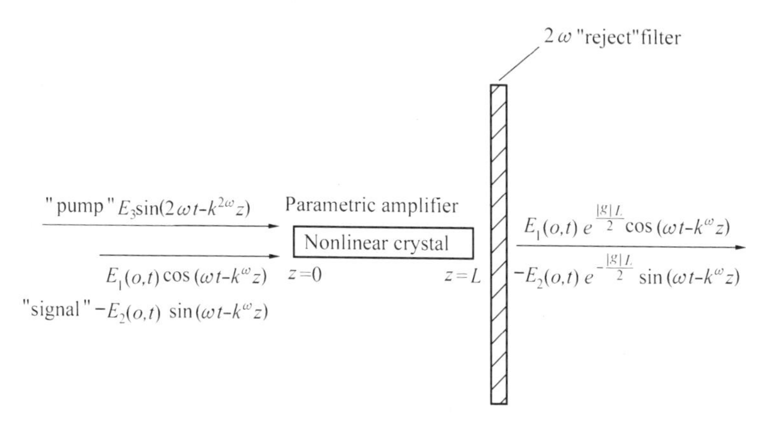

Figure 8.2 A degenerate parametric amplifier (pump frequency equal to twice the signal frequency) used in generating squeezed fields.

is

$$E_1(L,t)\equiv\Delta E_1(0,t)\exp\left(\frac{|g|L}{2}\right)$$
$$E_2(L,t)=\Delta E_2(0,t)\exp\left(-\frac{|g|L}{2}\right) \qquad (8.3.13)$$

so that

$$e_{\text{out}}(t)=\Delta E_1(0,t)\exp\left(\frac{|g|L}{2}\right)\cos\omega t$$
$$-\Delta E_2(0,t)\exp\left(\frac{-|g|L}{2}\right)\sin\omega t$$

corresponding to a field with a zero mean but with squeezed vacuum fluctuations — the so-called *squeezed vacuum*. The input (circle) and output (ellipse) fluctuation in this case are depicted by Figure 8.3(b) for a parametric gain $\exp(|g|L)=4$.

Experimental Demonstration of Squeezing

The experimental setup used often to demonstrate squeezing [5] is shown in Figure 8.4. The output of an optical parametric amplifier at ω is combined in a balanced homodyne receiver with the strong local

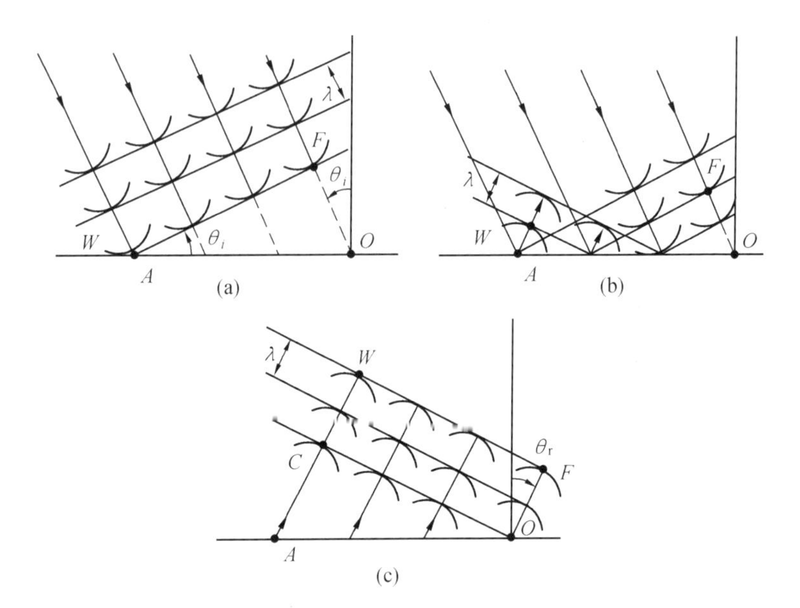

Figure 8.3 (a) The input field to a degenerate parametric amplifier shown a Figure 8.2 and the squeezed output fields for the case $\exp(|g|L/2) = 2$(6 db squeezing.) (b) Same as (a) except that the input field is zero.

oscillator, also of frequency ω, which is *coherent* with that of the pump field at 2ω. (Usually these two fields are derived from the same master laser oscillator at ω). The two combined fields, whose complex amplitudes are $\mathscr{E}_1 = 1/\sqrt{2}[E(t) - \mathscr{E}_L(t)]$ and $\mathscr{E}_2 = 1/\sqrt{2}[E(t) + \mathscr{E}_L(t)]$ are detected, respectively, by photodetectors D_1 and D_2. The resulting currents i_1 and i_2 are subtracted from each other. The net current $i_2 - i_1$ is fed to a spectrum analyzer that displays the spectral density of $(i_2 - i_1)$, a quantity, which according to (10.2.5) and (10.2.7), is proportional to $\langle(i_2 - i_2)^2\rangle$. The detected photocurrents are given according to (11.1.2) by

$$i_1 = t\mathscr{E}_1(t)\mathscr{E}_1^*(t) = \frac{t}{2}(E(t) - \mathscr{E}_L(t))(\mathrm{E}^*(\mathrm{t}) - \mathscr{E}_L^*(t))$$

$$i_2 = \frac{t}{2}(E(t) + \mathscr{E}_L(t))(E^*(t) + \mathscr{E}_L^*(t)) \qquad (8.3.14)$$

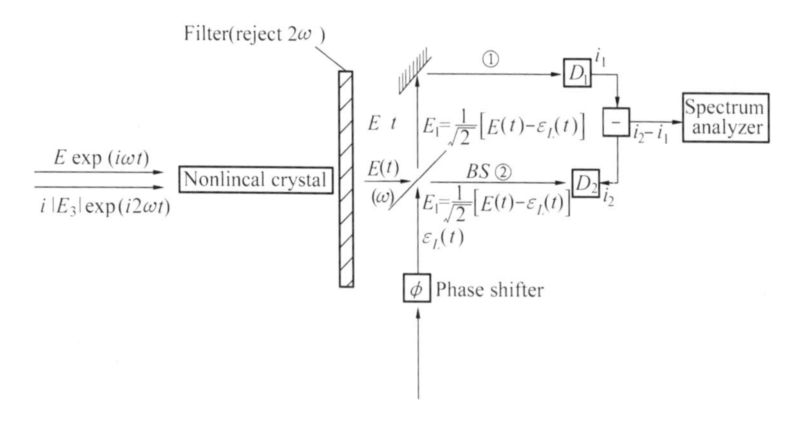

Figure 8.4 A balanced homodyne receiver for measuring the squeezing of the vacuum fluctuations of an electromagnetic field. The squeezing is caused by degenerate parametric amplification. Only the complex amplitudes are marked in the homodyne receiver section since the factor exp($i\omega t$) has been dropped.

where t is a detector constant. The result of the electronic subtraction is thus

$$i_2 - i_1 = t[\mathscr{E}_L^*(t)E(t) + \mathrm{c.c.}] \tag{8.3.15}$$

This result illustrates the raison d' être for the balanced homodyne receiver. Since the output ($i_2 - i_1$) contains only mixed("signal" × local oscillator) product terms, fluctuations $\Delta\mathscr{E}_L(t)$ of the local oscillator field, which lead to "large" terms $\mathscr{E}_L\Delta\mathscr{E}_L^*(t)$ in the photocurrents, i_1 and i_2, cancel out in the current subtraction, These terms would, in the case of a single detector receiver. mask the signal term $\mathscr{E}_L^* E(t)$.

In most of the recent experiments [4,5] demonstrating squeezing, there exists no input to the parametric amplifier. In this case , the output of the parametric amplifier is given by Equation (8.3.9) with $E_{10}(0) = E_{20}(0) = 0$

$$E(t) = \Delta E_1(0,t)\exp\left(\frac{|g|L}{2}\right) + i\Delta E_2(0,t)\exp\left(\frac{-|g|L}{2}\right) \tag{8.3.16}$$

i.e., the squeezed vacuum field, The complex amplitude of the local oscillator field at the beam splitter is taken as the sum of the average field plus a fluctuation term. The fluctuation may be due to basic quantum causes or any other cause. A phase factor $\exp(i\phi)$ accounts for the phase shifter

$$\mathscr{E}_L(t) = [\mathscr{E}_{LO} + \Delta\mathscr{E}_L(t)]\exp(i\phi) \qquad (8.3.17)$$

$\mathscr{E}_{LO}$ can, without any loss of generality, be taken as a real number. Substituting the last two equations for $\mathscr{E}_L(t)$ and $E(t)$ in (8.3.15) and neglecting the terms involving $\Delta\mathscr{E}_L(t)$, since $\Delta\mathscr{E}_L(t) \ll \mathscr{E}_{LO}$ results in

$$(i_2 - i_1) = 2t\mathscr{E}_{LO}\left(\Delta E_1(0,t)\exp\left(\frac{|g|L}{2}\right)\cos\phi - \Delta E_2(0,t)\exp\left(\frac{-|g|L}{2}\right)\sin\phi\right) \qquad (8.3.18)$$

Since both $\langle\Delta E_1(t)\rangle$ and $\langle\Delta E_1(t)\rangle$ are zero, the time-averaged $(i_2 - i_1)$, the quantity that normally will be registered by a sensitive ammeter, is zero, This problem is avoided by squaring $(i_2 - i_1)$, This is accomplished usually [5] by the spectrum analyzer, which displays the spectral density $S_f(\Omega)$ of the input $f(t)$ where

$$\langle f^2(t)\rangle = \int_0^\infty S_f(\Omega)\,\mathrm{d}\Omega$$

as discussed in Section 10.2. In our case, the input is $f(t) = i_2(t) - i_1(t)$, so that the output of the spectrum analyzer is proportional, in the case of a constant $S_{(i_2 - i_1)}(\Omega)$, to

$$S_{i_2 - i_1}(\Omega) \propto \langle(i_2 - i_1)^2\rangle = 4t^2\mathscr{E}_{LO}^2[\langle(\Delta E_1(0,t))^2\rangle\exp(|g|L)\cos^2\phi + \langle(\Delta E_2(0,t))^2\rangle\exp(-|g|L)\sin^2\phi] = t^2\mathscr{E}_{LO}^2A^2[\exp(|g|L)\cos^2\phi + \exp(-|g|L)\sin^2\phi] \qquad (8.3.19)$$

where we used (8.1.6) and $\langle\Delta E_1(t)\Delta E_2(t)\rangle = 0$.

A typical result of such an experiment is shown in Figure 8.5. The observed dependence of the photocurrent fluctuations on the phase ϕ of local oscillator field is in agreement with (8.3.19) and constitutes a dramatic verification of squeezing of the vacuum fluctuations. The dashed horizontal line is the result when the optical parametric amplifier is blocked.

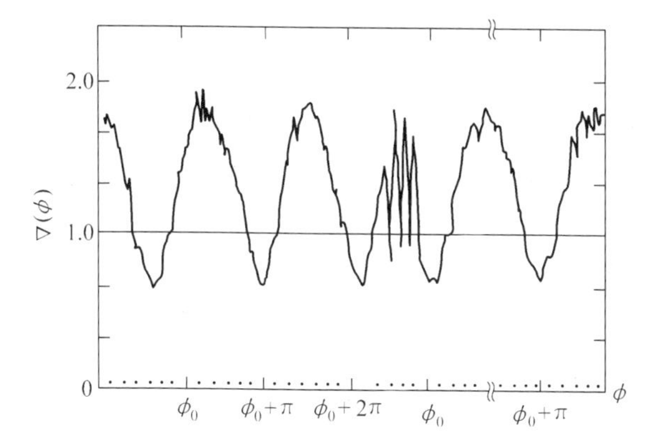

Figure 8.5 Measurement demonstrating the phase dependence of the quantum fluctuations in a squeezed state. The squeezing is achieved by degenerate parametric amplification. The phase dependence of the rms voltage from balanced homodyne receiver is displayed vs. the local oscillator phase ϕ. The noise voltage is centered on $v = 1.8$ MHz. With the parametric amplifier blocked. $|g| = 0$. the output is given by the dashed horizontal line with no ϕ dependence. The dips represent 50 percent of the electronic noise power relative to that of unsqueezed vacuum (i.e., $|g| = 0$) input[5].

It is instructive to view the squeezed states by plotting in Figure 8.6, the actual sinusoidal optical fields corresponding to six representative points inside the uncertainty ellipse of Figure 8.1(e). We recall that each such point represents a possible realization of the field (complex) amplitude. The case of no squeezing ($s = 0$) is shown in Figure 8.6(a), while a squeezed state with $x = 2.5$, $s = 4$, $\beta = 0$ is shown in (b). We note that in the squeezed state we trade an increase in the accuracy of measuring the frequency (or phase) for an increased amplitude fluctuation in qualitative agreement with (8.2.15). The vacuum state ($\langle E \rangle = 0$) is shown in (c), which contains plots of representative points from Figure 8.1(c), while the squeezed vacuum,

Figure 8.1(f), is shown in (d).

Another type of squeezing, "number squeezing," which results in photocurrent noise level below that of shot noise, can be exhibited in semiconductor diode lasers [6]. This squeezing results when the injection current to the lasers is highly constant and/or when proper, frequency dependent feedback is employed [7]. Such lasers may find practical uses in communication [8] and atomic measurements [9], for example, Experimental data showing such squeezing is shown in Figure 8.7.

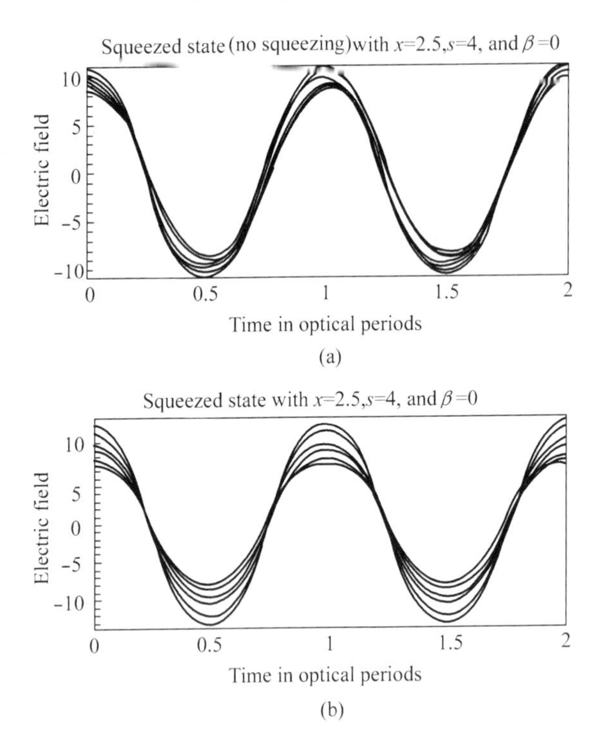

Figure 8.6 Representation of: (a) unsqueezed electric field, (b) squeezed ($s = 4$) state, (c) unsqueezed "vacuum" field, and (d) squeezed ($s = 4$) vacuum fed.

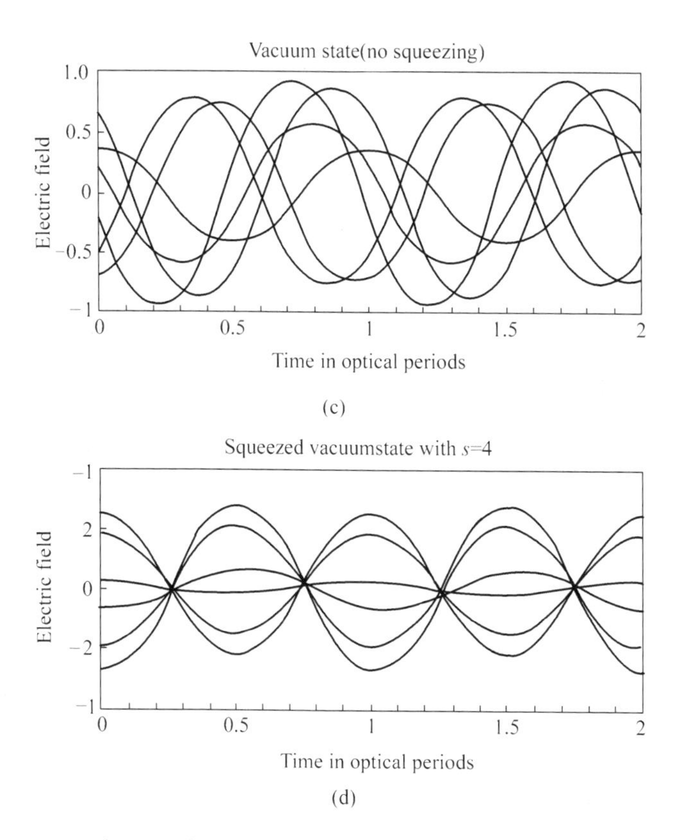

Figure 8.6 (*Continued*)

In conclusion it is worthwhile to remind ourselves that the classical treatment of this chapter seems to do a good job in representing the results of the rigorous quantum treatment — but only up to second-order electric field products. If we were to ask some more difficult questions, say those involving expectation values of the electric field raised to third power or higher, the classical approach fails.

A comprehensive review of the topic quantum noise in optics is found in Reference [10].

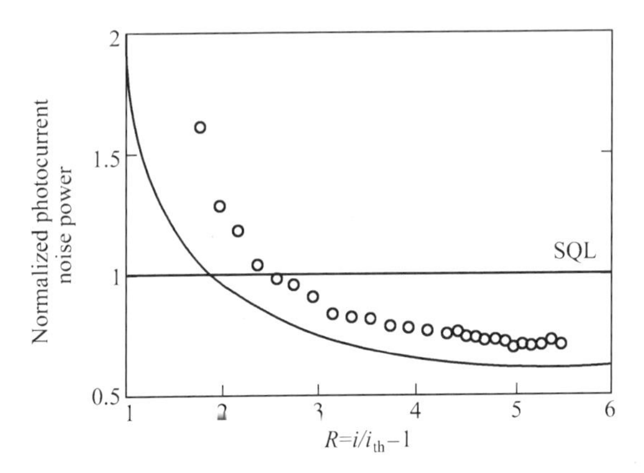

Figure 8.7 The normalized photocurrent noise current spectral density at υ = 29 MHz as a function of the injection current to a semiconductor laser. i_{th} is the oscillation threshold current. SQL — the standard quantum limit — is the level corresponding to shot noise. (After Reference [7].)

References

1. See, for example, A. Yariv, *Quantum Electronics*, 4th ed. Wiley, 1988, p. 13.
2. Glauber, R. J., "Coherent and incoherent states of radiation fields," *Phys*, *Rev*. 131:2776, 1963.
3. Louisell, W. H., *Radiation and Noise in Quantum Electronics*, New York: McGraw-Hill, 1964.
4. Yuen, H. P., and J. H. Shapiro, "Generation and detection of two-photon coherent states in degenerate four wave mixing," *Opt*. *Lett*. 4: 334, 1979.
5. Wu, L., J. H. Kimble, J. L. Hall, and H. Wu, "Generation of squeezed states by parametric down conversion," *Phys*, *Rev*, *Lett*. 57:2520, 1986.
6. Yamamoto, Y., S. Machida, and O. Nilsson, "Amplitude squeezing

in a pump-noise-suppressed laser oscillator," *Phys, Rev. A* 34: 4025-4042, 1986.

7. Kitching, J., A. Yariv, and Y. Shevy, "Room temperature generation of squeezed light from a semiconductor laser with weak optical feedback," *Phys. Rev. Lett.* 74, 1995.
8. Saleh, B. E. A., and M. C. Teich, "Information transmission with number-squeezed light," *Proc. IEEE* 80:451-460, 1992. Also by the same authors, *Fundamentals of Photonics*, New York: Wiley, pp. 414—416.
9. Wieman, C. E., and L. Hollberg, "Using diode lasers for atomic physics," *Rev. Sci. Instrum.* 62:1-20, 1991.

New Words and Expressions

ammeter	*n*. 电表
amplitude	*n*. [物]振幅
attenuation	*v*. 变薄,变细,稀释,冲淡
cluster	*v*. 聚合成
coherent	*adj*. 相干的
consistent	*adj*. 相容的,一致的
couterpart	*n*. 对应物
crystal	*n*. 晶体
decree	*v*. 决定,判定,颁布
dielectric	*n*. 电介质,绝缘体
	adj. 非传导性的
electromagnetic field	*n*. 电磁场
elegant	*adj*. [口]上品的,第一流的
elliptical	*adj*. 椭圆的,省略的
ensemble	*n*. 全体,整体,总效果
fluctuation	*n*. 波动,起伏
forego	*v*. (在位置时间或程度方面)走在……之前
formalism	*n*. 形式主义

graphical	*adj*. 绘成图画似的
homodyne	*adj*. [物]零差的,零拍的
incident	*adj*. 入射的
increment	*n*. 增加,增量
joule	*n*. [物]焦耳
mean	*n*. 平均数
momentum	*n*. 动力,要素,[物]动量
monochromatic	*adj*. [物]单色的,单频的
nonlinear	*adj*. 非线性的
oscillator	*n*. 震荡器
parametric complification	*n*. 参数放大
particle	*n*. 粒子,质点
phase shift	*n*. 位相漂移
phasor	*n*. 相位复矢量,相图,彩色信息矢量,相量
photocurrent	*n*. 光电流
photodetector	*n*. [电子]光电探测器
photoelectron	*n*. 光电子
photo-generated electron	*n*. 光生电子
photon	*n*. [物]光子
Poissonian	泊松
projection	*n*. 发射
quadrature	*n*. 求积,求积分
quantum	*n*. 量,额,[物]量子,量子论
quantum mechanics	*n*. 量子力学
quantum optics	*n*. 量子光学
refraction	*n*. 折射,折光
resonator	*n*. 谐振器,共振腔
scenario	*n*. 某一特定情节,方案
sinusoidal	*n*. 正弦分布的
spectral	*adj*. 分光谱的

spectrum	*n*. 光,光谱,频谱
spectrum analyzer	*n*. 光谱分析仪
splitter	*n*. 分离器
squeeze	*n*. 压榨,挤
	v. 压榨,挤,挤榨
temporal	*adj*. 暂时的,现世的,世俗的
	n. 世间万物
vacuum	*n*. 真空
	adj. 利用真空的
vector	*n*. [数]向量,矢量,带菌者
	v. 无线电导引
vectorial	*adj*. 向量的,矢量的
vibration	*n*. 震动,摆动

NOTES

1. Some of the most important and elegant phenomena related to optical waves and their detection can only be explained using the extension of the formalism of quantum mechanics to optics, i. e., "quantum optics."

关于光波以及在探测光波过程中所发现的一些极其重要而又微妙的现象,只能用量子力学在光学方面的延展方式即量子光学来解释。

2. This is the so-called zero point vibration energy of the mode.

这就是所谓的能量模式零点摆动。

3. We have thus demonstrated that the electronic shot noise in the photocurrent can be attributed to the quantum field fluctuations.

如此我们就证明了光电流中的电子噪声对量子场的波动具有贡献。

4. We can, however, appreciate many of the consequences and even obtain numerically correct results for the important scenarios by accepting just are basic consequence of quantum mechanics: that of the

uncertainty principle.

然而,对于很多重要的现象来说,我们能通过理解量子力学的基本结论即不确定原理的结论来得到很多结果甚至能获得正确的数字结果。

5. It is possible to take a coherent optical field, such as the output of a laser, and reduce the fluctuation of one of its quadrature components, say, ΔE_1, at the expense of ΔE_2, or vice versa.

我们能够得到一个连续的光场,例如激光的输出,同时我们能够减小它的起伏噪声中的一个正交分量,比如 ΔE_1,这是建立在另一个分量 ΔE_2 增大的基础上,反之亦然。

6. The output of an optical parametric amplifier at ω is combined in a balanced homodyne receiver with the strong local oscillator, also of frequency ω, which is coherent with that of the pump field at 2ω.

工作频率为 ω 的光学参数放大器的输出和另一个频率也是 ω 的本地震荡器的输出结合在一起,就形成了稳定的零差信号输出,这个震荡器也可用于频率为 2ω 的泵浦场。

PART FOUR
NEW OPTOELECTRONIC TECHNOLOGY

9

Holography and Optical Data Storage

9.1 Introduction

This chapter takes up the basic concepts and some key applications of the remarkable field of holography[1-8]. This field traces its beginning to a 1948 paper by D. Gabor [1] and to a major improvement by Leith and Upatnieks [2] who solved a major problem of the original proposal. Holography is an imaging technique in which the recording is accomplished by an interference in the recording medium of two, usually mutually coherent waves: the image-bearing "picture wave" and the, usually, plane wave or spherical, "reference" wave. The intensity pattern due to this interference "burns" itself into the volume (or surface) of the recording medium in a proportionate manner by modifying its index of refraction or gain. This pattern — the hologram — clearly contains both phase and amplitude information of the "picture." Viewing (reconstruction) is achieved when a wave identical and usually from the same direction as the original reference wave is incident on the hologram. The wave created by the diffraction of this wave from the hologram is identical in all essential respects to the original picture wave so that the viewer perceives the three-dimen-sional aspects of the originally photographed objects.

In volume holograms, it is possible to store a large number of pictures and to view each one of them selectively, with negligible crosstalk from the other pictures.

9.2 The Mathematical Basis of Holography[①]

Figure 9.1 illustrates the experimental setup used in making a simple hologram. A plane-parallel light beam illuminates the object whose hologram is desired. Part of the same beam is reflected from a mirror (at this point we refer to it as the reference beam) and is made to interfere within the volume of the photosensitive medium with the beam reflected diffusely from the object (object beam). The photosensitive medium is then developed and forms the hologram.

The image reconstruction process is illustrated in Figure 9.2. It is performed by illuminating the hologram with the same wavelength laser beam and in the same relative orientation that existed between the reference beam and the photosensitive medium when the hologram was made. An observer facing the far side (B) of the hologram will now see a three-dimensional image occupying the same spatial position as the original object. The image is, ideally, indistinguishable from the direct image of the laser-illuminated object.

The Holographic Process Viewed as Bragg Diffraction

To illustrate the basic process involved in holographic wavefront reconstruction, consider the simple case in which the two beams reaching

---

① Chapters 7 and 8 deal with the more advanced topic of dynamic holography in nonlinear optical media. The treatment of this section is mostly kinetic. It tells us when and how things happen but does not address the magnitudes involoved. It is meant as an introduction to the major concepts of holography.

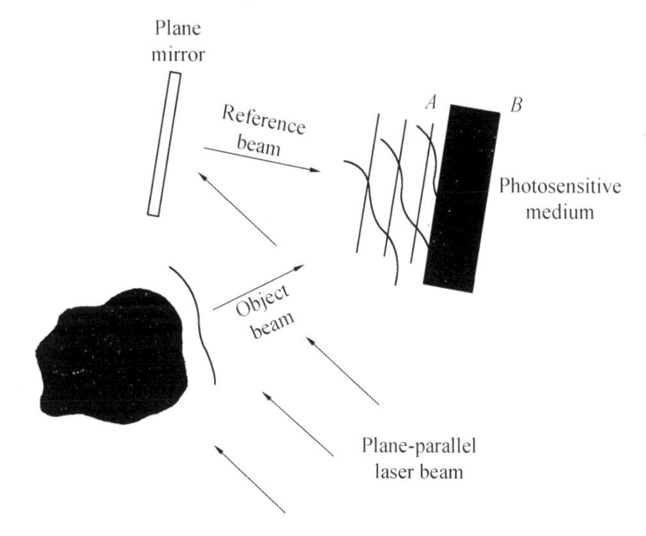

Figure 9.1 A hologram of an object can be made by exposing a photosensitive medium at the same time to coherent light. which is reflected diffusely from the object, and a planeparallel reference beam. which is part of the same beam that is used to illuminate the object.

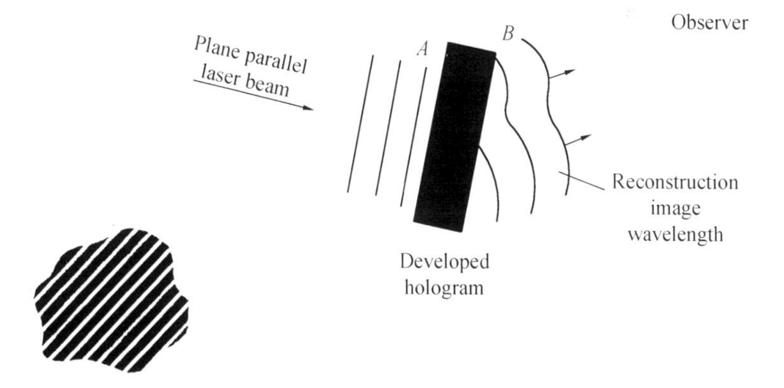

Figure 9.2 Wavefront reconstruction of the original image is usually achieved by illuminating the hologram with a laser beam of the same wavelength and relative orientation as the reference beam making it. An observer on the far side (B) sees a virtual image occupying the same space as the original subject.

the photosensitive medium in Figure 9.1 are plane waves. The situation is depicted in Figure 9.3. We choose the z axis as the direction of the bisector of the angle formed between the two propagation directions k_1 and k_2 of the reference and object plane waves inside the photosensitive layer. The x axis is contained in the plane of the paper. The electric fields of the two beams are taken as

$$e_{\text{object}}(r, t) = E_1 e^{i(k_1 \cdot r - \omega t)}$$

$$e_{\text{reference}}(\mathrm{r}, t) = E_2 e^{i(k_2 \cdot r - \omega t)} \tag{9.2.1}$$

From Figure 9.3 and the fact that $|k_1| = |k_2| = k$, we have

$$k_1 = a_x k \sin\theta + \mathrm{a}_z k \cos\theta$$

$$k_2 = -a_x k \sin\theta + \mathrm{a}_z k \cos\theta \tag{9.2.2}$$

where $k = 2\pi/\lambda$, and $\mathrm{a_x}$ and $\mathrm{a_z}$ are unit vectors parallel to x and a_x respectively.

The total complex field amplitude is the sum of the complex amplitudes of the two beams, which, using (9.2.1) and (9.2.2), can be written as

$$E(x, z) = E_1 e^{ik(x\sin\theta + z\cos)} + E_2 e^{ik(-x\sin\theta + z\cos\theta)} \tag{9.2.3}$$

If the photosensitive medium were a photographic emulsion, the exposure to the two beams and subsequent development would result in silver atoms developed out at each point in the emulsion in direct proportion to the time average of the square of the optical field. The density of silver in the developed hologram is thus proportional to $E(x, z)\ E^*(x, z)$, which, using(9.2.3), assuming $E_1$and E_2 real, becomes

$$E(x, z)\ E^*(x, z) = E_1^2 + E_2^2 + 2E_1E_2\cos(2kx\sin\theta) \tag{9.2.4}$$

The hologram is thus seen to consist of a sinusoidal modulation of the silver density. The planes x = constant(that is, planes containing the bisector and normal to the plane of Figure9.3) correspond to equidensity planes. The distance between two adjacent peaks of this spatial modulation pattern is, according to (9.1.4),

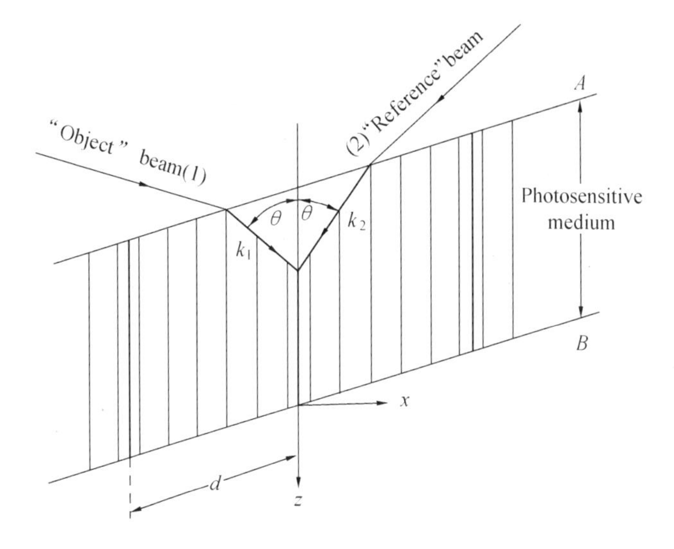

Figure 9.3 A sinusoidal "diffraction grating," produced by the interference of two plane waves inside a photographic emulsion. The density of black lines represents the exposure and hence the silver-atom density. The z direction is chosen as that of the bisector of the angle formed between the directions of propagation inside the photographic emulsion. It is not necessarily perpendicular to the surface of the hologram.

$$d = \frac{\pi}{k \sin \theta} = \frac{\lambda / n}{2 \sin \theta} \tag{9.2.5}$$

In the process of wavefront reconstruction, the hologram is illuminated with a coherent laser beam. Since the hologram consists of a three-dimensional sinusoidal diffraction grating, the situation is directly analogous to the diffraction of light from sound waves, which was analyzed in Section 12.1. Applying the results of Bragg diffraction and denoting the wavelength of the light used in reconstruction (that is, in viewing the hologram) as λ_R, a diffracted beam exists only when the Bragg condition (12.1.4)

$$2d \sin \theta_{\mathrm{B}} = \frac{\lambda_R}{n_R} \tag{9.2.6}$$

is fulfilled, where θ_B is the angle of incidence and of diffraction as shown in Figure 9.4 and n_R is the index of refraction. Substituting for d its value according to (9.2.5), we obtain

$$\sin\theta_B = \left(\frac{n}{n_R}\right)\frac{\lambda_R}{\lambda}\sin\theta \tag{9.2.7}$$

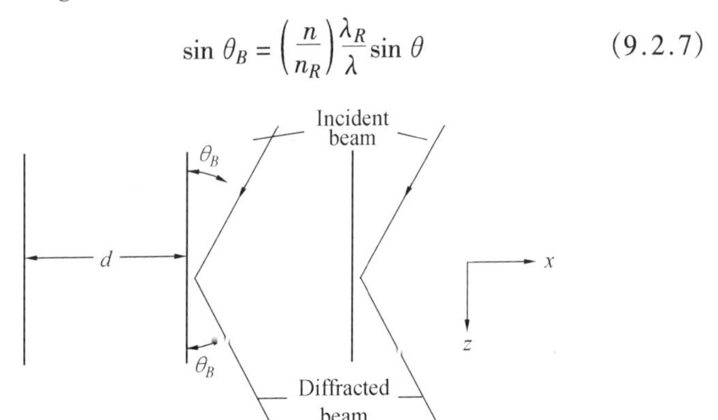

Figure 9.4 Bragg diffraction from a sinusoidal volume grating. The grating periodic distance d is the distance in which the grating structure repeats itself. In the case of a hologram we may consider the vertical lines in the figure as an edge-on view of planes of maximum silver density.

In the special case when $\lambda_R = \lambda$— that is to say, when the hologram is viewed with the same laser wavelength as that used in producing it — we have

$$\theta_B = \theta$$

so that wavefront reconstruction (that is, diffraction) results only when the beam used to view the hologram is incident on the diffracting planes at the same angle as the beam used to make the hologram. The diffracted beam emerges along the same direction (k_1) as the original "object" beam, thus constituting a reconstruction of the latter.

We can view the complex beam reflected from the object toward the photographic emulsion when the hologram is made, as consisting of a "bundle" of plane waves each having a slightly different direction. Each one of these waves interferes with the reference beam, creating, after development, its own diffraction grating, which is displaced slightly in angle from that of the other gratings. During reconstruction the illuminating laser beam is chosen so as to nearly satisfy the Bragg condition (9.2.6) for these gratings. Each grating gives rise to a diffracted beam along the same direction as that of the object plane wave that produced it, so the total field on the far side of the hologram (B) is identical to that of the object field.

Basic Holography Formalism

The point of view introduced above, according to which a hologram may be viewed as a volume diffraction grating, is extremely useful in demonstrating the basic physical principles. A slightly different approach is to take the total field incident on the photosensitive medium as

$$A(r) = A_1(r) + A_2(r) \qquad (9.2.8)$$

where $A_1(r)$ may represent the complex amplitude of the diffusely reflected wave from the object while $A_2(r)$ is the complex amplitude of the reference beam. $A_2(r)$ is not necessarily limited to plane waves and may correspond to more complex wavefronts.

The intensity of the total radiation field can be taken, as in (9.2.4), to be proportional to

$$AA^* = A_1A_1^* + A_2A_2^* + A_1A_2^* + A_1^*A_2 \qquad (9.2.9)$$

The first term $A_1A_1^*$ is the intensity I_1 of the light arriving from the object. If the object is a diffuse reflector, its unfocused intensity I_1 can be regarded as essentially uniform over the hologram's volume. $A_2A_2^*$ is the intensity I_2 of the reference beam. The change in the amplitude transmittance of the hologram ΔT can be taken as proportional to the exposure density so that

$$\Delta T \propto I_1 + I_2 + A_1A_2^* + A_1^*A_2$$

The reconstruction is performed by illuminating the hologram with the reference beam A_2 in the same relative orientation as that used during the exposure. Limiting ourselves to the portion of the transmitted wave modified by the exposure. we have

$$R = A_2 \Delta T \infty (I_1 + I_2) A_2 + A_1^* A_2 A_2 + I_2 A_1 \qquad (9.2.10)$$

Ther first term corresponds to a wavefront proportional to the reference beam. The second term, not being proportional to A_1, may be regarded as undesirable "noise." Since I_2 is a constant, the third term $I_2 A_1$ corresponds to a transmitted wave that is proportional to A_1 and is thus a reconstruction of the object wavefront.

9.3 The Coupled Wave Analysis of Volumeholograms

In this section, we wil extend the qualitative dynamic arguments of Section 9. 1 and obtain analytic expressions that will be useful in analyzing certain holographic applications. We will start with a description of the recording process, followed by a coupled wave analysis of the reconstruction of the hologram. We will simplify the problem by limiting it to two plane waves. One, A_1, is the "picture" field; A_{2r} is the reference wave. The results can be extended to more complex picture fields.

The total field during the recording phase is thus taken as

$$E(r) = Re[(A_1 e^{-ik_1' \cdot r} + A_{2r} e^{-ik_2' \cdot r}) e^{i\omega t}]$$

where the subscript r denotes the reference wave. We assume that the index of refraction (rather than the absorption) of the holographic medium changes by an amount that is proportional to the optical intensity $I(r)$

$$I(r) \infty [(A_1 e^{-ik_1' \cdot r} + A_{2r} e^{-k'_2 \cdot r})(c.c.)] =$$
$$|A_1|^2 + |A_{2r}|^2 + A_1 A_{2r}^* e^{-i(k_1' - k_2') \cdot r} +$$
$$A_1^* A_{2r} e^{i(k_1' - k_2') \cdot r} \qquad (9.3.1)$$

We can thus take the index of refraction distrubution of a hologram as

$$n(r) = n_0 + n_1 \cos(K \cdot r + \phi) \qquad (9.3.2)$$
$$n_1 \infty |A_1 A_{2r}| \infty \sqrt{I_1 I_2}$$
$$K = k_2' - k_1' \qquad (9.3.3)$$

A graphical demonstration of $n_1(r)$—the hologram —is shown in Figure 9.5.

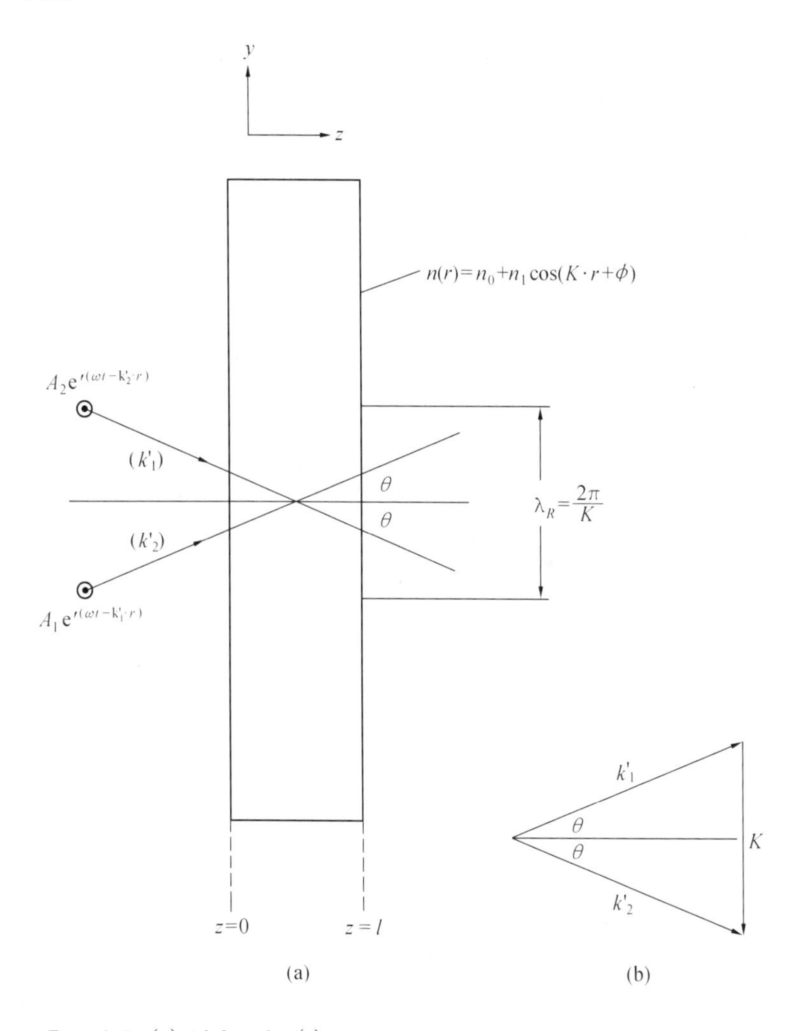

Figure 9.5 (a) A holograph $n(r)$ is recorded in a photosensitive medium by the standing wave pattern produced by two coherent beams that intersect in the medium. The hologram grating vector is $K=k_2'-k_1'$. (b) The Bragg condition diagram $K=k_2-k_1$.

During the reconstruction of the hologram, it is illuminated by a reference wave $A_2 e^{-ik_2 \cdot r}$ propagating along k_2. Our task is to obtain an expression for the "picture" wave A_1 that results. We know in advance that if the Bragg condition $k_2 - k_1 = K$ is satisfied. where k_1 is the propagation vector of the diffracted wave, then the interaction involves exclusively waves A_1 and A_2 so that we can write for the total field in a long medium hologram

$$E(r) = \frac{1}{2} A_1(r) e^{-ik_1 \cdot r} + \frac{1}{2} A_2(r) e^{-ik_2 \cdot r} + c.c. \quad (9.3.4)$$

These two waves "see" a medium with a modulated index of refraction, the hologram, given by (9.3.2). The total field thus obeys the Helmholtz equation

$$\nabla^2 E(r) + \omega^2 \mu \in (r) E = 0 \quad (9.3.5)$$

$$\in (r) = \in_0 n^2(r) \cong \in_0 [n_0^2 + (n_0 n_1 e^{-i(K \cdot r + \phi)} + c.c.)] \quad (9.3.6)$$

Substituting(9.3.4) and (9.3.6) in (9.3.5) leads to

$$\begin{aligned} &\frac{1}{2}\left(-2ik_1 \frac{dA_1}{dr_1} - \kappa_1^2 A_1\right) e^{-ik_1 \cdot r} + cc. + \\ &\frac{1}{2}\left(-2ik_2 \frac{dA_2}{dr_2} - \kappa_2^2 A_2\right) e^{-ik_2 \cdot r} + c.c. + \\ &\omega^2 \mu \in_0 [n_0^2 + (n_0 n_1 e^{-i\phi} e^{-iK \cdot r} + c.c.)] \times \\ &\left[\frac{A_1}{2} e^{-ik_1 \cdot r} + \frac{A_2}{2} e^{-ik_2 \cdot r} + c.c.\right] = 0 \end{aligned} \quad (9.3.7)$$

where we neglected

$$\frac{d^2 A}{dr^2} \ll k \frac{dA}{dr}$$

We observe by inspection that spatially cumulative exchange of power takes place when the (Bragg) condition

$$k_2 - k_1 = K \quad (9.3.8)$$

is satisfied[1]. Keeping only synchronous terms (terms with similar exponents) and recalling that in an isotropic medium $k_1 = k_2 = \omega\sqrt{\mu\epsilon_0}\, n_0$ helps us simplify (9.3.7) to

$$\cos\theta\frac{dA_1}{dz} = -\frac{\alpha}{2}A_1 + i\frac{\pi n_1}{\lambda}e^{i\phi}A_2e^{i(k_1-k_2+K)\cdot r}$$
$$\cos\theta\frac{dA_2}{dz} = -\frac{\alpha}{2}A_2 + i\frac{\pi n_1}{\lambda}e^{-i\phi}A_1e^{-i(k_1-k_2+K)\cdot r} \qquad (9.3.9)$$

where loss terms $-(\alpha/2)A_{1,2}$ were added phenomenologically to account for absorption, and $\lambda = 2\pi/(\omega\sqrt{\mu\epsilon_0})$ is the free space wavelength. 2θ is the angle between k_1 and k_2, and z is the distance measured along the bisector, so that $z = r_{1,2}\cos\theta$. Expressing the amplitudes in terms of magnitudes and phases by using the definition $A_j \equiv \sqrt{I_j}exp(-i\phi_j)$ leads to (in what follows, we take $k_1 - k_2 + K = 0$, i.e., the Bragg condition is satisfied)

$$\cos\theta\frac{dI_1}{dz} = -\alpha I_1 - \frac{2\pi n_1}{\lambda}\sqrt{I_1I_2}\sin(\phi_1 - \phi_2 + \phi)$$
$$\cos\theta\frac{dI_2}{dz} = -\alpha I_2 + \frac{2\pi n_1}{\lambda}\sqrt{I_1I_2}\sin(\phi_1 - \phi_2 + \phi) \qquad (9.3.10)$$

Note that the coupling at point **r** depends on the local phase $\Psi \equiv (\phi_1 - \phi_2 + \phi)$. If the phase $\Psi = \pm\pi/2$, the exchange is maximum. The case $\Psi = \pm\pi/2$ corresponds, according to Equations(9.3.4) and (9.3.5), to a grating that is displaced by a quarter period with respect to the intensity interference pattern of waves 1 and 2. In the most common scenario, a single wave, say 1, is incident on the grating, and wave 2 is the diffracted wave. In this case, it follows from the second equation of (9.3.10) that wave 2 is generated with a phase $\phi_2 = \phi_1 + \phi + \pi/2$, i.e., $\Psi = -\pi/2$, which results, according to the second equation of (9.3.10), in a *maximum* positive value for the power exchange dI_1/dz.

---

[1] When condition (9.3.8) is not satisfied the power exchange reverses sign every "coherence length" $L_c \equiv \pi/(|k_2 - k_2 - K|)$ amd averages out to near zero over distances $\gg L_c$.

The solution of (9.3.10) in the case of $\Psi = +\pi/2$ and $I_2(0) = 0$ becomes [9]

$$I_1(z) = I_1(0)\mathrm{e}^{-\left(\frac{\alpha z}{\cos\theta}\right)}\cos^2\left(\frac{\pi n_1 z}{\lambda\cos\theta}\right)$$

$$I_2(z) = I_1(0)\mathrm{e}^{-\left(\frac{\alpha z}{\cos\theta}\right)}\sin^2\left(\frac{\pi n_1 z}{\lambda\cos\theta}\right) \qquad (9.3.11)$$

so that in a grating of length l, the diffraction efficiency is

$$\eta = \frac{I_2(l)}{I_1(0)} = \exp\left(-\frac{\alpha l}{\cos\theta}\right)\sin^2\left(\frac{\pi n_1 l}{\lambda\cos\theta}\right) \qquad (9.3.12)$$

This formula is very useful in interpreting a large variety of experimental data involving fixed volume gratings and holograms.

Multihologram Recording and Readout — Crosstalk

It is possible in a volume hologram to record simultaneously a large number of holograms. The fundamental reason is that in a large volume, with dimensions $\gg L_1$, a reconstructed image results only when the Bragg condition, $k_2 - k_1 = K$, is satisfied (see footnote 2). Here k_2 and k_1 are the propagation vectors of the incident and diffracted wave during the reconstruction, and K represents the hologram, as in (9.3.2). If the $\boldsymbol{K}$ vectors representing the different holograms are sufficiently different, it is possible to read one specific hologram with negligible contributions — crosstalk — from the others , since, for all the other holograms, the Bragg condition is strongly violated. To consider this problem quantitatively, consider the situation depicted in Figure 9. 6, where two holograms $n_1(r)$ and $n_2(r)$ are recorded in the same volume using two different reference directions $k_2^{(1)}$ and $k_2^{(2)}$ but the same picture wave direction k_1,

$$n_1^{(1)}(r) \propto \sqrt{I_1^{(1)} I_2^{(1)}}\sin(K^{(1)}\cdot r + \phi^{(1)})$$

$$n_1^{(2)}(r) \propto \sqrt{I_1^{(2)} I_2^{(2)}}\sin(K^{(2)}\cdot r + \phi^{(2)}) \qquad (9.3.13)$$

where

$$K^{(1)} = k_2^{(1)} - k_1$$

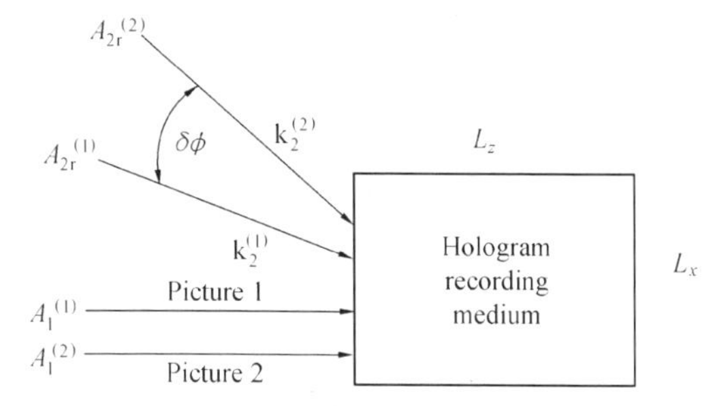

Figure 9.6 Two pictures $A_1^{(1)}$ and $A_1^{(2)}$ are recorded, separately as volume holograms each with its angularly unique reference wave. In practice, we can record many pictures in the same volume. This procedure can be repeated, resulting in the recording of hundreds (or even thousands) of holograms in the same volume.

$$K^{(2)} = k_2^{(2)} - k_1 \tag{9.3.14}$$

If we wish to reconstruct picture 1, we illuminate the hologram with the corresponding reference wave $k_2^{(1)}$ (i. e., the same reference wave used to record it), as discussed above. This reference wave will encounter in the crystal, not only the desired hologram $n_1^{(2)}(r)$ but also hologram $n_1^{(2)}(r)$. Any light scattered from hologram $n_1^{(2)}(r)$ in the direction of k_1 thus constitutes (noisy) crosstalk, which degrades the information contents of picture 1. This crosstalk places a fundamental limit on the number of holograms and their stored information contents: To quantify this argument, we will derive an expression for the power radiated along k_1 due to the undesirable scattering of the reference beam employed($k_2^{(1)}$) off the "wrong" hologram of picture 2 – $n_1^{(2)}(r)$. The equations describing this process were derived in (9.3.9) and are reproduced here for the incident (A_2) and the diffracted (A_1) beams

$$\frac{dA_1}{dz} = i\frac{\pi n_1^{(2)}}{\lambda\cos\theta}A_2 e^{i(k_1 - k_2^{(1)} + k_2^{(2)} - k_1)\cdot r}$$

$$\frac{dA_2}{dz} = i\frac{\pi n_1^{(2)}}{\lambda\cos\theta}A_1 e^{-i(k_1 - k_2^{(1)} + k_2^{(2)} - k_1)\cdot r} \tag{9.3.15}$$

where the grating vector $K^{(2)} = k_2^{(2)} - k_1$ is that of hologram 2 and we took $\phi = 0$. The direction k_1 is, according to Figure 9.6, the same for both $n_1^{(1)}(r)$ and $n_1^{(2)}(r)$, since the "picture" direction is the same for all the recorded holograms. We rewrite Equations (9.3.15) as

$$\frac{dA_1}{dz} = ikA_2 e^{-i\delta z}$$

$$\frac{dA_2}{dz} = ikA_1 e^{-i\delta z}$$

$$k = \frac{\pi n_1^{(2)}}{\lambda \cos\theta} \qquad \delta \equiv k_{2z}^{(2)} - k_{2z}^{(1)} \tag{9.3.16}$$

where, for the sake of simplicity, we assumed no optical losses ($\alpha = 0$).

We note that if we take $k_2^{(1)} = k_2^{(2)}$, Equations (9.3.16) lead to the solution (9.3.11) for the Bragg matched condition. Here, however, we are interested in the crosstalk case, $k_2^{(1)} \neq k_2^{(2)}$. In this case, $\delta \neq 0$ and the formal solution of (9.3.16) at the output of the hologram $z = L$ is

$$A_2(L) = A_2(0)e^{-i\delta L}\left[\cos(sL) + \frac{i\delta}{s}\sin(sL)\right]$$

$$A_1(L) = iA_2(0)e^{i\delta L}\frac{k}{s}\sin(sL) \tag{9.3.17}$$

$$s^2 = \kappa^2 + \delta_2^2 \tag{9.3.18}$$

where we used the boundary condition $A_1(0) = 0$ so that only the reference wave $A_2(0)$ is present at the input. The output field $A_1(L)$ constitutes our crosstalk. The fraction of the incident power thus scattered is given by (9.3.17) as

$$\left|\frac{A_1(L)}{A_2(0)}\right|^2 \approx (\kappa L)^2 \frac{\sin^2(\delta L)}{(\delta L)^2}, \delta \gg \kappa \tag{9.3.19}$$

We can thus reduce this power to an acceptable level by using a sufficiently large δ. Using the definition $\delta = k_{2z}^{(2)} - k_{2z}^{(1)}$, we find that by employing a sufficiently large angular separation $\delta\phi$ between the two reference waves $k_2^{(1)}$ and $k_2^{(2)}$, we can reduce the crosstalk. Choosing, somewhat arbitrarily, the location of the second zero of (9.3.19) as a measure of the necessary selectivity, we obtain

$$\delta = k_{2z}^{(2)} - k_{2z}^{(1)} = K_z^{(2)} - K_z^{(1)} \geq \frac{2\pi}{L} \tag{9.3.20}$$

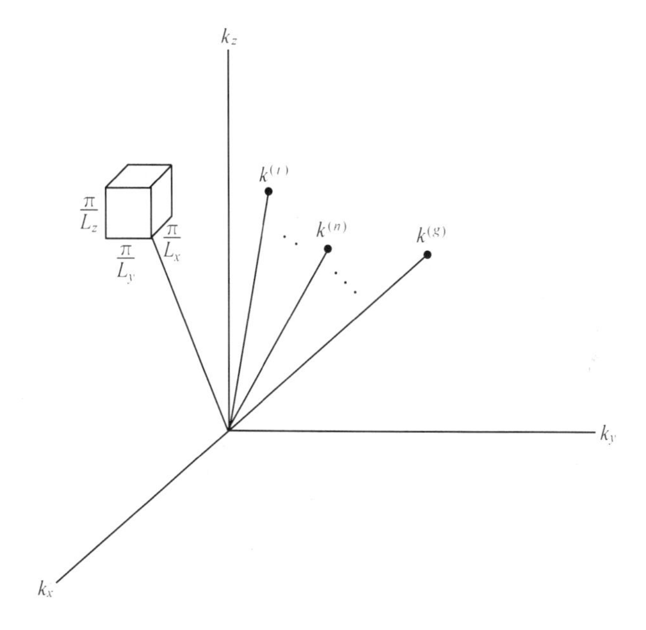

Figure 9.7 K space representation of holograms. Each hologram, $n_1^{(i)}(r) \propto \sin(k_1 \cdot r) + \phi^{(1)}$, is represented by a fuzzy volume centered on K_1. The fuzziness reflects the finite dimension L_x, L_y, L_z of the hologram. A plane wave hologram, such as, g, n, and t, is represented ideally as a point in K space. The small "fuzzy" volume shown represents the spread in K due to the finite volume of the hologram. The unit volume in K space associated with a single hologram is shown.

The strategy for volumetric data storage is thus to fill K space with $K^{(1)}$ vectors, each representing a single hologram, which are separated by

$$\Delta K_x \geqq \frac{2\pi}{L_x}, \Delta K_y \geqq \frac{2\pi}{L_z}, \Delta K_z \geqq \frac{2\pi}{L_z}$$

where L_x, L_y, L_z are the dimensions of the hologram.

We may thus need allocate to each hologram a volume

$$d^3 K = \frac{8\pi^3}{L_x L_y L_z} = \frac{8\pi^3}{V_{\text{holog}}} \tag{9.3.21}$$

in K space in order to avoid crosstalk. The total number of holograms that can be stored thus corresponds to packing K space with nonoverlapping holograms whose total number is

$$N_{\text{holog}} = \frac{\text{Total volume in K space}}{\text{Volume per hologram}} \approx \left(\frac{1}{2}\right)\frac{4\pi k^3 V_{\text{holog}}}{3 \times 8\pi^3} = \left(\frac{2\pi}{3}\right)\frac{V_{\text{holog}}}{\lambda^3}$$

where we took $|K| \sim k$. (Since $K = k_2 - k_2$, $|K|$ can vary between zero and $2k$, so on the average $|K| \sim k$.) The factor of 1/2 accounts for the fact that holograms with K and $-K$ are not independent. λ is the wavelength in the recording medium.

In our example of plane wave holograms, each hologram carries one bit of information. The hologram either exists (a "1") or does not (a "0"), so that the number of stored bits is equal to the number of holograms. The total number of bits that can be stored is thus

$$N_{\text{bits}} \sim \frac{V_{\text{holog}}}{\lambda^3} \tag{9.3.22}$$

According to (9.3.22), if we use $\lambda = 1\ \mu\text{m}$, the number of bits that can be stored exceeds $10^{12}/\text{cm}^3$. This storage density is intriguingly large and helps explain the interest in holographic data storage. Figure 9.8 shows a diagram of the experimental system used in recording some 1,000 angle-multiplexed diagrams in a $LiNbO_3$ crystal.

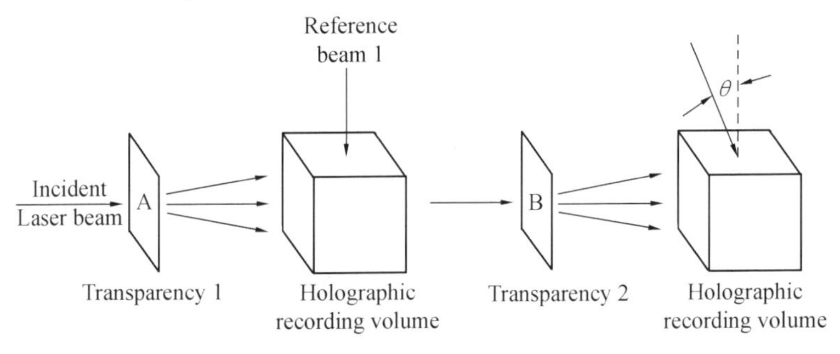

Figure 9.8 Generic schematic diagram of the optical setup for angular multiplexing of volume holograms. Two holograms are shown recorded in the same photosensitive volume but with each hologram using a different direction reference beam.

Wavelength Multiplexing(11)

Another method for multiplexing numerous holograms uses the geometry demonstrated in Figure 9.9. Here a given hologram is recorded with two oppositely traveling waves of the same wavelength. Each hologram is recorded with a different wavelength. The K space representation of this method is shown in Figure 9.9(a). The reconstruction of a given picture, say i, is accomplished by illuminating the hologram with the reference wave at λ_i, used to record it, as shown in Figure 9.9(b). This method of recording is called *wavelength multiplexing*. It makes the best use of K space and minimizes crosstalk, which results when the information contents of the holograms causes them to spread beyond their nominal K space address and encroach on the territory of adjacent holograms [11]. Figure 9.10 shows the reflection vs. λ of a large number of wavelength multiplexed holograms.

Crosstalk in Data-Bearing Holograms(12)

Up to this point we considered only plane wave holograms. We showed that in a recording volume with dimensions $L_i(i=1,2,3)$, such holograms needed to be separated in K by $\Delta K_i = 2\pi/L_i$ to avoid crosstalk. If the picture wave, prior to reaching the crystal, passes through a transparency, or a spatial light modulator, it is modulated spatially and can no longer be represented by a simple plane wave. If the wave incident on the transparency is taken as $A_1\exp(-ikz)$ and the transmittance of the transparency as $t(x,y)$, then the wave incident on the holographic medium can be taken as

$$E_{\text{picture}} = A_1 e^{-ikz} t(x,y) = A_1 \iint \exp[i(k_x x + k_y y) + i\sqrt{k^2 - k_x^2 - k_y^2}z]\tilde{t}(k_x, k_y)\mathrm{d}k_x \mathrm{d}k_y \quad (9.3.23)$$

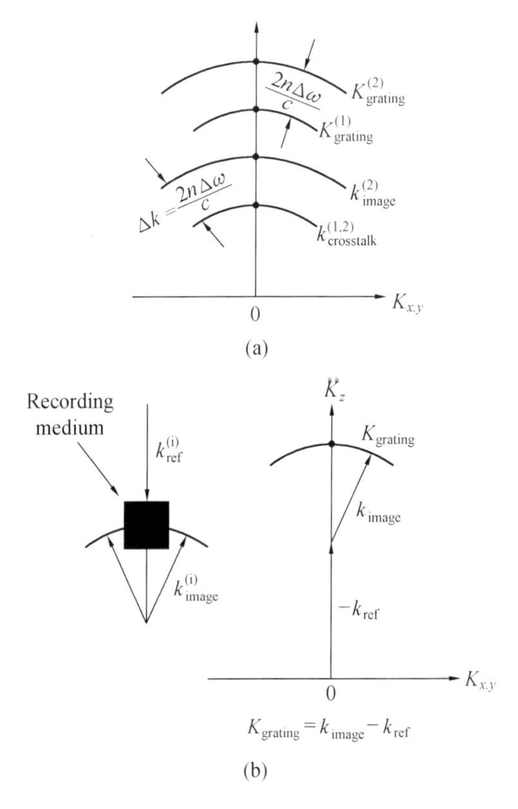

Figure 9.9 Holographic data storage with wavelength multiplexing. (a) The black dots on the K_z axis correspond to "blank" (no information) holograms. The transverse quasi-circular curves correspond to the loci of the tip of the $K^{(1)}$ vector (recorded with λ_i) when the holograms bear pictorial information. (b) A construction illustrating how the angular fanning of the $k^{(i)}_{\text{image}}$ due to stored information spreads the loci of $K^{(1)}$. (After Reference [11].)

where

$$t(x, y) = \iint \tilde{t}(k_x, k_y)\exp[i(k_x x + k_y y)]\mathrm{d}k_x \mathrm{d}k_y$$

with $\tilde{t}(k_x, k_y)$ the spatial Fourier transform of $t(x, y)$.

According to (9.3.23), the picture wave is no longer a single plane wave but a continuum of such waves, each with an amplitude $\tilde{t}(k_x, k_y)$ $dk_x dk_y$ propagating along the direction

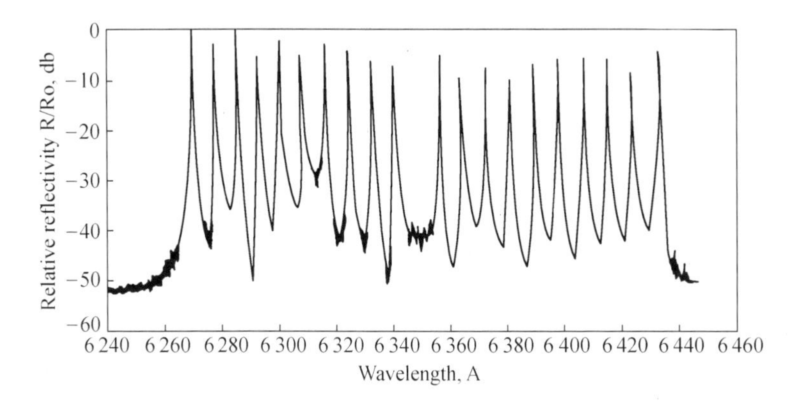

Figure 9.10 Diffraction(reflection) efficiency vs. λ in the case of 20 wavelength multiplexed hologram [Courtesy V. Leyva, G. Rakuljic-Accuwave Corp.]

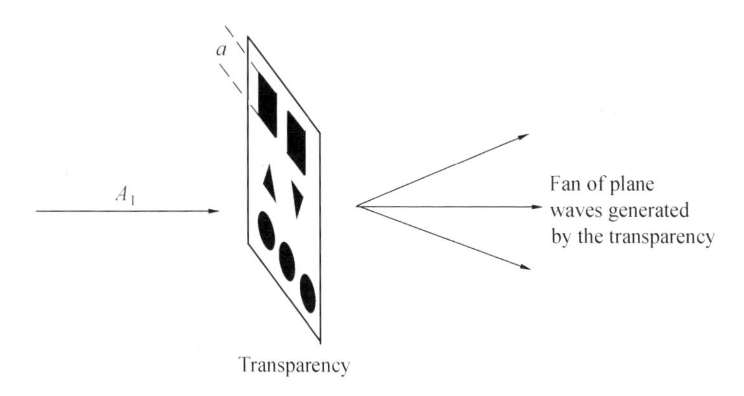

Figure 9.11 Passage of a plane wave through a transparency generates a continuum of plane waves on the output size.

as illustrated in Figure 9.11. If we plot the $\boldsymbol{K}$ vector of this hologram in $\boldsymbol{K}$ space as in Figure 9.9(b), the tip of the $\boldsymbol{K}$ vector now occupies a volume rather than a point. If the smallest feature size in the transparency is a, then we expect $\tilde{t}(k_x, k_y)$ to have appreciable value up to ~ $(k_x)_{\max} \sim \pi/a$, $(k_y)_{\max} \sim \pi/a$, so that, in general, the volume in $\boldsymbol{K}$

space needed to avoid overlap of holograms is now no longer $8\pi^3/L_xL_yL_z$ as in the case of plane wave (no information) holograms but becomes π^3/a^3. This reduces the number of holograms that can be recorded with negligible crosstalk from $\sim L_xL_yL_z/\lambda^3$ (see 9.3.22) to $\sim a^3/\lambda^3$. If we now associate a single bit with the volume a^3, the volume of the smallest feature, we find that the total number of bits that can be recorded and read with little crosstalk is

$$N_{\text{bits}} = N_{\text{holog}} \times \text{bits per hologram} =$$

$$\left(\frac{a}{\lambda}\right)^3\left(\frac{V}{a^3}\right) = \left(\frac{V}{\lambda}\right)^3$$

($V = L_xL_yL_z$ is the hologram volume)

which is the same as the result of Equation (9.3.22) for the number of bits that can be stored in the case of (one bit each) plane wave holograms. This is just one manifestation of the fact that the basic limit $\sim V/\lambda^3$ to holographic data storage is insensitive to the spatial modulation format, i.e., single-bit plane holograms or holograms that store multibit pages.

Problems

9.1 Show that if a planar hologram is made using a wavelength λ but is reconstructed with a wavelength λ_R, the reconstructed image is magnified by a factor of λ_R/λ with respect to the original object.

9.2 A monochromaticwave $A(r)$ is propagating essentially in the $+z$ direction. Show that, if in any plane z we replace A by its complex conjugate (but leave the factor $\exp(i\omega t)$ unchanged), the result is a new wave propagating in the $-z$ direction, but possessing everywhere wavefronts (i.e., loci of constant phase) identical to those of the original wave. [Hint: The expansion(1.6.12) may be useful.]

9.3

a. Using the notation of section 9.2, show that if the hologram is illuminated with a plane wave A_2^* is real—that is, A_1^* instead of A_1.

b. Show that the reconstructed image A_2^* instead of A_2 the reconstructed image is A_1^* actually converges to an image. [Hint: Consider what happens to a bundle of rays originally emanating from a point on the object.]

c. Show that the reconstructed image A_1 observed when the hologram is illuminated by A_2 is virtual; that is, rays corresponding to a given image point do not cross unless imaged by a lens.

9.4 Consider the problem of making a hologram in which the reference and object beams are incident on the emulsion from two opposite sides. Draw the equidensity planes for the case where the beams are nearly antiparallel. Show that the viewing (reconstructing) of this beam is performed in the reflection mode: that is, the viewer faces the side of the emulsion that is illuminated by the beam.

9.5 Show that in an infinitely thin hologram both virtual and real images can be reconstructed simultaneously. [Hint: Consider the problems of light scattering from a surface grating (as opposed to a volume grating).]

9.6 Calculate the reconstruction angle sensitivity $d\theta_B/d\lambda_R$ for transmission holograms (as described in the text) and in reflection holograms (as described in Problem 9.4). θ_B is the Bragg angle, and λ_R is the wavelength used in reconstruction. Show that $d\theta_B/d\lambda_R$ is much larger in the case of the transmission hologram. Which hologram will yield better results when illuminated by white light?

9.7 Plot the locus in **K** space of holograms that contain pictorial information with dimensions a. (The holograms were recorded with plane wave passing through a transparency with feature size ~ a.)

a. Plot for the case of angular multiplexing.

b. Plot for the case of wavelength multiplexing.

c. Show that in case (a) the preferred geometry for minimizing crowding in $\boldsymbol{K}$ space is one where the reference and picture waves are at 90° to each other.

9.8 Calculate the wavelength selectivity of a hologram recorded with two oppositely traveling plane waves ($+z$ and $-z$) of wavelength λ. Specifically, plot the power reflectivity $R(\lambda) = |r(\lambda)|^2$ for a wave incident along the z direction.

References

1. Gabor. D. Microscopy by reconstructed wavefronts. *Proc. Roy. Soc.* (*London*). 1949, 197(A): 454
2. Leith E N, J Upatnieks. Wavefront reconstruction with diffused illumination and three-dimensional objects. *J. Opt. Soc. Am.* 1964, 54: 1295
3. Collier R J. Some current views on wavefront reconstruction. *IEEE Spectrum*, 1966, 3: 67
4. Stroke G W. *An Introduction to Coherent Optics and Holography*. 2nd ed. New York: Academic, 1969
5. DeVelis J B. G O Reynolds. *Theory and Applications of Holography*, Reading, MA: Addison-Wesley, 1967
6. Smith. H M. *Principles of Holography*. New York: Interscience, 1969
7. Goodman J W. *Introduction to Fourier Optics*. New York: McGraw-Hill, 1968
8. Yu T S F. *Introduction to Diffraction Information Processing and Holography* Cambridge. MA: MIT, 1973
9. Kogelnik H. Coupled wave theory for thick hologram gratings. *Bell Syst. Tech. J* 1969, 48: 2909
10. Mork H F. Angle multiplexed storage of 5 000 holograms in lithium

niobate. Opt. Lett., 1993, 18:915

11. Rakuljic G, Leyva V, and Yariv, A. Optical data storage by using wave-length multiplexed volume holograms, *Opt. Lett.*, 1992, 17: 1473

12. Yariv A. Interpage and interpixel crosstalk in wavelength multiplexed holograms. *Opt. Lett.* 1993, 18:652

New Words and Expressions

analogous *adj*. 类似的,相似的

arbitrarily *adv*. 任意地

bisector *n*. [数]平分线

boundary condition *n*. 边界条件

Bragg 布拉格,英国物理学家

Bundle *n*. 包,束
v. 捆扎

continuum *n*. 连续统一体

crosstalk *n*. 色度亮度干扰

cumulate *adj*. 堆积的,累积的

diffraction *n*. 衍射

diffraction grating *n*. 衍射光栅

dynamic *adj*. 动态的

emulsion *n*. [摄]感光乳剂

equidensity *n*. 等密度

fanning *n*. 铺开,展开

fuzzy *adj*. 模糊的,失真的

hologram *n*. 全息图,全息摄影

holography *n*. 全息术,全息摄影术

intensity *n*. 强度,亮度

isotropic *adj*. 各向同性的

kinetic	*adj*. 运动的,动力学的
manifestation	*n*. 显示,表现,证明
multiplexing	*n*. 多路技术
mutually	*adv*. 互相地
negligible	*adj*. 可以忽略的
optical intensity	*n*. 光强
optical loss	*n*. 光损耗
perpendicular	*n*. 垂线
	adj. 垂直的,正交的
phase	*n*. 相,相位
	v.定相
photosensitive	*adj*. [物]感光性的,光敏的
plane wave	*n*. 平面波
propagation	*n*. (声波、电磁辐射等)传播
proportionate	*v*.成比例
	adj. 成比例的
qualitative	*adj*. 定性的
radiate	*v*. 发光,放射,辐射
	adj. 辐射状的
readout	*n*.读出器,读出
respectively	*adv*. 分别地,各自地
sinusoidal modulation	*n*. 正弦调制器
synchronous	*adj*. [物]同步的
transparency	*n*.透明,透明度,幻灯片
wavefront	*n*. 波阵面
wavelength	*n*. [物]波长

NOTES

1. If the photosensitive medium were a photographic emulsion, the exposure to the two beams and subsequent development would result in silver atoms developed out at each point in the emulsion in direct proportion to the time average of the square of the optical field.

如果光敏介质是照相感光乳剂,那么两束光线重叠曝露后,接下来的变化将导致银原子在感光乳剂上的每一个点都能按正方形光场的平均时间准确的按比例分布。

2. Each one of these waves interferes with the reference beam, creating, after development, its own diffraction grating, which is displaced slightly in angle from that of the other gratings.

与参考波干涉的每个波都会产生各自的衍射光栅,每个光栅都与另一个光栅有一个很小的角度偏移。

3. If the ***K*** vectors representing the different holograms are sufficiently different, it is possible to read one specific hologram with negligible contributions —— crosstalk ——from the others, since, for all the other holograms, the Bragg condition is strongly violated.

如果用以表现不同全息图的 ***K*** 矢量完全不同,就有可能通过比较某个详细而又精确的全息图与其他全息图样得到色度亮度的微小差别,因为对所有其他全息图来说,布拉格条件是完全不同的。

4. It makes the best use of ***K*** space and minimizes crosstalk, which results when the information contents of the holograms cause them to spread beyond their nominal K space address and encroach on the territory of adjacent holograms.

它充分的利用了 ***K*** 空间,并使色度亮度干扰达到最小。当全息图所包含的信息内容过多而导致它超越了它们名义上的 K 空间地址,使相邻全息图区域重叠。

PART FIVE
PAPER EXAMPLES

10

Paper Examples

Designing an Optical Disk Lens without Analytical Definition of Aspheric Surfaces

1. Introduction

Aspherics are applied more and more widely in modern optical systems as the use of aspheric surfaces not only reduces the number and cost of elements used in an optical system but also allows better correction of wave-front aberrations and even design configurations that are not possible with all-spherical systems. Thus aspherics have been playing an indispensable role in some critical cases, e. g., compact optical disk storage.

The conventional approach of designing an appropriate profile of the aspheric surface uses automatic optical design software based on ray-tracing aberration calculations and optimization algorithms, among which the most widely used is the damped least-squares method.[1] In addition, many other methods have been introduced in efforts to develop an optical design technique, for example, the adaptive correction method[2] and the simultaneous nonlinear inequalities method.[3] Despite the distinctive aspects of each method, many common properties can be found in all these methods, one of which is that all the surfaces involved in the optical system must be described analytically with some characteristic parameters. For instance, to describe an aspheric surfaceproperly, the surface curvature, the conic constant, and a series of aspheric coefficients have to be applied. On the one hand, because the aspheric surface definition used must be able to approximate the ideal surface that provides the full necessary correction to the wave-front aberrations, it is desirable that the aspheric coefficient order be as high as possible. On the other hand, owing to the principle that the speed of optimization is roughly proportional to the square of the number of variables used, it is desirable that the aspheric representation define the proper surface with a minimum of terms. Thus the dilemma arises Moreover, though the standard aspheric surface can effectively control aberrations in most optical systems, it falls short in representing highly aspheric shapes that compensate. For a high degree of wave-front asphericity as the surface normal becomes perpendicular to the optical axis. Although people have been trying to devise new approaches to describe aspheric surfaces better,[4] there could hardly be a versatile analytic definition for all cases. Many other methods have been developed with different techniques and concepts. Schulz suggested an algorithm[5] capable of designing an m thorder aplanatic lens on the basis of expansion of the wave aberration. This method needs two stages in the design procedure, and the whole algorithm is quite complicated. Benitez and Mi ñ ano developed a method to form an ultrahigh-numerical-aperture imaging concentrator.[6] In this method, two symmetrical fields ($\pm\ \alpha°$) of rays in the object side are coupled with another two in the image side. Their design turns out to be aberration free in fields of $\pm\ \alpha°$, but the axial performance is not well guaranteed, so the design is not suitable for cases such as optical disk systems that required both axial stigmatism and offaxial aplanatism. Considering such disadvantages of those techniques, we developed a simpler method to design aspheric surfaces, applying discretization instead of analytic definition; thus an arbitrary aspheric profile of the surfaces can be automatically solved point by point.

2. Principle of the Method

Here we give an example of an optical disk lens to illustrate the basic principle of this new method. Figure 1 shows the general configuration of the objective lens in an optical disk pickup head. Typically, the incident light is collimated, and it is focused onto the plane carrying bit information by the objective lens, which is usually designed as a singlet for the purpose of reducing cost. Thus the lens has to be biaspherical to achieve diffraction-limited performance.[7] People have been used to using conventional optical design software to calculate the proper profile for the lens. But we intend to show that with our new method the same purpose can easily be attained. Furthermore, as we illustrate below, results derived by our method are better than those of conventional methods.

In general, an optical system is rotationally symmetrical; thus we consider only the case in the tangential plane.[6,8,9] Furthermore, our solving domain is restrained to the y-positive portion, namely, $0 \leqslant y \leqslant R$, where R is the entrance pupil radius. Let S^1 and S^2 denote surfaces 1 and 2 of the lens; discretization is implemented on their generatrix, and we get point sequences $\{S_j^1\}$ and $\{S_j^2\}$; here the superscript denotes the surface index and the subscript denotes the point index. In addition, though surface 3 (S^3) is a flat surface rather than an asphere, we also need to discretize it into point sequence $\{S_j^3\}$, which is of benefit for derivation. Once the coordinates of $\{S_j^1\}$ and $\{S_j^2\}$ are determined, so are the desired profiles of the aspheres. Therefore our goals are to seek

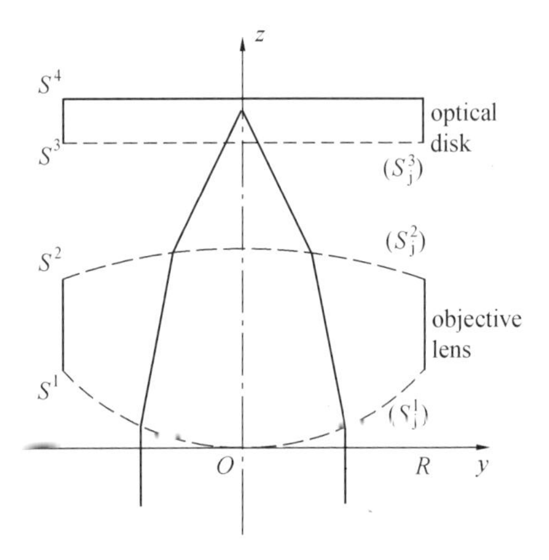

Figure 1 Configuration of the optical disk system. The collimated light is focused onto an optical disk by the objective lens. Dashed curves, discretized surfaces.

the requirements that $\{S_j^1\}$ and $\{S_j^2\}$ have to meet to include both stigmatism on axis and aplanatism in the $\alpha°$ field and then eventually to get them resolved. For convenience, we prescribe coordinates of those relevant points in Table 1, and, next, we set up the relationship among them. We assume points $S_{j-2}^{1,2}$, $S_{j-1}^{1,2} \in \{S_j^{1,2}\}$ are known or already determined.

Table 1. Coordinates of the Relevant Points

Point	Coordinates	Point	Coordinates
S_j^1	(y_j^1, z_j^1)	A_{+1}^1	(ξ_j, η_j)
S_{j-1}^1	(y_{j-1}^1, z_{j-1}^1)		
S_{j-2}^1	(y_{j-2}^1, z_{j-2}^1)		
S_j^2	(y_j^2, z_j^2)		
S_{j-1}^2	(y_{j-1}^2, z_{j-1}^2)		
S_{j-2}^2	(y_{j-2}^2, z_{j-2}^2)		
S_j^3	$(v_j, t + WD)$	A_{+1}^3	$(\lambda^j, t + WD)$
P_0	$(0, t + WD + td)$	P_+	$(f' \tan\alpha, t + WD + td)$

A. Requirements To Be Met in the Case of a 0° Field

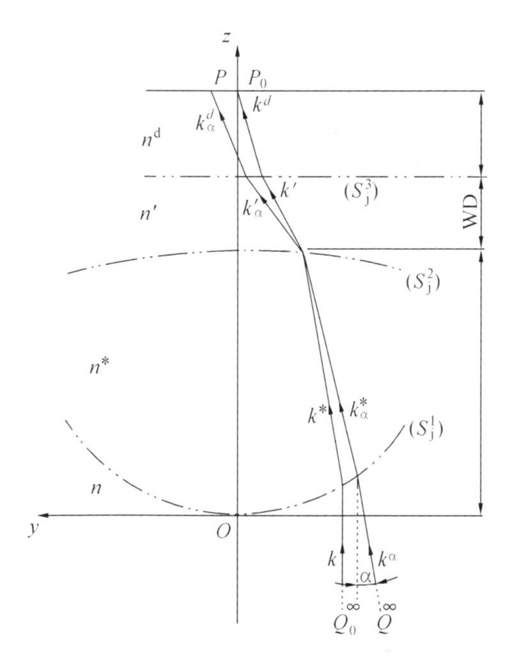

Figure 2 Schematic plot of the ray paths for the optical system. WD, working distance.

We join Q_0, S_j^1, S_j^2, S_j^3, and P_0 with a solid intersecting line, as shown in Fig 2, which denotes the path that the 0° field ray passes along, where Q_0 is the object point at infinite and P_0 is the ideal Gaussian image point on the focal plane. It is clear that this line should satisfy Snell's refraction law at points S_j^1 and S_j^2. Assuming $\boldsymbol{T}$, $\boldsymbol{T}^*$, and $\boldsymbol{T}'$ are tangential vectors at S_j^1, S_j^2, and S_j^3, respectively, we get

$$(n\boldsymbol{k} - n^*\boldsymbol{k}^*)\cdot\boldsymbol{T} = 0, \tag{1}$$

$$(n^*\boldsymbol{k}^* - n'\boldsymbol{k}')\cdot\boldsymbol{T}^* = 0, \tag{2}$$

$$(n'\boldsymbol{k}' - n^d\boldsymbol{k}^d)\cdot\boldsymbol{T}' = 0, \tag{3}$$

Components of vectors $\boldsymbol{k}$, $\boldsymbol{k}^*$, $\boldsymbol{k}'$, $\boldsymbol{k}^d$, $\boldsymbol{T}$, $\boldsymbol{T}^*$, and $\boldsymbol{T}'$ are shown in Table 2.

Table 2. Components of Vectors k, k^*, k', k^d, T, T^*, and T'

Vectors	Components[a]		
$\boldsymbol{k}$	0	0	1
$\boldsymbol{k}^*$	0	$\cos\left[\tan^{-1}\left(\frac{z_j^2 - z_j^1}{y_j^2 - y_j^1}\right)\right]$	$\sin\left[\tan^{-1}\left(\frac{z_j^2 - z_j^1}{y_j^2 - y_j^1}\right)\right]$
$\boldsymbol{k}'$	0	$\cos\left[\tan^{-1}\left(\frac{t + WD - z_j^2}{v_j - y_j^2}\right)\right]$	$\sin\left[\tan^{-1}\left(\frac{t + WD - z_j^2}{v_j - y_j^2}\right)\right]$
$\boldsymbol{k}^{\mathrm{d}}$	0	$\cos\left[\tan^{-1}\left(\frac{t_d}{-v_j}\right)\right]$	$\sin\left[\tan^{-1}\left(\frac{t_d}{-v_j}\right)\right]$
$\boldsymbol{T}$	0	$\cos\left[\tan^{-1}\left(\frac{dz}{dy}\bigg\vert_{S_j^1}\right)\right]$	$\sin\left[\tan^{-1}\left(\frac{dz}{dy}\bigg\vert_{S_j^1}\right)\right]$
$\boldsymbol{T}^*$	0	$\cos\left[\tan^{-1}\left(\frac{dz}{dy}\bigg\vert_{S_j^2}\right)\right]$	$\sin\left[\tan^{-1}\left(\frac{dz}{dy}\bigg\vert_{S_j^2}\right)\right]$
$\mathbf{T}'$	0	1	0

"WD, working distance."

If we replace the differential with the difference, namely,

$$\frac{dz}{dy}\bigg\vert_{S_j^1} = \frac{\Delta z}{\Delta y}\bigg\vert_{S_j^1} = \frac{z_j^1 - z_{j-1}^1}{y_j^1 - y_{j-1}^1}, \tag{4}$$

$$\frac{dz}{dy}\bigg\vert_{S_j^2} = \frac{\Delta z}{\Delta y}\bigg\vert_{S_j^2} = \frac{z_j^2 - z_{j-1}^2}{y_j^2 - y_{j-1}^2}, \tag{5}$$

then Eqs. (1) - (3) can be rewritten into a form that contains only known quantities.

B. Requirements To Be Met in the Case of $\alpha°$ Field

In this case a special ray is needed for derivation. This ray meets point S_j^2 on surface S^2 and it intersects S^1 at A_{+1}^1. Points A_{+1}^1 and S_j^1 do not coincide, and the former is merely an auxiliary point for derivation, whereas the latter is the solution goal. Similarly, the refraction law should be met at points A_{+1}^1 and S_j^2, namely,

$$(n\boldsymbol{k}_\alpha - n^*\boldsymbol{k}_\alpha^*)\cdot\boldsymbol{T}_\alpha = 0, \tag{6}$$

$$(n^*\boldsymbol{k}_\alpha^* - n'\boldsymbol{k}'_\alpha)\cdot\boldsymbol{T}_\alpha^* = 0, \tag{7}$$

$$(n'\boldsymbol{k}'_\alpha - n^d\boldsymbol{k}_\alpha^d)\cdot\boldsymbol{T}'_\alpha = 0. \tag{8}$$

Here $\boldsymbol{T}_\alpha$ and $\boldsymbol{T'}_\alpha$ are the tangential vectors at points A^1_{+1} and A^3_{+1}, respectively.

Components of vectors $\boldsymbol{k}_\alpha$, $\boldsymbol{k}^*_\alpha$, $\boldsymbol{k'}_\alpha$, $\boldsymbol{k}^d_\alpha$, $\boldsymbol{T}_\alpha$, $\boldsymbol{T}^*_\alpha$ and $\boldsymbol{T'}_\alpha$ are shown in Table 3.

Table 3. Components of Vectors $\boldsymbol{k}_\alpha$, $\boldsymbol{k}^*_\alpha$, $\boldsymbol{k'}_\alpha$, $\boldsymbol{k}^d_\alpha$, $\boldsymbol{T}_\alpha$, $\boldsymbol{T''}_\alpha$, $\boldsymbol{T}^*_\alpha$ and $\boldsymbol{T'}_\alpha$

Vectors	Components[a]		
$\boldsymbol{k}_\alpha$	0	$\sin\alpha$	$\cos\alpha$
$\boldsymbol{k}^*_\alpha$	0	$\cos\left[\tan^{-1}\left(\frac{z^2_j-\eta_i}{y^2_j-\xi_j}\right)\right]$	$\sin\left[\tan^{-1}\left(\frac{z^2_j-\eta_i}{y^2_j-\xi_j}\right)\right]$
$\boldsymbol{k}_d'$	0	$\cos\left[\tan^{-1}\left(\frac{t+WD-z^2_j}{\lambda_j-y^2_j}\right)\right]$	$\sin\left[\tan^{-1}\left(\frac{t+WD-z^2_j}{\lambda_j-y^2_j}\right)\right]$
$\boldsymbol{k}^d_\alpha$	0	$\cos\left[\tan^{-1}\left(\frac{t_d}{f'\sin\alpha-\lambda_j}\right)\right]$	$\sin\left[\tan^{-1}\left(\frac{t_d}{f'\sin\alpha-\lambda_j}\right)\right]$
$\boldsymbol{T}_\alpha$	0	$\cos\left[\tan^{-1}\left(\left.\frac{dz}{dy}\right\rvert_{A^1_{+1}}\right)\right]$	$\sin\left[\tan^{-1}\left(\left.\frac{dz}{dy}\right\rvert_{A^1_{+1}}\right)\right]$
$\boldsymbol{T}^*_\alpha$	0	$\cos\left[\tan^{-1}\left(\left.\frac{dz}{dy}\right\rvert_{S^2_j}\right)\right]$	$\sin\left[\tan^{-1}\left(\left.\frac{dz}{dy}\right\rvert_{S^2_j}\right)\right]$
$\boldsymbol{T'}_\alpha$	0	1	0

Then the differential at point S^2_j is replaced with the difference in the manner shown in Eq. (6), and the differential at point A^1_{+1} is predicted by linear interpolation:

$$\left.\frac{dz}{dy}\right\rvert_{A^1_{-1}} = L(\xi_j, z_y') = \left.\frac{\Delta z}{\Delta y}\right\rvert_{S^1_j} + \left.\frac{\Delta\left(\frac{\Delta z}{\Delta y}\right)}{\Delta y}\right\rvert_{S^1_j}(\xi_j - y^1_j)$$

$$= \frac{z^1_j - z^1_{j-1}}{y^1_j - y^1_{j-1}} + \frac{\frac{z^1_j - z^1_{j-1}}{y^1_j - y^1_{j-1}} - \frac{z^1_{j-1} - z^1_{j-2}}{y^1_{j-1} - y^1_{j-2}}}{y^1_j - y^1_{j-1}}(\xi_j - y^1_j) \quad (9)$$

Then Eqs. (6) – (8) can also be rewritten in a form containing only known quantities.

Now we have derived Eqs. (1) – (3) and (6) – (8) from Snell's law, and they are called the aspheric refractive equations, which define the requirements that the aspheric lens should meet. In addition, we intend to add one constrained equation that may make the definition more accurate. This equation is to define the relationship between the y coordinate and the z coordinate of the auxiliary point A^1_{+1} with a quadratic interpolation:

$$
\begin{aligned}
z(A^1_{+1}) &= \eta_j \\
&= L_2(\xi_j, z) \\
&= \frac{(\xi_j - y^1_{j-1})(\xi_j - y^1_j)}{(y^1_{j-2} - y^1_{j-1})(y^1_{j-2} - y^1_j)} z^1_{j-2} \\
&\quad + \frac{(\xi_j - y^1_{j-2})(\xi_j - y^1_j)}{(y^1_{j-1} - y^1_{j-2})(y^1_{j-1} - y^1_j)} z^1_{j-1} \\
&\quad + \frac{(\xi_j - y^1_{j-2})(\xi_j - y^1_{j-1})}{(y^1_j - y^1_{j-2})(y^1_j - y^1_{j-1})} z^1_j .
\end{aligned}
\tag{10}
$$

Obviously, Eq. (10) contains only known quantities, too.

Without loss of generality, we put a uniformly spaced grating on the generatrix of S^1; namely, the y coordinate of point S^1_j is determined as

$$
y^1_j = jh = j\frac{R}{N} \qquad (j = 0, 1, 2, \cdots, N), \tag{11}
$$

where h is the step and N is the sample number. Thus, we get a group of equations, i.e., Eqs. (1) – (3), (6) – (8), (10), and (11), which is called the aspheric complete equations (ACEs) and can be rewritten in the form

$$
\begin{aligned}
&\boldsymbol{F}_j(\boldsymbol{X}_j) = 0, \\
&\boldsymbol{F}_j = [f^1_j, f^2_j, f^3_j, f^4_j, f^5_j, f^6_j, f^7_j, f^8_j]^T, \\
&\boldsymbol{X}_j = [y^1_j, z^1_j, y^2_j, z^2_j, \xi_j, \eta_j, \upsilon_j, \lambda_j]^T .
\end{aligned}
\tag{12}
$$

Equations (12) show that if points $S^{1,2}_{j-2}, S^{1,2}_{j-1} \in \{S^{1,2}_j\}$ are known, then point $S^{1,2}_j$ can be determined when the ACE is solved. This means that if boundary conditions, namely, coordinates of points $S^{1,2}_0$ and $S^{1,2}_1$, can be given, the whole point sequence $\{S^{1,2}_j\}$ can be solved point by point, namely, in a procedure such as $S^{1,2}_2 \to S^{1,2}_3 \to \cdots \to S^{1,2}_N$, and then the desired aspheric profile is obtained.

3. Boundary Conditions and Solution Procedure of the Aspheric Complete Equation

As mentioned above, to determine the whole point sequence $\{S_j^{1,2}\}$, we have to prescribe appropriate boundary conditions for the ACE. Thus coordinates of points $S_0^{1,2}$ and $S_1^{1,2}$ must be given in advance. Fortunately, a domain close enough to the optical axis of a conventional spherical lens can be treated as a Gaussian domain, and such a domain of an appropriate spherical lens can be sampled and taken as the boundary conditions for the aspheric lens that is to be solved. Now the problem is how we can construct the spherical lens, namely, how we can determine the radii of its two surfaces, to meet the requirements of the optical system. According to the paraxial formula, if only the refraction index (n), the central thickness (t), the focal length (f'), and the working distance (WD) of the objective lens and the refraction index (n_d) and thickness(t_d) of the optical disk are given, the curvature of the first surface ($c_1 = 1/r_1$) of the appropriate spherical lens is determined as

$$c_1 = \frac{1}{t}\left(1 - \frac{WD + t_d/n_d}{f'}\right)\frac{n}{n-1} \tag{13}$$

With Eq. (13), curvature of the second surface ($c_2 = 1/r_2$) can be determined easily by application of the focal-length formula. Thus the two radii of the appropriate spheres are all determined. and then ray tracing is implemented on this spherical lens. First, we apply ray tracing with incident rays that are vertically collimated and spaced by $h = R/N$ as Eq. (11) shows, thus the intersection of the jth ray with surfaces 1,2, and 3 can be determined as $S_j^1(\tilde{y}_j^1, \tilde{Z}_j^1)$, $S_j^2(\tilde{y}_j^2, \tilde{z}_j^2)$, and $\tilde{S}_j^3(\tilde{y}_j^3, t + WD)$, respectively. Next, we apply inverse ray tracing and find a ray that also meets surface 2 at point $\tilde{S}_j^2$, and this ray meets surface 3 and 1 at points $\tilde{A}_{+1}^3(\tilde{y}_{+1}^3, t + WD + t_d)$ and $\tilde{A}_{+1}^1(\tilde{y}_{+1}^1, \tilde{z}_{+1}^1)$, respectively.

These quantities of the spherical lens are then taken as conditions for the aspheric lens, namely,

$$
\begin{aligned}
y_j^1 &= \tilde{y}_j^1, & \xi_j &= \tilde{y}_{+1}^1, \\
z_j^1 &= \tilde{z}_j^1, & \eta_j &= \tilde{z}_{+1}^1, \\
y_j^2 &= \tilde{y}_j^2, & \upsilon_j &= \tilde{y}^3, \\
z_j^2 &= \tilde{y}_j^2, & \lambda_j &= \tilde{z}_{+1}^3 \quad (j = 0,1).
\end{aligned}
$$

Because $S_0^{1,2}$ and $S_1^{1,2}$ are now all determined, we can solve the whole point sequence by applying the ACE as Eqs. (12) shows.

The typical methold to solve nonlinear equations such as the ACE is the Newton iterative algorithm. However, it is not convergent if the initial value of iteration is not given properly; therefore we apply the parametric Newton method instead of the conventional method, and the whole soultion procedure can be carried out as follows:

1. Use the paraxial formula to determine the radii of the appropriate spherical lens that meets requirements of the optical system.

2. Apply ray tracing to the appropriate spherical lens and determine $S_0^{1,2}$ and $S_1^{1,2}$.

3. Solve the jth ACE with the parametric Newton algorithm, i.e.,

a. $\boldsymbol{X}_j^{(0)} = \boldsymbol{X}_{j-1}, j = 2,3,\cdots,N$

b. $\boldsymbol{X}_j^{(k+1)} = \boldsymbol{X}_j^{(k)} + \Delta\boldsymbol{X}_j^{(k)}, k = 0,1,2,\cdots,$

$$\frac{\mathrm{d}\boldsymbol{F}_j}{\mathrm{d}\boldsymbol{X}}[\boldsymbol{X}_j^{(k)}]\Delta\boldsymbol{X}_j^{(k)} = -\omega\cdot\boldsymbol{F}_j[\boldsymbol{X}_j^{(k)}],$$

ω is selected to satisfy $||\boldsymbol{F}_j[\boldsymbol{X}_j^{(k+1)}]|| \leqslant ||\boldsymbol{F}_j[\boldsymbol{X}_j^{(k)}]||$.

c. If $||\Delta\boldsymbol{X}_j^{(k)}|| \leqslant \epsilon$, let $\boldsymbol{X}_j = \boldsymbol{X}_j^{(k)}$.

4. Increase j by 1 and return to step 3 until $j > N$.

4. Simulation Results

We have performed a computer simulation successfully by designing a biaspheric objective lens used in a digital versatile disk [(DVD), an optical disk format, whose capacity is usually 4.7 G] pickup head with N.A. = 0.6. The first-order design parameters are listed in Table 4, and the schematic setup is like that shown in Fig 2. To compare the performance of the lens from our method with that from the conventional method, we first design the lens by using conventional optical design software, ZEMAX version 5.0. In this design we try to account for both stigmatism on axis and aplanatism in the 0.5° field. The highest order of the aspheric coefficient we used is the twelfth, and lens A is obtained. Then we apply the new method in MATLAB, which is software for scientific calculation, containing many mathematic tool kits, to design the lens under the same conditions, with sample number $N = 25{,}000$, and this time we get lens B.

Table 4. First-Order Design Parameters of the Optical System

Parameters	Quantities
Focal length	3.36 mm
Entrance pupil diameter	4 mm
Refractive index of the lens (n^*)	1.49
Refractive index of the disk (n^d)	1.58
Central thickness of the lens	2.2 mm
Working distance	1.71 mm
Thickness of the disk	0.6 mm

Figure 3 shows the profile difference between lens A and lens B. We can see that A and B almost coincide in the portion for which pupil height y is small; when y increases, the two profiles deviate from each other, with the largest departure of tens of micrometers at the maximum pupil height. This shows that the new method is quite different from the conventional method in designing procedure and concept.

To compare the performance of lenses A and B. we present ray aberrations of lenses A and B in Figs 4 and 5, respectively. Note that the conventional vector algorithm of ray tracing for analytic surfaces is not suitable for the discretely defined surfaces here. This means we cannot

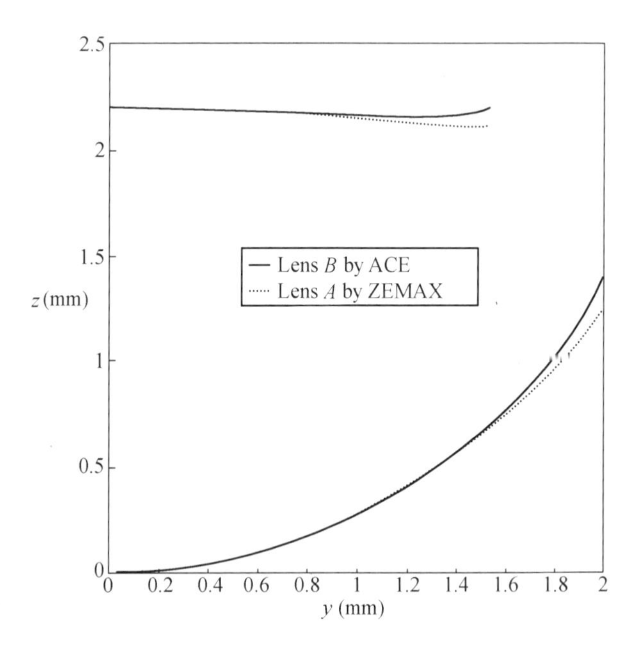

Figure 3. Profile difference between lens *A* designed with ZEMAX and lens *B* designed with the new method.

evaluate lens *B* by using current software such as ZEMAX. Therefore we have developed a modified vector algorithm for rigorous evaluation. It is based on the conventional method and amended to accommodate features of discretely defined surfaces. As the figures show, in the case of the 0° field, ray aberrations of lens *B* are smaller than those of lens *A* at normalized pupil height $PY(PX) < 0.7$; even in the domain where $PY(PX) > 0.7$, ray aberrations of lens B are quite comparable with those of lens *A*. In contrast, in the case of the 0.5° field, it is obvious that the tangential aberration of *A* is considerably larger than that of *B*, as the maximum of the former is 1.18 μm, which is 4 times larger than that of the latter, which is less than 0.3 μm. Thus we can say that the performance of lens *B* is better than that of lens *A* in general.

In addition, we evaluate the performance over the whole aperture of

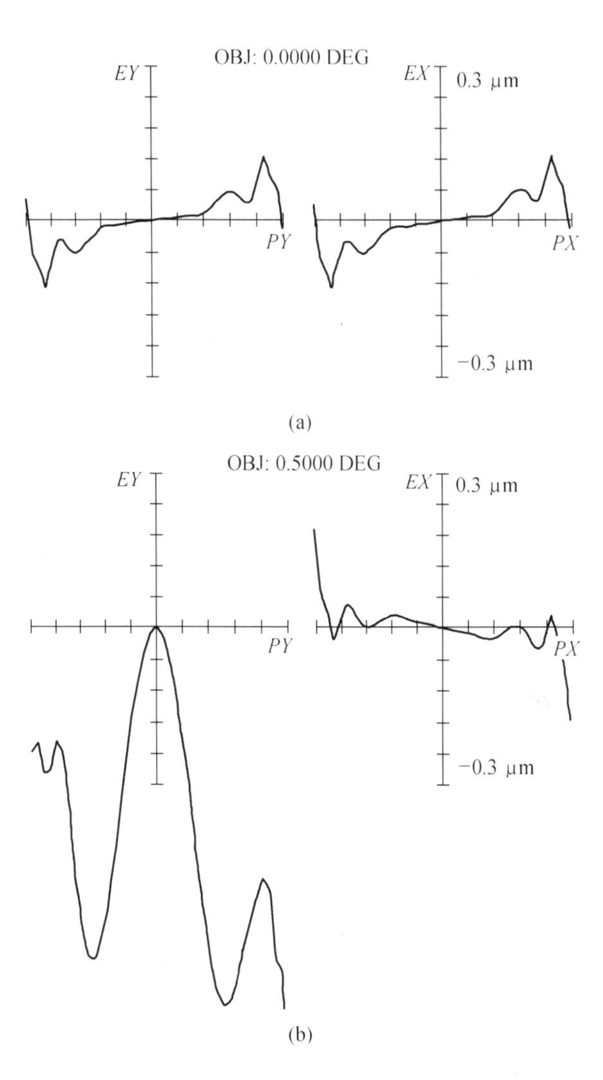

Figure 4. Ray aberrations of lens A designed with ZEMAX in the case of the (a) 0° field and (b) 0.5° field.

the lens with the spot diagram shown in Fig. 6, from which we can also say that lens *B* is a better result because of its better off-axis performance. The GEO spot sizes [a term used in ZEMAX software and

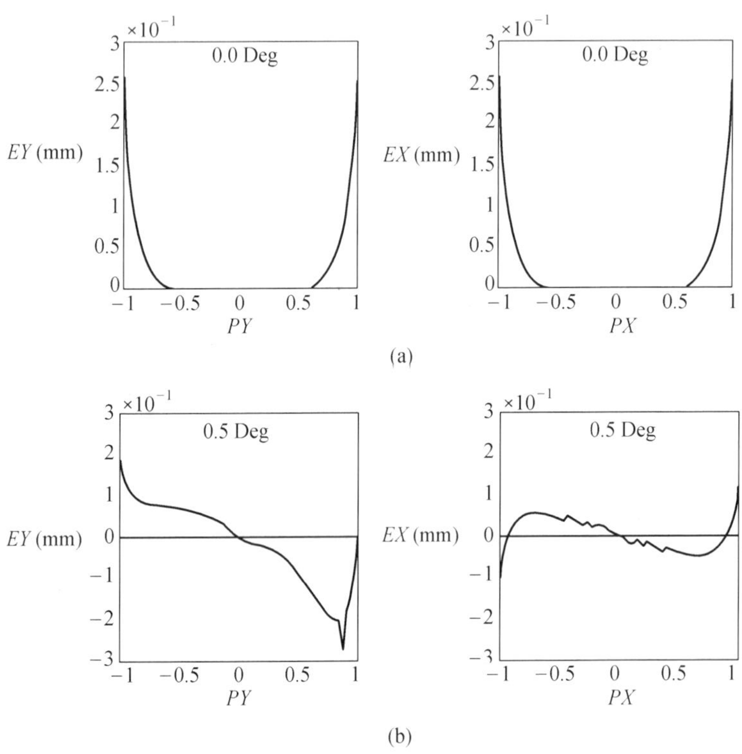

Figure 5 Ray aberrations of lens B designed with the new method in the case of the (a) 0° field and (b) 0.5° field.

defined as the radius of the circle that encloses all rays] of lenses *A* and *B* are 0.746 and 0.27 μm, respectively. When the rms spot radius is referred to, the ratio is 37%; the rms spot size of lens *A* is 0.224 μm and that of lens *B* is 0.083 μm.

There is an explanation of why our method is superior to the conventional one. We have tried to fit coordinates of the discrete points of both surfaces in lens *B* into the form of the analytic equation that is used in ZEMAX to describe aspheric surfaces, as[10]

$$z = \frac{cy^2}{1 + [1 - (1 + k)c^2y^2]^{1/2}} + \alpha_1 y^2 + \alpha_2 y^4 + \alpha_3 y^6 + \alpha_4 y^8 + \alpha_5 y^{10} + \alpha_6 y^{12}. \tag{14}$$

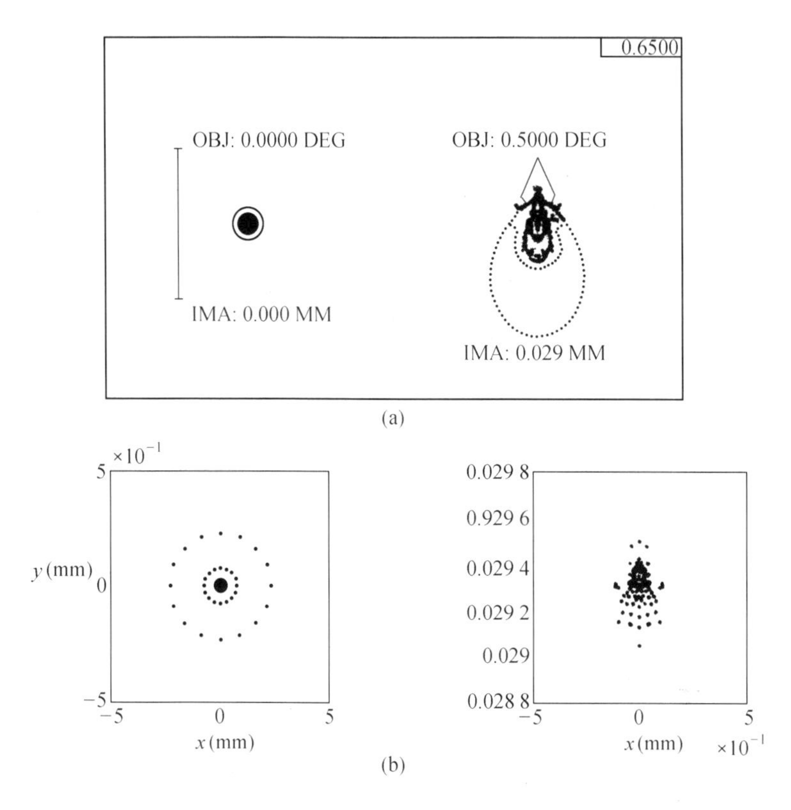

Figure 6. Spot diagram of the lens (a) designed with ZEMAX and (b) designed with the new method.

Thus we get the best-fit analytic aspheric surfaces of the discrete design, and differences between them are shown in Fig.7. There exists a maximum deviation of as much as 2 μm on surface 1. Thus if the highest aspheric order is only the twelfth, then the analytic surfaces cannot match the performance of the discrete ones, no matter what aspheric coefficients are applied in ZEMAX. If the maximum deviation of 0.1 μm is taken as the minimum requirement, according to our further calculations, the analytic surfaces are able to perform as well as the discrete ones only if the highest aspheric order is the twentieth. This sufficiently shows that our method is more efficient, flexible, and reliable, regardless of aspheric order. And we can predict that for those applications that need ultrahigh asphericity, the method we developed is preferable.

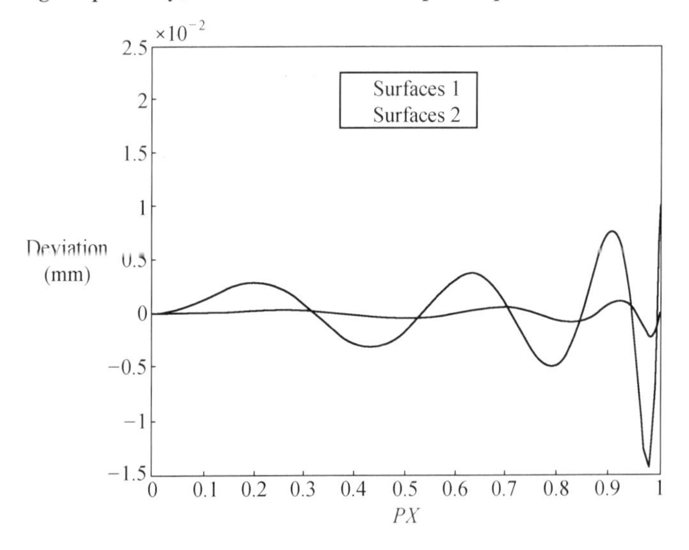

Figure 7. Deviation of the best-fit analytic aspheric surfaces with the discrete ones.

Moreover, another point is to be addressed. As our solving domain is constrained to $0 \leqslant y \leqslant R$, and trajectories of the α° field rays are defined only in that portion, it seems that a sharp focusing of the field is ensured only for the rays falling within that domain; however, it can be seen from the evaluation that those rays in the domain $-R \leqslant y \leqslant 0$ do not cause deterioration of the performance. That means fulfillment of aplantism over the whole aperture of the lens.

5. Conclusion

We have applied a new method, using discretization instead of analytic definition, to represent aspheric surfaces in designing a biaspheric objective lens for a DVD pickup head of the next generation. The aspheric complete equation is implemented to describe the requirements of the optical system. Because the solution procedure is performed point by point, we can solve an arbitrary aspheric profile that meets the design requirements for the optical system. Comparing results from the new method with those from ZEMAX, we can conclude that our method is more effective and reliable than conventional optical design software in designing aspheric lenses. In addition, it is obvious that the calculating speed is determined only by the sampling number N and that it is independent of the aspherical order; i.e., it will not take more time when the ideal aspheric profile is of a high degree of asphericity than when the ideal profile has little deformation form a sphere, unlike in conventional methods in which computational time increases greatly when we design deeply aspheric surfaces with higher and higher aspheric orders. Also, our method presents a new art and concept for optical design, and it is expected to work well in those cases that cannot be managed with conventional methods. Finally, we discuss techniques for manufacturing the discretely defined surfaces obtained by our method. The point sequence $\{S_j^m\}(m=1,2)$ we get is, in fact, the points on the generatrix of the lens ' s surface, whose spatial coordinates are determined; thus we can compile programs for the numerical control lathe and make the lathe tool follow the path of those coordinates to form the correct generatrix, so as to get the desired surface around the spindle of the lathe. Of course, when the number of points is large, an automatic interface program is needed between the computer and the numerical control machine to avoid inputting those data manually.

References

1. R R Shannon. *The Art and Science of Optical Design*. Cambridge: Cambridge U. Press, 1997.
2. E Glatzel, R Wilson. Adaptive automatic correction in optical design. Appl. Opt. 1968, 7: 265 ~ 276

3. T Suzuki and S Yonezawa. System of simultaneous nonlinear inequalities and automatic lens-design method. J. Opt. Soc. Am. 1966,56:677 ~ 683

4. S A Lerner and J. M. Sasian. Novel aspheric surfaces for optical design. *Novel Optical Systems Design and Optimization* Ⅲ. J. M. Sasian, ed., Proc. SPIE 4092, 2000,17 ~ 25

5. G Schulz. Higer order aplanatism. Opt. Commun, 1982,41:315 ~ 319

6. P Benftez and J C Miñano. Ultrahigh-numerical-aperture imaging concentrator. J. Opt. Soc. Am., 1977,4:1986 ~ 1997

7. J J M Braat, A Smid, M M B Wijinakker. Design and production technology of replicated aspheric objective lenses for optical disk systems. Appl. Opt., 1985,24:1853 ~ 1855

8. J C Miñano, P Benitez and J C González. RX: a nonimaging concentrator. Appl. Opt. 1995,34:2226 ~ 2235

9. J C Miñano and J C González. New method of design of nonimaging concentrators. Appl. Opt, 1992,31:3051 ~ 3060

10. *ZEMAX*. *Optical Design Program User Guide*. *Version* 5.0 (Focus Software, Tucson, Ariz., 1996).

Computer-aided Alignment of a Wide-field, Three-Mirror, Unobscured, High-resolution Sensor

Abstract

Santa Barbara Research Center has been exploring the technology related to the design, tolerance, and alignment of wide-field, all-reflecting sensors for multispectral earth observation. The goals of this study are to design an optical system with reduced fabrication risks, to-develop a detailed to lerance budget, and to demonstrate our ability to align the system to a tolerance of 0.05 waves rms, at 0.632 8 microns. The optical system is a three – mirror, unobscured telescope. It is telecentric and flat field over 15 degrees at F/4.5, and achieves diffraction-limited imagery at visible wavelengths.

A separate paper[1] describes the design and error budget approach for the telescope. This paper reports on the effort that led to the alignment of a scaled, prototype optical system. The approach used interferometric measurements of the wavefront at multiple field points and a computer alignment algorithm to define the rigid-body adjustments of the mirrors to achieve the alignment goal.

1. Background

Very high-quality imaging systems have been designed and built, as have wide-field, unobscured telescopes. However, no one has demonstrated the precision alignment of a high-quality, wide-field, unobscured telescope. This milestone is essential to demonstrate the total capability of designing and building such systems.

A multiyear research and development project was undertaken at

Santa Barbara Research Center to demonstrate all the key technologies required to design, tolerance, and build just such a telescope. The following design attributes describe the optical system:

- Three-mirror, unobscured design
- 1.0 meter EFL at F/4.5
- 0.6 × 15.0 degree field of view; telecentric and flat field
- Minimum number of aspheres to reduce fabrication difficulty
- Average performance across the field:

  0.007 mm rms spot size

  0.09 rms wavefront error at 0.632 8 microns

  0.92 MTF at 20 lp/mm

More information on the design and error budget for this system appears in reference 1.

The error budget analysis defined an alignment residual error of 0.055 waves rms at 0.632 8 microns across the field of view. To demonstrate our ability to achieve this goal, a prototype optical system was built. This prototype was scaled down to a 0.42 meter EFL from the nominal design to minimize fabrication costs. This scaled system had the same angular alignmet sensitivity and a reduced decenter sensitivity due to the scaling.

The mechanical structure was made of aluminum; the optics were single-point, diamond-turned aluminum mirrors. We recognized that the diamond-turned optics would not give us a good wavefront at visible wavelengths, but, in an alignment demonstration, only the low-order aberrations are important (even in nonsymmetric systems such as this). The characteristic ripple of diamond-turned optics would be seen as a fixed, background error, The low-order surface errors, however, would change the performance goal of our system. Therefore, a successful alignment would be judged relative to a "best possible" goal that was defined as the wavefront performance of the computer-optimized system in the presence of mirror fabrication errors.

2. Computer-aided Alignment

Computer algorithms to help align optical systems have been used successfully in past efforts.[2-5] The need for such an exotic approach is motivated by two reasons: the complexity of the optical system and the need for precise results. The two techniques most often applied are a geometric-based Hartmann test to evaluate the quality of a focused image, or interferometry to evaluate the quality of the system wavefront. The approach is analogous to an optimization cycle in optical design. The merit function is defined as the difference between the theoretical design performance and the as-measured performance. The sensitivity matrix is the change in the system performance as a function of rigid-body motions of the optics. The alignment motions are the results of a linear, least-squares solution.

Our need to achieve the 0.055 wave rms alignment error across a 15 degree field of view was best served using interferometry. Interferometry offers the following advantages over geometric-based approaches:

- The wavefront sampled at the exit pupil of the system does not suffer from the diffraction effects of an image point. This ultimately limits the accuracy of image-qualifying techniques.

- By sampling an entire wavefront, the effects of local slope errors are more accurately recorded. Local slope errors can be a problem with subaperture Hartmann samples.

- A wavefront map can be analyzed in terms of aberration coefficients. This is important because it allows the designer to isolate alignment aberrations from design-residual aberrations.

A team at the University of Arizona, Optical Sciences Center,[6] has reported success in modeling the computer-aided alignment of a wide-field IR sensor where a minimum effort was devoted to initial alignment. The use of a sufficiently robust algorithm was shown to converge to an acceptable solution. Our exercise, however, closely followed the constraints that might be imposed on mounted optics in a structure. The dynamic range of the alignment actuators were restricted to achieve the necessary precision of motion and rigidity. Also, the optics could not be allowed to "float" in an unconstrained space because of the potential for vignetting from the mounting and metering structure. Through a careful alignmet simulation exercise, it was determined how well the mirrors had to be mechanically aligned to ensure that the performance would converge within the imposed constraints. Our alignment approach combined an initial mechanical alignment with the precision of computer-aided alignment.

3. Mirror Fabrication and Test

The primary and tertiary mirrors are off-axis portions of rotationally-symmetric aspheres. The secondary mirror is symmetric about it's vertex. The mirrors were diamond turned at Hughes Optical Products, Inc. Three datum surfaces were built into the mirror substrates to locate the vertex axis of the eccentric aspheres.

Null lenses were used to characterize the surface figures of the primary and tertiary mirror. The primary mirror null lens was a two-element Offner null that provided correction to better than 0.1 wave P-V across the full parent. The tertiary mirror null corrector was a single negative element and corrected the parent asphere to better than 0.5 wave P-V. The secondary mirror had such a mild aspheric departure that it could be directly tested against a spherical reference. The fabrication and assembly tolerances for the null correctors were part of an error budget category separate form alignment, and all tolerances were easily met.

In the upper part of the figure are photographs of the actual components, while the lower part of the figure is a contour plot generated by a 25-term Zernike polynomial fit to the measured data. The ellipse represents the approximate size of the region of measured data inscribed in the circle that limits the Zernike terms.

This data was used to update the optical design model and to define a new alignment and performance goal. An optimization cycle was run using only rigid-body motions of the mirrors and was constrained to restrict beamwalk and line-of-sight variations. The result defined the best performance in the presence of the low-order errors, and the new offset positions defined for each mirror became the target for the mechanical alignment.

4. Mechanical Structuer

The mechanical structure used for the alignment demonstration has three main attributes:

- High stiffness-to-weight ratio
- Kinematic positioning of the mirrors
- "Stress-free" mounting and adjustment for primary and tertiary mirrors

The mounting and adjustment scheme evolved as part of an integrated design effort that balanced the needs for performance, mirror fabrication, and alignment flexibility. The number of precision degrees of freedom and the dynamic range of adjustment for the mirror mounts was minimized as a result of computer modeling of the alignment process.

It consists of a main bulkhead, on which the primary and tertiary mirrors are mounted, and a separate bulkhead for the secondary mirror. These bulkheads are secured to each other through a metering structure that avoids obstructing the light bundle.

Primary and tertiary mirrors are mounted at three points, and each mount constrains two degrees of motions, one radial and one tangential. One mounting point is located near the vertex of the optical surface. This allows tilt adjustments to be made about the mirror vertex, which greatly simplifies the alignment adjustments and modeling. The other two points are radially positioned at opposite "corners" of the mirror. The design of the primary and tertiary mirrors allowed the mounting fixtures to be identical in function and design.

The keys to alignment of primary and tertiary mirrors were their datum surfaces and the precision manufacture of the main bulkhead. The bulkhead's front face was machined flat and incorporates accurately positioned holes that secure tooling balls for positional references. The mechanical alignment was successful in positioning the two mirrors, with respect to their offset nominal settings, to within 0.002 inch and about 1.5 arc-min.

Secondary mirror alignment was aided by use of two reference marks. One was calibrated with respect to the main bulkhead, and the second identified the optical center of the secondary. An alignment telescope was used to extrapolate an axis between these two points.

The quality of the adjustment mechanisms, stability of the structure, and stress-free mounting were verified in practice. During alignment we achieved 0.04 wave P-V repeatability of aberration coefficients and we were able to accurately impart angular motions of less than 15 arc-seconds, which related to a linear motion of the adjustment screws of 0.000 4 inches.

5. Interferometric Text Stand

The test stand used for system-level interferometry is as important as the optical structure itself in terms of stability and ease of operation. Some of the interesting design features are:

• The input is a point source at the focal plane. The wavefront double-passes the system using an auto-collimating flat in "object space."

• The optical structure is held at an 8-degree angle to simulate the field-biased input.

• The auto-collimating flat pivots around the entrance pupil of the system. The image space is telecentric, so a linear travel can be used to sample the focal plane.

• A remote, phase modulator/objective assembly is used. This eliminates the need to support and move the entire interferometer.

• The modulator/objective assembly is on a three-axis, remotely-controlled precision positioner. The large linear stage allows us to traverse a 12-inch focal plane with micron resolution.

• The optical assembly, separate from the interferometer and motor controllers, can be shrouded from environmental fluctuations. The entire assembly is on an air-suspension table.

6. System-level Interferometry

Two years of design, tolerancing, and optical and mechanical fabrication and assembly, had culminated in a well-balanced, precision optical system. The moment of truth had come when we were faced with the task of aligning this system in the presence of a wavefront.

Though we had anticipated a poor-quality wavefront, this was a real challenge. The mechanical alignment achieved three waves peak-to-valley and easily bettered the capture range of the interferometer; nevertheless, we could not acquire data due to local phase discontinuities. It was then, to our relief, and to the credit of the software staff at Zygo Corporation,[7] that we were provided an experimental copy of a software package being developed specifically for handing phase discontinuities in the wavefront. This allowed us to pass the first hurdle-acquiring interferometric data.

At this point we had to determine the validity of the measurements. Two lengthy tests were conducted. One established the repeatability of the software algorithm (as well as the mechanical structure), and the second established the accuracy of the interpreted data.

The repeatability test consisted of taking a series of measurements of the same wavefront and examining the stability of the data. Each wavefront was sampled with approximately 25,000 data points. The Zygo algorithm identified areas of discontinuous phase and eliminated those regions. Then, the remaining data points (typically 15,000 to 20,000) were fit to a continuous surface. This surface, representing the double-pass OPD error of the system, was fit with a 15-term Zernike polynomial. Typical 1-sigma variations were:

Individual coefficients	0.002 to 0.02 waves zero-peak
P-V of data points	0.01 to 0.02 waves
rms of data points	0.002 to 0.005 waves rms
rms of residual	0.001 to 0.003 waves rms
Number of data points	75 to 200 points

As a measure of accuracy, we compared actual wavefront data to theoretical predictions, It would have been extremely difficult, and probably inconclusive, to attempt to model the actual, static, wavefront. The high-spatial frequency error alone would be impossible to model exactly, so we measured a change in the wavefront as a function of tilting one mirror. From sensitivity data, it was evident that astigmatism was the dominant error to expect when either the primary or tertiary mirror was tilted.

Primary mirror tilt was applied for convenience. A series of measurements was made where the mirror was tilted 30, 60, 90, 60, 30, and 0 arc-seconds. The results are shown in Figure 1, The banded linear plot shows the theoretical change in the astigmatism coefficient as a function of mirror tilt. Each band represents a 0.01 wave rms increment away from theoretical. The actual data points are plotted with ± 1-sigma error-bars for a series of three runs. These results show that the measured data tracked the theoretical plot to within an average value of 0.015 waves rms. The analyses for repeatability and accuracy indicate that even in the presence of a poor quality wavefront, the low-order wavefront data being measured can be believed.

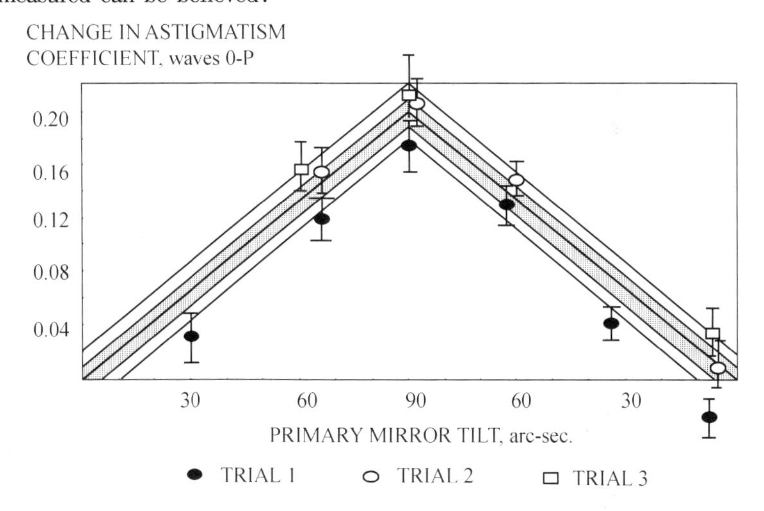

Figure 1. To demonstrate accuracy, measured changes in wavefront followed predicted values.

7. Alignment

The final task, and indeed the actual reason for all this activity, alignment of the telescope, could now be undertaken. The procedure is actually very straightforward:

1. The goal for the best aligned system had been established based on the best performance one can hope to achieve in the presence of errors on the mirror surfaces.

2. Five interferograms were taken across the field of view of the telescope. At each field point, five wavefront sets were measured . Each

set was fit to a 15 term Zernike polynomial, and the average of the five data sets was used to define the coefficients that describe the wavefront.

3. These coefficients were defined as INTerferogram files to Code-V.[8]

4. The ALignment option of Code-V was run using these interferogram files and the "goal" optical system as the target for the performance.

5. Various alignment solutions were attempted prior to each iteration to determine the minimum number of alignment motions that give maximum performance improvement.

The following sequence of figures shows that the alignment goal was met after four iterations. The "Y-axis" of the plots is the value for the rms wavefront error. The "X-axis" indicates the field position of the measurements. The measurements were limited to an 8-degree field of view due to vignetting in the test stand. Though the 15-degree field is not fully sampled, because of the quality of the model, the excellent agreement of measured data, the degree of control in alignment convergence and the excellent agreement to the theoretical goal, we feel that the full field of view would have been just as easily aligned.

Figures 9 through 11 show the rms of the low-order Zernike fit to the wavefront as a function of field position. These terms includes coma, astigmatism, and spherical aberration. Focus was not included because the variation of focus across the field is indicative of a tilted, flat focal plane, which is acceptable, The last figure, Figure 12, shows the rms error for the low-and high-order Zernike fit for iteration number four. This includes all the third-order terms as well as the fourth-order and some fifth-order expansion terms. Consistency between the low-order and high-order results supports the claim that one can align the system using only low-order terms.

The curve labeled "Goal" is the computer-predicted, best-aligned

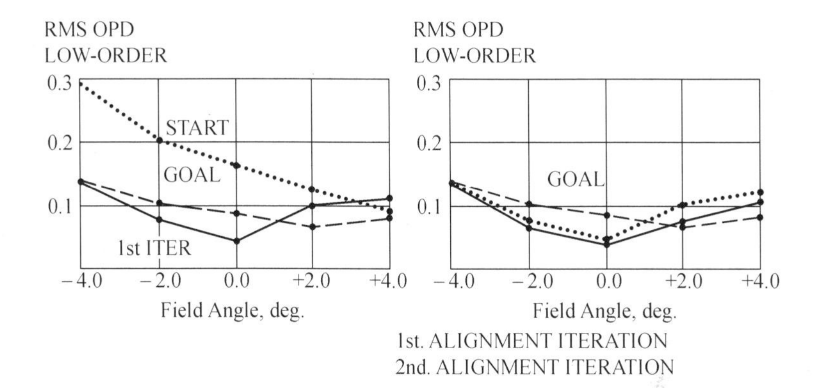

Figure 9 and 10. Plots show the measured results of the first two alignment iterations.

value. The width of the hatch marks is the 0.055 wave rms acceptance tolerance rss'd with the predicted performance value. The shaded line titled "Start," in Figure 9, is the value of the measured wavefront after mechanical alignment. The solid black line is the measured data after an alignment cycle had been completed. In Figures 10 through 12 the "Start" curve has been removed and successive iterations are compared to the previous one. Table 1 gives highlights of each alignment (See Page440).

Table 1. Highlights of Alignment Iterations

Iteration 1. This first iteration made the greatest progress. This was the result of applying *five simultaneous rigid-body motions to three optical components*.

Iteration 2. A 15 arc-second tilt of one mirror was predicted to make a small, controlled improvement. Results were as predicted, showing our ability to control the process to the order of 0.02 waves rms.

Iteration 3. A two-axis tilt was applied to the primary mirror. Though there was no statistically significant change of the rms wavefront error, there were changes in the astigmatism terms that were as expected: one went up while the other went down.

Iteration 4. The fourth and final iteration was attempted after a

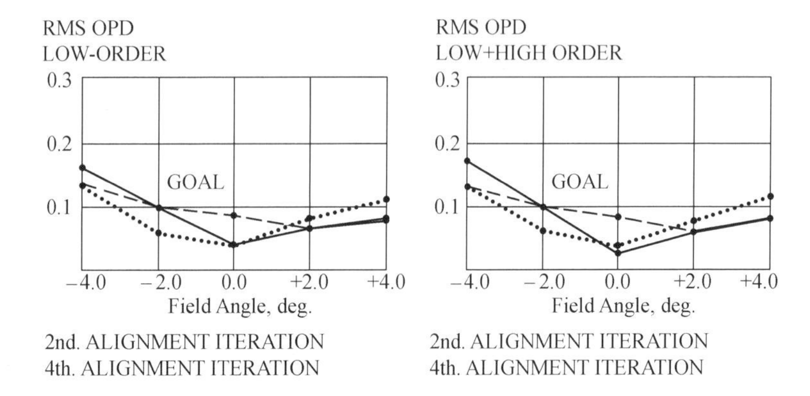

Figure 11 and 12. Successful conclusion of the alignment is verified whether one uses only low-order terms or low-plus high-order terms.

different field weighting factor was applied to the alignment solution, This brought the entire field, with the minor exception of one field point, to within, or better than the predicted goal. Control of the process was fine enough that the last point could have been recovered.

8. Summary

A three-mirror, unobscured, wide-field telescope with diffraction-limited performance has been designed and toleranced. To demonstrate our ability to assemble such an instrument, a scaled version of the telescope was built. An alignment demonstration was conducted that brought the optical system into alignment to the error-budgeted value of 0.055 waves rms at 0.632 8 microns.

The alignment approach used interferometric measurements of the system wavefront at five field points. This data was used in conjunction with a computer alignment algorithm to predict mirror adjustments. Prior to the actual alignment, tests were conducted to assure the repeatability and accuracy of the measurements of a wavefront that was adversely affected by residual high spatial frequency errors.

There are three key reasons for the success of this program:

1. From the beginning, the design, tolerances, structure, and alignment procedure were oriented to the computer-aided alignment approach. In this way the tolerances could be balanced throughout each stage of fabrication and alignment.

2. Eease of implementing actual alignment motions was a direct consequence of the quality of the mechanical structure and adjustment mechanisms. Fine, precise motions in a stable structure allowed us to control wavefront changes of a few-hundredths of a wave rms.

3. Before undertaking the actual alignment, care was taken to assure good correlation between the computer model and actual hardware.

The use of computer-aided alignment is essential to some problems (such as complex, high-performance systems), useful and cost-effective in others (such as rapid alignment of production optics), and probably unnecessary in many applications. For the procedure to be successful, the user must be willing to accept the up-front technical challenge of defining and verifying the appropriate test hardware and procedure.

9. Acknowledgements

The authors would like to acknowledge the important contributions of T.W. Tourville for the design of the optics structure and mounting and W.C. Cushman for the design of the test stand hardware. Also, this work would not have been successful without the commitment of SBRC management to the development of quality space sensors.

10. References

1. J W Figoski. Design and tolerance specification of a wide-field, three-mirror, unobscured, high-resolution sensor. SPIE Proc., vol. 1049, 1989

2. S G L Williams, et. al.. U.S. patents 4m471,447 and 4,471,448,

1984.

3. I M Egdall. Manufacture of a three-mirror wide-field optical system. Opt. Eng., 1985, 24(2): 285

4. W S Smith, et. al.. Interferometric alignment of multi-element optical systems. Proc. SPIE, 1980, 251: 5

5. M J Fahniger. Alignment of a full aperture system test of a cassegrain telescope. Proc. SPIE, 1980, 251: 5

6. H J Jeong, G N Lawrence. Auto-alignment of a three-mirror, off-axis telescope; reverse optimization. Proc. SPIE, 1988, 966

7. C A Krajewski, Zygo Corporation, Laurel Brook Road, Middlefield, CT, 06455.

8. Code-V is a registered trademark of the Optical Research Associates, 550 North Rosemead Blvd, Pasadena, CA, 91107.

Wide Field-of-view Three Mirror Telescope Designed for Improved Manufacturability

Abstract

A new form of three mirror anastigmat (TMA) has emerged in which the primary and tertiary mirrors share a common vertex. This allows the two mirror surfaces to be machined on the same substrate. The literature to date has shown that reimaging forms have been restricted to a moderate field-of-vies(< 3°). This illustrates design forms which achieve excellent performance over a wide field-of-view (3° × 7°).

1. Introduction

Three Mirror Anastigmat (TMA) telescope designs have been studied for over 15 years by a variety of researchers. Early designs required interferometric assisted alignment in which wavefront measurements were made and analyzed to determine the appropriate mirrors adjustments required[1,2]. As diamond turning technology and techniques improved, the ability to diamond machine individual mirrors and the support structure allowed the assembly of TMA telescopes without interferometric alignment[3,4].

Recently, a new form of TMA has been introduced in which the primary and tertiary mirrors share a common vertex[5]. This design feature has facilitated improved manufacturability by relaxing the allowed machine tolerances between the secondary and primary/tertiary pair. To further examine the advantages of this design form a study was undertaken to develop a Wide Field of View (defined as exceeding 3 or 4 degrees after Cook[6]) Common Substrate Reimaging TMA Telescope.

2. Design Requirements

A reimaging TMA design was selected to provide an accessible entrance pupil, 100% cold stop efficiency and high stray light rejection capability. The telescope performance specifications are summarized in Table 1.

Table 1 Reimaging TMA Design Requirements

Parameter	Requirement
Spectral Response	MWIR (3.0 to 5.0 μm)
Cold Stop Efficiency	100 percent
Blur Spot Performance	Near Diffraction Limit
F/number	2.8
Field of View	7 × 3 degrees
Aperture Size	3 inch

3. Optical Design

The resulting TMA optical design operated off axis and decentered in both aperture and field. Figure 1.0 is a layout of the TMA design in the yz plane. Figure 2 illustrates the design and the mirror parents, dearly identifying the common primary/tertiary vertex design feature. The design is similar to the type IIIb generic concave-convex-concave configuration defined by Korsch[7] in that the primary and secondary produce a physical accessible intermediate image where a field stop can be placed. The tertiary mirror then relays the intermediate image to the detector plane. In addition the entrance pupil which lies approximately 100 mm in front of the primary is imaged onto the cold aperture stop to meet the 100% cold stop efficiency requirement.

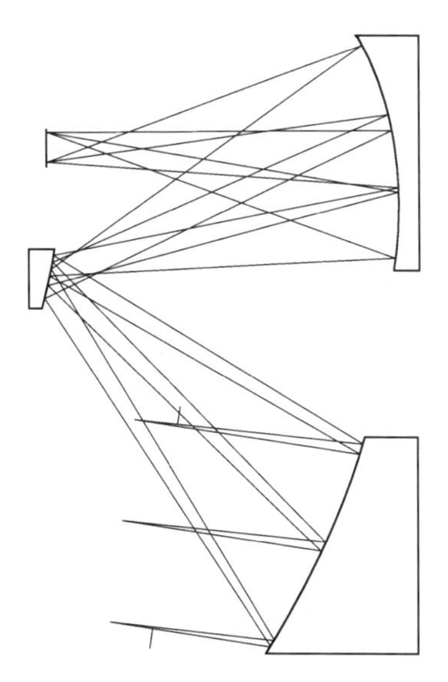

Figure 1 TMA Layout

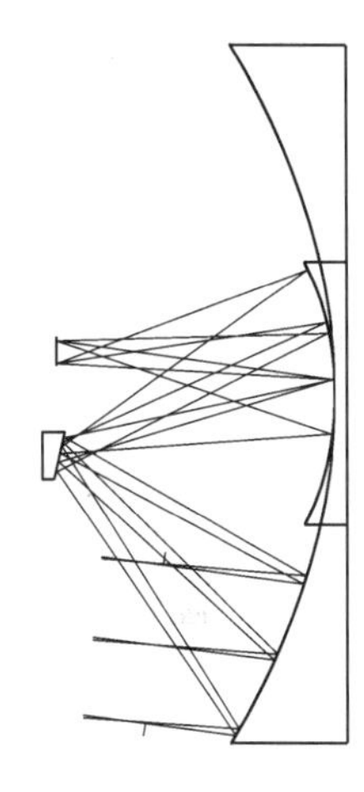

Figure 2 TMA Layout with Parents

The TMA has an ellipsoidal primary, a hyperbolic secondary and an ellipsoidal tertiary. Each of these conics have sixth and eighth order aspheric deformations which were employed to control the residual aberrations over the FOV. The secondary was also decentered with respect to the opto-mechanical axis defined by the common primary tertiary to enhance system performance.

4. Optical Performance

Since this system has bilateral symmetry the analysis task was simplified by performing the evaluation over field points covering only one half of the rectangular format. The geometric blur spot performance is exhibited in Figure 3 where the spots show excellent symmetry for a design of this form. Figure 4 illustrates the diffraction MTF curves out to 30 lp/mm. The performance approaches the diffraction limit and will provide excellent imagery.

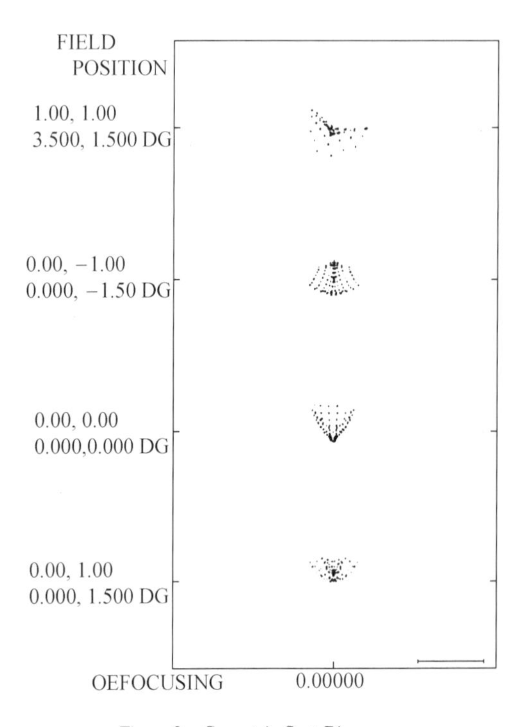

Figure 3 Geometric Spot Diagrams

Distortion is illustrated in Figure 5 which exhibits the anticipated "barrel" type effect characterized by the horizontal lines. Keystoning is also evident from the figure. Both of these effects were anticipated for the off axis design form and could be reduced somewhat by sacrificing blur spot performance and by reducing the field offset angle.

Constraints were devised during optimization to ensure that adequate clearance existed between the primary and tertiary mirror to allow both elements to be turned on a common substrate. The resulting primary/tertiary layout is depicted schematically in Figure 6.

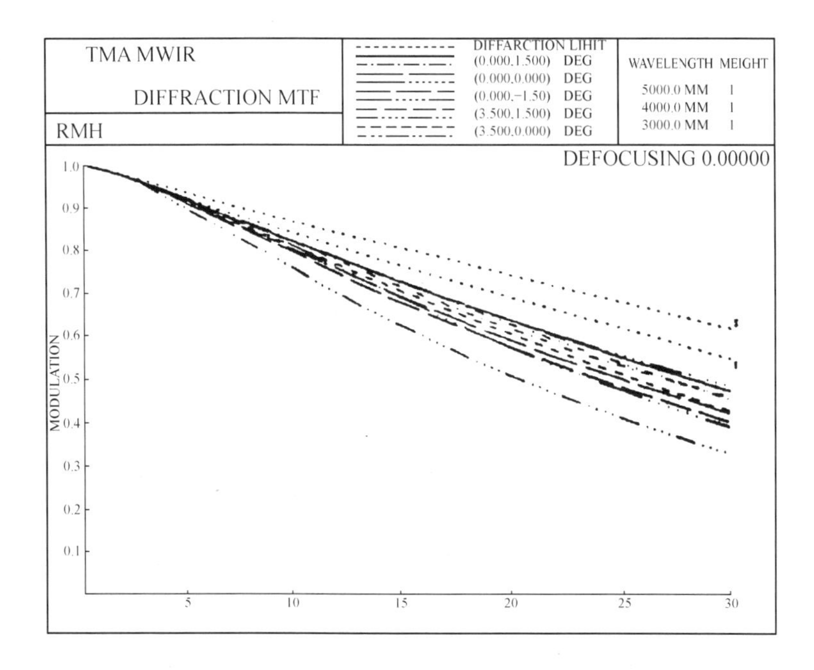

Figure 4 Representative Modulation Transfer Function Curves

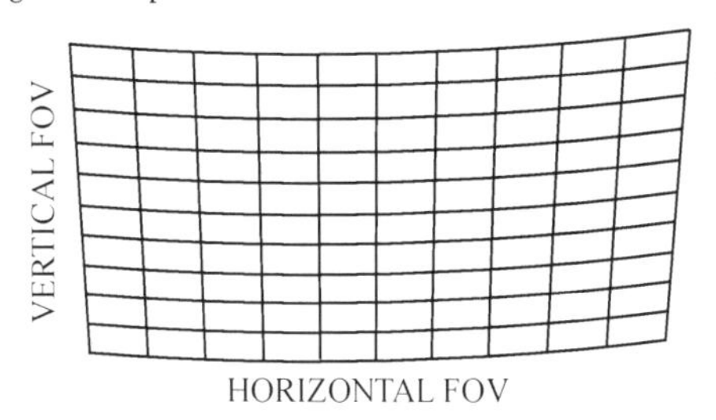

Figure 5 Predicted TMA Distortion

5. Telescope Tolerance and Manufacturability Assessment

Utilizing the design described in the previous section a manufacturability assessment was performed to investigate further the advantages and limitations of the common substrate design form.

A tolerance analysis was conducted using CODE V's real ray based

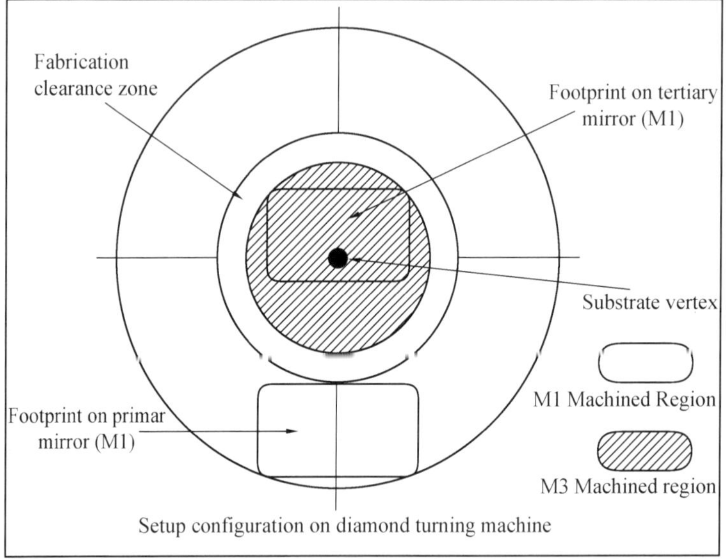

Figure 6 Schematic Primary/Tertiary Substrate Layout

tolerance option TOR in which the sensor focus was employed as the only compensator. Manufacturable tolerances were predicted with less than a 10% drop in the MTF performance anticipated. For comparative purposes the design was modelled two ways:

(a) Using the assumption that the mirrors were mounted independently

(b) Using the assumption that the primary and tertiary were "linked" as a common substrate (the tolerances between the two mirrors defined by the diamond turning process).

Table 2 illustrates the results of a preliminary tolerance study which shows that significant tolerance relaxation can be achieved utilizing the common substrate design. Note that the common substrate decenter tolerances are 2.5 to 3.0 times looser and the tip/tilt tolerances are even less stringent. Production cost savings associated with the relaxed tolerances are generally characterized as an improvement in yield and a reduction in the required machining time for each part. As a minimum the combined primary/tertiary mirrors will eliminate one machine set-up. Since set-ups can be costly a savings of up to 25% percent could be realized[9].

Table 2 Relative Tolerance Relaxation Achieved with Common Substrate Design

Mirror	Parameter	Independent Substrate Model	Common Substrate Model
Primary	Decenter	± .02 mm	± .05
	Translation	± .04 mm	not applicable
	Tip/Tilt	± 20 mrad	± .50
	Clocking	± .20 mrad	± .50
	Sag	± .025 mm	± .025
Secondary	Decenter	± .02 mm	± .07
	Translation	± .04 mm	± .05
	Tip/Tilt	± .20 mrad	± 1.0
	Clocking	± .20 mrad	± 1.0
	Sag	± .02 mm	± .02
Tertiary	Decenter	± .02 mm	linked
	Translation	± .04 mm	linked
	Tip/Tilt	± .20 mrad	linked
	Clocking	± .20 mrad	linked
	Sag	± .01	± .01

To further examine the impact of the common substrate design a tradeoff was performed to examine the impact of tilt, decenter and surface profile on the overall telescope performance. The results of this tradeoff are illustrated in Table 3. The blur spot diameter is an average value derived from the 80% energy capture blur diameter at six different points spread across the telescope field of view. As anticipated the conic surface profile is essential to ensuring high performance. The addition of higher order aspheric deformations and a tilted/decentered secondary does not have a strong impact upon the average blur diameter but does result in a more symmetrical blur spot performance across the field. To achieve strict diffraction limited performance, tilted and decentered primary, secondary and tertiary mirrors are required. Consequently manufacturing advantages of the common substrate design form are lost.

Table 3 Average Blur Spot Diameter Versus Mirror Tilt, Decenter & Surface Profile

Design	Sphere	Conic	Asphere	Decenter	Tilt	Average Blur Diameter (μm)
TMA-DES51	S	···	PT	S	S	66
TMA-CON32	···	PT	···	S	···	38
TMA-CON3	···	PST	···	S	S	36
TMA-DES54	···	S	PST	S	S	34
TMA-NOM	···	···	PST	S	···	35
TMA-NOM1	···	···	PST	S	S	35
TMA-DES44	···	···	PST	PST (1)	PST (1)	31

The results of table 3 were examined through focus as illustrated by Figure 7. Assuming an allowed "as built 50 μm depth of focus requirement" the more complex TMA-DES44 is seen to offer only a 10% performance improvement over the TMA-NOM1 common substrate design. Since this small performance gain is dependent on a cost premium of 25% to 50%, the optical designer should carefully examine the system requirements before selecting a decoupled design.

(1) Primary and Secondary Decoupled

(2) Diffraction limited blur = 2.44 × 4 μm × F/# = 27.3 μm

P—Primary S—Secondary T—Tertiary

An important limitation imposed upon the common substrate design is the maximum diameter of the part which can be diamond machined. The maximum part diameter which can be turned, depends upon the machine employed For example:

Moore M-18 Aspheric Generator [4]

Over rotary table 13 inches

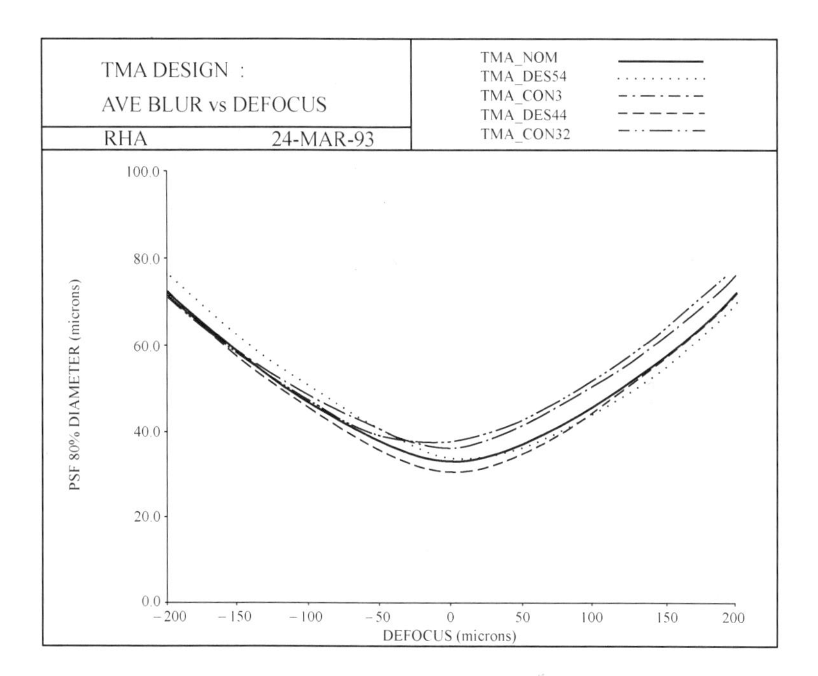

Figure 7 Average TMA Blur Spot Diameter versus Defocus

no rotary table 22 inches

with risers 33 inches

Rank Pneumo Nanoform 600 [8]

Swing capacity 24 inches

Scaling the existing design, the required substrate diameter was calculated as a function of aperture diameter. Table 4 illustrates the results. Based upon existing machines, as illustrated above the common substrate TMA design is limited to apertures of 5 to 7 inches.

Table 4 Substrate Parent Diameter Versus Telescope Aperture

Aperture Diameter	Primary/Tertiary Substrate Diameter
7 inches	34 inches
6 inches	27 inches
3 inches	13.5 inches

To validate the analysis reviewed in this paper a prototype reimaging

TMA Telescope with the primary and tertiary mirrors sharing a common substrate is currently being fabricated. Figure 8 illustrates a preliminary mechanical layout of the TMA telescope. Precision diamond turned mounting surfaces are added to the base and near the center of the large substrate. These mounting surfaces allow the secondary support structure to be fastened directly to the primary/tertiary substrate. The secondary is supported by a precision cylinder which has diamond turned mounting surfaces at its mating end. The secondary is held on a mounting plate, which is diamond turned in an integral fashion with the mirror itself. In this manner the telescope is comprised of only three pieces which can simply be bolted together.

6. Conclusions

A wide field of view common vertex reimaging TMA was designed to achieve near diffraction limited performance over a 3 × 7 degree field of view. This design form offers potential manufacturing benefits by combining the primary, and tertiary onto a single substrate. Compared with a conventional TMA with independent mirrors, a preliminary tolerance analysis shows that the primary/tertiary common substrate design increases the decenter tolerances 2.5 to 3 times while the tip/ tilt tolerances are even less stringent. Based upon current diamond turning technology this design form can also be scaled to yield apertures of 5 to 7 inches depending upon the FOV.

7. References

1. J W Figoski, T E Shrode, G F Moore. Computer-aided. alignment of a wide-field, three mirror, unobscured, high resolution sensor. SPIE, 1989,049:167 ~ 177
2. I M Egdare. Manufacture of a three mirror wide field optical system. Optical Eng., 1985,24(2):285 ~ 289

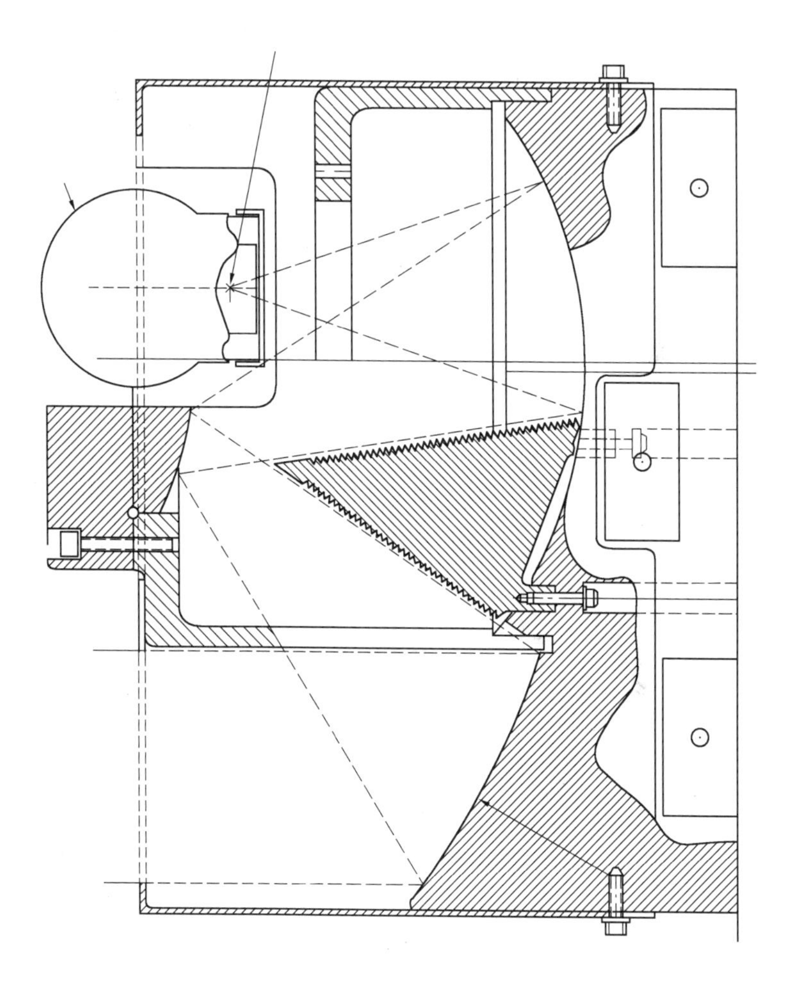

Figure 8 Preliminary Opto-mechanical Layout for a Prototype TMA Telescope

3. D Morrison. Design and manufacturing considerations for the integration of mounting and alignment surfaces with diamond turned optics. SPIE, 1988, 966: 219 ~ 227

4. D J Erickson, R A Johnston, A B Hull. Optimization of the optomechanical interface employing diamond machining in a concurrent engineering environment. SPIE, 1992, CR43: 329 ~ 366

5. G Y Chan, & K Preston. Advanced Manufacturability Three Mirror Telescope. SPIE, 1992, 1690: 49 ~ 55

6. L G Cook. The last three mirror anastigmat? SPIE, 1992, CR41: 310 ~ 324

7. D G Korsch. Reflective Optics. Academic Press, 1991

8. Rank Pneumo Product Brochure for the Nanoform 600

9. Private Communication, Robert Clark, OFC Diamond Turning Division